3.95

MATERIALS SCIENCE

G K Narula
Indian Institute of Technology
Delhi

K S Narula
Pusa Polytechnic
Delhi

V K Gupta
Delhi College of Engineering
Delhi

Tata McGraw-Hill Publishing Company Limited
NEW DELHI

McGraw-Hill Offices

New Delhi New York St Louis San Francisco Auckland Bogotá Guatemala
Hamburg Lisbon London Madrid Mexico Milan Montréal
Panama Paris San Juan São Paulo Singapore Sydney
Tokyo Toronto

© 1988, TATA McGRAW-HILL PUBLISHING COMPANY LIMITED

First Reprint 1992

RLRYRRDERXQXX

No part of this publication can be reproduced in any form or by any means without the prior written permission of the publishers

This edition can be exported from India only by the publishers,
Tata McGraw-Hill Publishing Company Limited

Published by Tata McGraw-Hill Publishing Company Limited
4/12 Asaf Ali Road, New Delhi 110 002, and printed at
A.P. Offset, Naveen Shahdra, Delhi 110 032

FOREWORD

I congratulate the authors for having brought out a really good book on materials science. The style of the book is very clear and easy to read. The plan adopted by the authors is well conceived. A reasonably wide coverage, in sufficient depth, has been attempted, which will prove the book useful for the first course in engineering materials in various Indian universities, AMIE courses and Engineering institutions. There is no doubt that the book is of great use to students and all those who wish to undergo a general course in materials science.

I have no doubt that this venture of the authors will be a success, and will also encourage them in future.

PROF. PKC PILLAI
Department of Physics
Indian Institute of Technology
Delhi

PREFACE

Modern science and technology is dependent on materials whose properties can be controlled to suit different applications. Polycrystalline materials, dielectrics, single crystals having controlled purity and perfection, thin films of different materials, polymers, glasses and superconductors, etc., are the building blocks of modern science and technology. It is often said that the rate of growth of technology is hindered by the limited availability of materials with the desired properties. The multidisciplinary area of materials science has the responsibility of development of materials. The interdisciplinary field of materials science has assumed for itself the responsibility to discover and control properties of materials for fundamental research, as well as for applications.

The purpose of this book is to present the basic principles necessary for understanding the nature and properties of engineering materials, and to make clear the significance of these principles in engineering applications. The book requires knowledge of general physics, general chemistry, and some engineering.

Materials science has become a very important subject as an interdisciplinary course in most of the universities in our country. We have written this book for a first course in materials science at the undergraduate level in engineering institutes. The students of applied sciences would also find this book useful. It is hoped that this book will be extremely useful to the students preparing for the degree examination of MSc (Physics), BE (Mechanical/Metallurgy/Production Engineering) of all Indian universities, and for the diploma examinations conducted by the various Boards of Technical Education and also for the AMIE and UPSC examinations. It can also serve as a source book for practising engineers.

The book begins with an introduction designed to orient the reader in the study of engineering science and to understand the properties of materials. The succeeding chapters are concerned with the atom, structure, properties, processing and application of materials in an integrated manner. Discussions of industrial processes and topics such as deformations, alloy systems, heat treatment, electrical and magnetic properties of materials, polymers and composites are presented in a concise manner. A detailed account of service behaviour (corrosion), mechanical properties, photoelectric effects, powder metallurgy and electron theory of metals is included to cover up the syllabi of the engineering institutions. In all chapters topics which are of great use to the readers are included to bring the text uptodate.

Review questions appear at the end of the chapters in the book. These are intended to test the readers comprehension of the principles covered in the section. If the reader cannot answer the questions he should restudy the section. It will also give an idea of the examination standard maintained by the various examination bodies.

We wish to express our gratitude for the inspiration and guidance received from Prof. PKC Pillai (of IIT Delhi) and Dr RN Chatterjee (of Delhi College of Engineering).

Mr Anokh Singh aided greatly during the initial planning of the book and subsequently read the various drafts of the manuscript. Mr Arora skillfully converted the preliminary sketches into illustrations that convey much of the message of the book.

We acknowledge the excellent cooperation received from our colleagues, especially Mr DS Sital, Mr KL Arora, Mr RK Kapoor and Mr JK Gupta in the preparation of this text.

We are also thankful to M/s Tata McGraw-Hill Publishing Company, for their untiring efforts in bringing out the book with excellent printing and nice get-up within the shortest possible period.

We would welcome suggestions for the improvement of the book.

GK NARULA
KS NARULA
VK GUPTA

CONTENTS

Foreword iii
Preface v

1. Introduction 1

1.1 What is materials science *1*   1.2 Classification of materials *1*   1.3 Engineering requirements of materials *2*   1.4 Important properties of engineering materials *2*   1.5 Material structure *4*

2. Atomic Structure and Electronic Configuration 5

2.1 Introduction *5*   2.2 Atomic structure *5*   2.3 The electron *5*   2.4 Properties of cathode rays *6*   2.5 Electron as universal constituent of all matter *6*   2.6 Protons-positive rays *7*   2.7 Nucleus *8*   2.8 Atomic number *8*   2.9 Atomic weight *8*   2.10 Isotopes *8*   2.11 Isobars *9*   2.12 Avagadro's number *9*   2.13 Atomic models *9*   2.14 Vector model *21*   2.15 Quantum numbers *22*   2.16 Pauli's exclusion principle *24*   2.17 Wave mechanics *25*   2.18 The periodic table *25*

3. Crystal Geometry and Structure 28

3.1 Introduction *28*   3.2 Crystal structure *28*   3.3 Space lattice *29*   3.4 Unit cell *30*   3.5 Crystal systems *31*   3.6 Atomic packing *31*   3.7 Co-ordination number *34*   3.8 Crystal structures for metallic elements *34*   3.9 Body-centred cubic structure (BCC) *34*   3.10 Face-centred cubic structure *35*   3.11 Close-packed hexagonal structures *35*   3.12 Crystal symmetry *35*   3.13 Atomic radius *36*   3.14 Atomic packing factor (APF) *38*   3.15 Lattice planes and directions (Miller Indices) *40*   3.16 Important features of miller indices *43*   3.17 Interplanar spacings *43*

4. Bonds in Solids   48

4.1 Introduction *48*   4.2 Types of bonds *48*
4.3 Mechanism of bond formation *49*   4.4 Ionic bond *51*
4.5 Covalent bond *54*   4.6 Metallic bond *56*
4.7 Comparison of bonds *59*   4.8 Secondary bonds *60*
4.9 Mixed bonds *60*   4.10 Chemical bonding and properties of solid materials *61*   4.11 Chemical bonding and the periodic table *62*

5. Crystal Imperfection   64

5.1 Introduction *64*   5.2 Types of imperfections *64*
5.3 Frank-Read source *71*   5.4 Some salient points in relation to the theory of dislocation *72*

6. X-rays   76

6.1 Introduction *76*   6.2 Discovery of X-rays *76*
6.3 Production of X-rays *76* 6.4 Properties of X-rays *78*
6.5 Scattered X-rays *79*   6.6 Characteristic X-rays *79*
6.7 Beta-rays (β-rays) *80*   6.8 Origin of X-rays *80*
6.9 Wave nature of X-rays or X-ray diffraction *81*
6.10 Bragg's Law *82*   6.11 Intensity measurement of X-rays *84*   6.12 Bragg's conditions for X-ray diffraction *84*   6.13 Bragg' X-ray spectrometer *85*
6.14 Mosley's Law *87* 6.15 Importance of Mosley's Law *88*
6.16 Bragg's law and crystal structure *88*   6.17 Practical X-ray diffraction *89*   6.18 Laue method *90*
6.19 Rotating-crystal method *91*   6.20 Powder crystal method *92*   6.21 Application of X-rays *93*
6.22 X-ray absorption *94*

7. Diffusion in Solids   97

7.1 Introduction *97*   7.2 Application of diffusion *97*
7.3 Classification of diffusion *9.7*   7.4 Diffusion mechanisms *98*   7.5 Vacancy mechanism *98*   7.6 The interstitial mechanism *99*   7.7 Direct interchange mechanisms *99*   7.8 Diffusion coefficient: Fick's Law *100*
7.9 Fick's Second Law—Time dependence *102*   7.10 Factors affecting the diffusion coefficient *102*   7.11 Self-diffusion *103*   7.12 Inter-diffusion *103*   7.13 Diffusion couple *103*   7.14 Diffusion with constant concentration (case-hardening) *104*   7.15 Diffusion in oxides and ionic crystals *104*   7.16 Grain-boundary and surface diffusion *105*   7.17 Activation energy for diffusion *106*

8. Deformation of Materials — 108

8.1 Introduction *108* 8.2 Elastic deformation *109*
8.3 Plastic deformation *110* 8.4 Deformation in polycrystalline metals *119* 8.5 Work hardening *121*
8.6 Season cracking *122* 8.7 Bauschinger effect *122*
8.8 Yield point phenomenon and related effects *123*
8.9 Cold-working *126* 8.10 Preferred orientation *126*
8.11 Recovery, recrystallisation and grain growth *127*
8.12 Hot-working *131*

9. Theory of Alloys: Constitution and Equilibrium Diagrams — 134

9.1 Introduction *134* 9.2 Basic terms *134*
9.3 Solid solutions *135* 9.4 Hume-Rothery's rules *137*
9.5 Intermediate compounds or intermediate phases *138*
9.6 Phase diagrams *139* 9.7 The phase rule/Gibb's phase rule *140* 9.8 Time-temperature cooling curves *141*
9.9 Construction of a phase diagram *142* 9.10 The lever rule *142* 9.11 Equilibrium diagram of a binary system in which the two metals are completely soluble in the liquid and solid states *144* 9.12 Eutectic systems *146*
9.13 Eutectoid system *150* 9.14 Peritectic and peritectoid system *151* 9.15 Ternary equilibrium diagram *152*

10. Phase Transformations — 154

10.1 Introduction *154* 10.2 Rate of transformation *154*
10.3 Mechanism of phase transformation or nucleation *155*
10.4 Homogeneous nucleation *157* 10.5 Heterogeneous nucleation *157* 10.6 Nucleation and growth *157*
10.7 Applications of phase transformations *158*
10.8 Micro-constituents of iron-carbon system *162*
10.9 The allotropy of iron *163* 10.10 Iron-carbon equilibrium diagram *164* 10.11 Formation of austenite *170*
10.12 Isothermal transformation—TTT diagram *172*
10.13 Transformation of austenite upon continuous cooling *174* 10.14 Martensitic transformation *175*

11. Heat Treatment — 176

11.1 Introduction *176* 11.2 Heat-treatment processes *177*
11.3 Annealing *177* 11.4 Normalising *181* 11.5 Hardening *182* 11.6 Tempering *190* 11.7 Sub-zero treatment of steel *192* 11.8 Case hardening *192* 11.9 Surface hardening *197* 11.10 Heat-treatment furnaces *200*

11.11 Temperature measurement or pyrometry *204*
11.12 Defects in the heat treatment of steel *205*

12. Organic Materials   209

12.1 Introduction *209*   12.2 Polymers *209*   12.3 Mechanisms of polymerization *210*   12.4 Additions in polymers *212*   12.5 Polymer structures *213*   12.6 Plastics *214*   12.7 Elastomers and rubbers *217*   12.8 Fibres and filaments *219*   12.9 Composite materials *220*   12.10 Single crystals *222*   12.11 Agglomerated structures *222*   12.12 Protective coatings *223*

13. Electrical and Magnetic Properties of Materials   227

13.1 Introduction *227*   13.2 Resistivity *227*   13.3 Conductivity *229*   13.4 Semiconductors *233*   13.5 Classification of semiconductors on the basis of fermi level and fermi energy *236*   13.6 pn Junction as rectifier *237*   13.7 Transistors *239*   13.8 Insulators *240*   13.9 Dielectrics *240*   13.10 Ferroelectricity *244*   13.11 Electrostriction *246*   13.12 Piezoelectricity *247*   13.13 Uses of dielectrics *248*   13.14 Magnetic properties *248*   13.15 Classification of magnetic materials *250*   13.16 Magnetostriction *256*   13.17 Soft magnetic materials *258*   13.18 Hard magnetic materials *259*

14. Corrosion   262

14.1 Introduction *262*   14.2 Factors influencing corrosion *262*   14.3 Types of corrosion *263*   14.4 Specific types of corrosion *270*   14.5 Control and prevention of corrosion *274*

15. Mechanical Properties   281

15.1 Introduction *281*   15.2 Definition of common terms used in mechanical properties *281*   15.3 Fundamental mechanical properties *287*   15.4 Creep *292*   15.5 Mechanical tests *296*   15.6 Models of fracture *316*   15.7 Factors affecting mechanical properties *324*   15.8 Technological properties of metals *326*

16. Photoelectric Effect   330

16.1 Introduction *330*   16.2 Experimental study of the photoelectric effect *330*   16.3 Characteristics of the photoelectric effect *332* 16.4 Laws of photoelectric emission *333* 16.5 Einstein's photoelectric emission *334* 16.6 Millikan's verification of Einstein's equation *335* 16.7 Photoelectric cells *337* 16.8 Applications of photoelectric cells *343*

17. Powder Metallurgy 347

17.1 Introduction *347* 17.2 Applications of powder metallurgy *347* 17.3 Advantages of powder metallurgy *348* 17.4 Disadvantages and limitations *349* 17.5 Design considerations for powder metallurgy *349* 17.6 Process description *350* 17.7 Secondary operations *357* 17.8 Powder metallurgy and fabrication *359*

18. Electron Theory of Metals 366

18.1 Introduction *366* 18.2 Metallic bonding *366* 18.3 Drude and Lorentz theory *367* 18.4 Sommerfield free/electron theory *367* 18.5 Electron energies in a metal *369* 18.6 Zone theory of solids *371* 18.7 Factors affecting electrical resistance of materials *376*

Appendices 377

Index 386

1
INTRODUCTION

1.1 WHAT IS MATERIALS SCIENCE

Materials science is a branch of science that reveals the many diverse factors in materials. Materials comprise a wide range of metals and non-metals which must be processed to form the end product. Information and knowledge about their different properties is very important; else manufacturing processes could become complex and costly. Materials science investigates relationships existing between the structures of materials and their properties and concerns the inter-disciplinary study of materials for engineering and practical purposes. It is based on solid state physics and the chemistry of the structure of materials.

1.2 CLASSIFICATION OF MATERIALS

Common engineering materials that come within the scope of materials science may be classified into one of the following three types:
   (i) Metals (ferrous and non-ferrous)
   (ii) Ceramics
   (iii) Organic polymers

Metals are element substances that readily give up electrons to form metallic bonds and conduct electricity. When two or more pure metals are melted together to form a new metal whose properties are quite different from those of the original metals, it is called an alloy.

Metals possess specific properties like plasticity and strength. Some favourable characteristics of metals are lustre, hardness, resistance to corrosion, thermal and electrical conductivity, malleability, stiffness, the property of magnetism, etc. These properties are due to two factors:
   (i) The atoms of which the metals are composed and
   (ii) The way in which these atoms are distributed in the space lattice.

1.2.1 Ceramics

These are materials consisting of phases. These are compounds of metallic and nonmetallic elements. Common ceramics are rocks, minerals, glass, glass fibre, metallic compounds, fired clays, abrasives, etc.

1.2.2 Organic Polymers

These materials are carbon compounds. They are solids composed of long molecular chains. Their structures are fairly complex and three dimensional. Common organic materials are plastics and synthetic rubbers.

Table 1.1 *Some Important Groupings of Materials*

Material Group	Important Characteristics	Common Examples of Engineering Use
1. Metals and alloys	Lustre, hardness, resistance to corrosion, thermal and electric conductivity, malleability, stiffness and the property of magnetism.	Iron and steels, aluminium, copper, zinc, magnesium, brass, bronze, Invar, super alloy, conductors, etc.
2. Ceramics	Thermal resistance, brittleness, hardness, opaqueness to light, electrical insulation abrasiveness, resistance to corrosion and high temperature strength	Silica, glass, cement, concrete, refractories, silicon carbide, boron nitride abrasives, ferrites, insulators, garnets, etc.
3. Organic polymers	Soft, light in weight, dimensionally unstable, poor conductors of electricity and heat, ductile, combustible, low thermal resistance.	Plastics: PVC, PTFE, polyethylene, polycarbonate Natural: and synthetic fibres: rubber, leather, cotton, nylon terylene
		Other uses: explosives, refrigerants, insulators, lubricants, detergents, fuels, vitamins, medicines for surface treatment, adhesives, Fibre—reinforced plastics, etc.

1.3 ENGINEERING REQUIREMENTS OF MATERIALS

Engineering components should be made of materials whose properties suit the application they are required for. This will permit the component to perform its function successfully. The main engineering requirements are economic, fabrication and service requirements.

A knowledge of materials science helps select the right engineering material for practical application under a wide range of conditions.

1.4 IMPORTANT PROPERTIES OF ENGINEERING MATERIALS

The important properties of engineering materials are:

(i) Mechanical
(ii) Electrical
(iii) Thermal
(iv) Physical
(v) Chemical
(vi) Magnetic
(vii) Optical

1.4.1 Mechanical Properties

These include the characteristics of the material that indicate its behaviour when external forces are applied. Important mechanical properties are strength, ductility, toughness, stiffness, malleability, plasticity, elasticity, hardness, brittleness, creep constant and fatigue.

1.4.2 Electrical Properties

These signify the ability of the material to resist the flow of an electric current. These include conductivity, dielectric strength and resistivity.

1.4.3 Thermal Properties

A sound knowledge of the thermal properties is required to know the response of the material to thermal changes. Thus, materials can be selected to withstand fluctuating and high temperatures. Common thermal properties are specific heat, thermal expansion and conductivity.

1.4.4 Physical Properties

These are required to evaluate the condition of the material without any external force acting on it. Dimensions, density, porosity and structure are included in the physical properties.

1.4.5 Chemical Properties

The tendency of the material to combine with other substances, its reactivity, solubility, and effects like corrosion, chemical composition, acidity alkalinity, etc., called are its chemical properties.

1.4.6 Magnetic Properties

Permeability, hysteresis and coercive force are the magnetic properties which need consideration, especially for materials used for transformers, etc.

1.4.7 Optical Properties

Colour, light transmission and light reflection depend upon the refractive index, reflectivity and absorption co-efficient of the material used for optical work.

1.5 MATERIAL STRUCTURE

The *internal structure* of a material can be observed with a degree of magnification. The details of the structure disclosed at a certain level of magnification are different from the details obtained with other levels of magnification. Material structure can be classified as: macrostructure, microstructure, sub-structure, crystal structure, electronic structure and nuclear structure.

1.5.1 Macrostructure

This is seen by low-power magnification or the naked eye. It deals with the size, shape and atomic arrangement in a crystalline material. Macrostructure may be observed directly on a fracture surface or on a forging specimen. Internal flaws open up under applied stress. Much expense can be saved by rejecting such defective materials at an early stage.

1.5.2 Microstructure

Full information regarding structure cannot be obtained without the examination of prepared sections of the specimen. The structure is observed under an optical microscope at magnifications ranging from $20\times$ to $2000\times$.

1.5.3 Substructure

When crystal imperfections such as dislocation in a structure are to be studied, a special microscope having much higher magnification than an optical microscope is used. Electron microscope with magnifications up to $100,000 \times$ are used for this.

1.5.4 Crystal Structure

This shows the atomic arrangement within a crystal. Electron microscopes are used only for surface studies in crystalline solids as, otherwise, the regularity of the internal structure will be affected due to charging of particles by electron beams. Electron diffraction method and X-rays are commonly used for studying crystal structure.

1.5.5 Electronic Structure

This refers to the electrons in the outermost shells of individual atoms that form the solid. Spectroscopy is used to study this.

1.5.6 Nuclear Structure

This is determined by nuclear spectroscopic method. Mossbauer studies and magnetic resonance (NMR) are the common techniques used for this.

2
ATOMIC STRUCTURE AND ELECTRONIC CONFIGURATION

2.1 INTRODUCTION

Scientific research of recent times has shown that the atom is not the smallest indivisible particle of matter. Scientists like Sir J J Thomson, Rutherford, Neils Bohr and many others have found that the atom consists of smaller particles called *electrons*, *protons* and *neutrons*. The electrons and protons are electrical in nature, whereas neutrons are neutral.

2.2 ATOMIC STRUCTURE

The first recorded speculations as to whether matter is continuous or composed of discrete particles were made by Greek philosophers during 500-428 BC. The question was reopened following the experimental discoveries of the gas laws, the kinetic theory of gases and the law of chemical combination. This was followed in 1807 by J Dalton's laws of multiple proportions according to which, when two elements combine to form different compounds, for a fixed weight of one element the weight of the other is in the ratio of small integers. These laws were explained by Dalton in 1808 when he made the hypothesis that elements are composed of discrete atoms. This was the beginning of the atomic model. With the explanations put forward by J J Thomson, Rutherford and Bohr, we will try to find out more about the atom and the nucleus. Rutherford and his colleagues did marvellous work to show that the mass of the atom is concentrated in the nucleus. The atom essentially has an electrical structure, having a diameter of the order of 10^{-10} metre or 1 Å unit and is made up of smaller particles, the principle ones being electrons, protons and neutrons.

The atom is considered to be made up of a heavy nucleus, consisting of protons and neutrons, surrounded by highly structured configurations of electrons.

2.3 THE ELECTRON

The first experimental evidence that electrical charge was not infinitely divisible, but existed in discrete units was obtained by M Faraday in his experiments on the laws of electrolysis in 1833. He found that a given quantity of electricity always liberated the same mass of a given substance, and that this mass was proportional to the equivalent weight of the substance; where the equivalent weight is defined as the atomic (or molecular) weight divided by the valency.

When electrodes are placed in a gas at normal atmospheric pressure, no current passes and the gas acts as an insulator. When the electric field is increased to above 3 to 4 MV/m sparking takes place. In contrast, at low pressures, a steady current can be maintained in a gas. At pressures of about 1 mm of mercury, the discharge is accompanied by the emission of light; but at still lower pressures a dark region forms near the cathode. If, under these low pressure conditions, a small hole is made in the anode, a green glow is observed on the glass wall of the discharge tube. The rays of light appear to travel in straight lines from the surface of the cathode and move away from it. These are called cathode rays since they start from the cathode.

2.4 PROPERTIES OF CATHODE RAYS

The properties of these rays were studied by W Crookes who showed that the rays,

  (i) travelled in straight lines
 (ii) cast shadows
(iii) carried sufficient momentum to set in motion a light paddle wheel
(iv) carried negative charge (by collecting the charge on electrometer or by the deflection of a magnet)
 (v) can induce some chemical reactions, excite fluorescence on certain substances, and possess high kinetic energy.

The properties of the cathode rays were best explained by JJ Thomson by his hypothesis that the rays consist of a stream of particles, each of mass m and charge $-e$, originating at the cathode of the discharge tube. These particles are called *electrons*. He determined the specific charge (e/m), the ratio between the charge and mass, and accepted its value as -1.76×10^{11} coulombs per kilogram.

2.5 ELECTRON AS UNIVERSAL CONSTITUENT OF ALL MATTER

This is considered true for the following reasons:

  (i) The electrons emitted in cathode rays exhibit identical features in different modes of cathode ray discharge tubes.
 (ii) The electrons obtained from any source are the same and have identical effects when used for different applications, i.e., for electromagnetic waves (radio and television), producing fluorescence effects in exposed metals, and in X-ray applications.

(iii) Electrons obtained from any source possess the same charge 1.602×10^{-19} coulomb, same effective radius 4.6×10^{-6} nomometer, same mass 9.1×10^{-31} kg, and a rest mass equivalent energy of 0.51 MeV. All the matter consists of atoms as the universal constituents where electrons form the structure of the atom and thus electrons are considered the universal constituent of all matter.

The electron exists in shells surrounding the nucleus. It carries negative electrical charge equal in magnitude to the positive charge of a proton and can be considered to be a minute particle or an energy wave. The arrangement of electrons around the nucleus is similar to the arrangement of planets in the solar system. The motion of an electron around the nucleus is also limited to discrete orbits or quantum states, each of which represents a discrete level of energy—an integral multiple of quanta of fundamental energy units.

2.6 PROTONS—POSITIVE RAYS

It has been established from the above statements that an atom contains negatively charged particles (electrons). However, as the atom is electrically neutral it must contain some positive charge to neutralize the electrons. During the latter part of the nineteenth century Goldstien designed a special discharge tube with which he discovered certain new rays he called *canal rays*. The name is derived from the fact that the rays, travelling in straight lines through a vaccum tube in a direction opposite to that of the cathode rays, pass through and emerge from a canal or hole in the cathode. These rays are now commonly known as positive rays.

These rays possess positive charge and knock out electrons from the atoms and molecules of gas in the discharge tube. This process is called *ionization*. This collision is of great significance in determining the atomic structure of the element.

It was also observed that positive rays can be deflected by magnetic and electrical fields like cathode rays, but in the opposite direction due to their positive charge.

Positive rays have been found to be charged atoms of different weights, therefore their charge to mass ratio is not constant. The unit of positive charge is the same as that of the electron but of the opposite sign, i.e., $+1.6 \times 10^{-19}$ coulomb.

The positively charged basic nuclear particles that coexist with neutrons in the nucleus are called protons. A proton has a unit mass of 1.673×10^{-27} kg. The mass of each proton is equal to that of one atom of hydrogen. Each proton carries a unit positive charge. Neutrons are particles 1.008 times heavier than protons but carrying no charge at all. The mass of each neutron is very nearly the same as that of the proton.

$$\text{Neutron} = \text{Proton} + \text{Electron}$$

2.7 NUCLEUS

In 1911, Rutherford made use of the alpha particles emitted by a radioactive source to explore the structure of the atom. By passing a stream of particles through a film of gold foil and measuring the angles through which the beam scattered, he was able to conclude that most of the mass of the gold-atom resided in a volume called the nucleus, which carried a positive charge. The diameter of the nucleus is found to be 1/10,000 of the diameter of the atom. The diameter of atom is measured in Angstrom units (1 Å = 10^{-8} cm).

Since the atom is electrically neutral, the number of protons in the nucleus must be equal to the number of electrons in the surrounding orbits.

The nucleus is normally composed of protons and neutrons and both these particles make up the mass of the atom.

2.8 ATOMIC NUMBER

The number of protons contained in the nucleus of a normal atom is known as the atomic number. It is designated by the letter Z. In an electrically neutral atom there must, therefore, be Z electrons. In iron there are 26 protons and 26 balancing electrons, therefore its atomic number is 26. This number specifies the position of the element in the periodic table of elements.

2.9 ATOMIC WEIGHT

The atomic weight of an element is the average relative weight of its atom as compared to the weight of one atom of oxygen which is taken to be 16.

The mass number is equal to the sum of the number of protons plus neutrons. This should not be confused with the atomic weight of the element. The mass number of an element is always a whole number and its value is very close to the atomic weight.

Most elements have fractional atomic weights due to the existence of different isotopes of the same element. Atomic weight can also be expressed in terms of atomic mass unit (amu). Hydrogen has one mass unit and Carbon 12 amu respectively.

2.10 ISOTOPES

All atoms having different atomic weights but belonging to the same element are called isotopes. The external structures of all isotopes are identical. Therefore, isotopes are atoms of different weight belonging to the same chemical element and having the same atomic number. For example, chlorine has two isotopes, of Cl_{35} and Cl_{37}, which are available in the ratio of 3:1. Their average atomic weight is

$$\frac{37 \times 3 + 35 \times 1}{4} = 35.48$$

2.11 ISOBARS

Atoms with the same mass but belonging to different chemical elements are called isobars. The first pair of isobars is argon and calcium.

2.12 AVAGADRO'S NUMBER

This is the number of atoms in a gramme-atom of any substance. It is a universal constant. It is designated as N and its value is 6.02×10^{23} mol^{-1}.

A mol or a gramme-molecule of a compound is the amount of mass in grammes that equals the molecular weight, which in turn is the sum of the atomic weights of the atoms which make up the molecule.

2.13 ATOMIC MODELS

We have established the existence of some elementary particles that are more fundamental than chemical elements. Electrons, for example, are common to all elements and are a common building block of all matter. In order to study the extra-nuclear electronic structure, mainly with the help of positive rays and mass spectroscopy, several theories have been put forward over the years after obtaining quantitative measurements on electrons and positive rays. These theories about atomic structure are also known as atomic models. Some of the important atomic models as proposed by Sir J J Thomson, Rutherford, and Bohr are discussed below.

2.13.1 Thomson Atom

The electron was first clearly identified as an elementary particle by J J Thomson in 1897. He was able to conclude that,

(i) The electron is a constituent of all matter

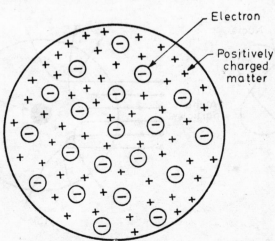

Fig. 2.1 *Thomson's plum-pudding model*

10 *Materials Science*

(ii) It has a negative charge of 1.602×10^{-19} coulomb and a mass of 9.1085×10^{-31} kilogramme at rest.

(iii) Electrons were distributed along with positive charges in the atom, like plums in a pudding (Fig. 2.1).

Thomson also computed the total number of electrons in an atom by several methods of spectroscopy.

His proof for the existence of electrons was an essential prerequisite for the subsequent theories on the structure of atoms. By using different materials for the filament wire, he established that the same value is always obtained for the charge to mass (e/m) ratio of the electron, thus supporting the statement that the electron is a constituent of all matter.

Thomson's atomic model however, could not explain the spectral series, and it was given up with the development of Rutherford's scattering experiments on alpha particles.

2.13.2 Rutherford's Nuclear Atomic Models

In 1911 Rutherford by his differential scattering experiment, proposed a new atomic model. His experiments on alpha particle scattering established the incorrectness of the uniform distribution of positive charge in a sphere of atomic dimensions in the Thomson atomic model. He suggested that all the mass and positive charges of an atom are concentrated at the centre of the atom in a nucleus of very small dimensions. He further suggested that the internal structure of the atom could be assessed by the bombardment of matter with high speed alpha particles emitted from a radioactive substance.

According to the Rutherford model, an atom consists of a positive nucleus with the electrons moving very rapidly around different orbits, as shown in the Fig. 2.2 (a). It was assumed that the force of attraction bet-

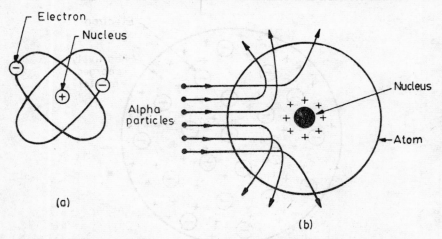

Fig. 2.2 (a) *Rutherford's model*   (b) *Scattering of alpha particles at different angles at the atomic nucleus*

ween the electrons and the nucleus was balanced by the centrifugal force attained by the electrons due to their motion.

Rutherford's model explains alpha particle scattering in the following way.

The forces between electrons and alpha particles may be neglected, but an alpha particle passing near the centre of an atom is subjected to an increasingly large coulomb repulsion as the separation between the two particles decreases. Because the atom is mainly empty space, many of the alpha particles go through the foil with practically no deviation. But an alpha particle that passes close to the nucleus, experiences a very large force that is exerted by the massive positive core and, is deflected through a large angle in a single encounter (Fig. 2.2 (b)).

The significant interaction, then is between the doubly charged alpha particle ($+2e$) and the positive core having an integral number Z of positive charges of electronic magnitude $+Ze$. Rutherford assumed that these two charges acted on each other with a coulomb force which, in this case was repulsive. The equation for this force is given by:

$$F = \frac{1}{4\pi\epsilon_0} \frac{2e\,Ze}{r^2} = \frac{2\,Ze^2}{4\pi\epsilon_0\,r^2}$$

where ϵ_0 is the permittivity of free space,

$$\left(\frac{1}{4\pi\epsilon_0} = 9 \times 10^9 \text{ MKS units.}\right)$$

This force is inversely proportional to the square of the distance between the bodies.

If the electron moves in a circular orbit of radius r with a constant linear velocity v, it will be subjected to two forces, one acting inwards and the other outwards. Acting inwards, will be the force of electrostatic attraction described by Coulomb's law:

$$F = \frac{q_1\,q_2}{4\pi\epsilon_0\,r^2}$$

where q_1 and q_2 are the positive and negative charges, each in this case having the value of charge e on the electron.

2.13.3 Limitations of the Rutherford Model

In the Rutherford model (Fig. 2.3) the electrons move in the coulomb field of the nucleus in orbits, like a planetary system. A particle moving on a curved trajectory is accelerating and an accelerating charged particle radiates electromagnetic waves and loses energy. Newton's law of motion and Maxwell's electromagnetic equations, if applied to the Rutherford atom, show that in a time of the order 10^{-10} seconds, all energy of the atom would be radiated away and the electrons would collapse into the nucleus.

This is clearly contrary to the laws of classical mechanics and it is concluded that Rutherford's nuclear atomic model is defective.

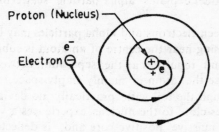

Fig. 2.3 *Spiralling of an electron into the nucleus as it radiates energy due to its acceleration*

2.13.4 Bohr Atomic Model

In 1913, Neil Bohr proposed a model of the hydrogen atom which could successfully explain the spectrum of the hydrogen atom and gave a model for all other atoms.

Bohr accepted the Rutherford model of the atom. He retained the small positively charged nucleus of the atom and proposed, in addition, that there were electrons moving in orbits around the nucleus. In the case of hydrogen, Bohr proposed that the nucleus consists of one proton with one electron revolving about it. He assumed that the electron was a particle which orbited the nucleus along a circular path.

The electrons are assumed to be in a definite planetary system (in circular orbits) of fixed energy. These stationary states are known as energy levels since there is a definite value of potential energy associated with each orbit. More than one energy level is considered to be possible for any electron of the atom.

The allowed energy levels can then be determined by the condition that the angular momentum of the electron moving in a circular orbit can take one of the values $L = nh/2\pi$ where n is a positive integer equal to 1, 2, 3, ..., and h is the Planck's constant.

An electron can either emit or absorb energy when making a transition from one possible orbit to another.

2.13.5 Postulates of Bohr's Theory

(i) In the case of the hydrogen atom, Bohr introduced a restrictive condition which is known as the first Bohr postulate. Bohr's first postulate is that only those orbits occur for which the angular momenta of the planetary electron are integral multiples of $h/2\pi$

(ii) Bohr's second break with classical Physics is contained in his second postulate which states that no electron radiates energy as long as it remains in one of the orbital energy states. Radiation occurs only when an electron goes from a higher energy state to a lower one. The energy of the quantum of radiation, hf, is equal to the energy difference of the states.

Let the quantum $n = n_2$ represent a higher energy state and $n = n_1$ a lower energy state. Then the second Bohr postulate can be written as

$$hf = E_{n_2} - E_{n_1}$$

$hf =$ The energy difference ejected from the atom in the form of light waves of energy called a *photon*. Here then, is the origin of light waves from the atom. Hence, light is not emitted by an electron when it is moving in one of its fixed orbits, but only when it jumps from one orbit to another.

(iii) To keep the electron in its orbit and prevent it from spiralling toward the nucleus or away from it to escape, Bohr next assumed that the centripetal force is equal to the inward electrostatic force. He assumed that the electron was a particle which orbited the nucleus along a circular path at a distance r from it (Fig. 2.4 (a)).

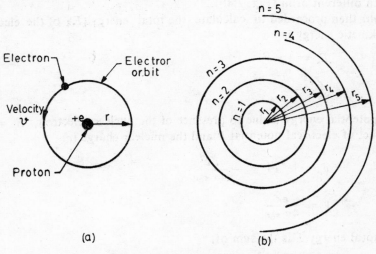

Fig. 2.4 *Bohr's model of the hydrogen atom with circular orbits*

2.13.6 Bohr's Theory of Hydrogen Atom

Starting with what should be the simplest of all atoms, Bohr assumed that a hydrogen atom ($Z = 1$) consists of a nucleus with one positive charge e and a single electron of charge $-e$ revolving around it in a circular orbit of radius r. Because it is 1836 times heavier than electron, the nucleus could be assumed at rest.

The centripetal force acting on the electron of mass m is given be mv^2/r where v is the velocity of the electron in the orbit. The electrostatic force of attraction between the proton and electron is F, where

$$F = \frac{Ze^2}{4\pi \epsilon_o r^2}$$

where $\dfrac{1}{4\pi \epsilon_0} = 9 \times 10^9$ MKS units

$Z = 1$ for hydrogen

According to Newton's third law of motion,

$$\frac{mv^2}{r} = \frac{Ze^2}{4\pi \epsilon_0 r^2} \tag{1}$$

Orbits with the angular momentum of the orbiting electrons equal to $nh/2\pi$ should be occurring on the assumption of the quantum hypothesis, i.e., the angular momentum of the electron should be an integral multiple of $h/2$. Thus, the product of the linear momentum mv and r is given by

$$mvr = \frac{nh}{2\pi} \tag{2}$$

Where $n = 1, 2, 3 \ldots \infty$ is an integer, each value of which is associated with a different orbit (Fig. 2.4b).

Bohr then proceeded to calculate the total energy (E) of the electron. The kinetic energy is

$$E_k = \tfrac{1}{2} mv^2$$

or $\qquad\qquad = \dfrac{Ze^2}{8\pi \epsilon_0 r} \tag{3}$

The potential energy is due to presence of the nuclear electron, $E_p =$ the product of electrical potential V and the nuclear charge (e)

$\therefore \qquad E_p = -eV = -\dfrac{Ze^2}{4\pi \epsilon_0 r} \tag{4}$

$\because \qquad V = \dfrac{eZ}{4\pi \epsilon_0 r}$

The total energy E is the sum of,

$$E = E_k + E_p = \frac{1}{2} mv^2 - \frac{Ze^2}{4\pi \epsilon_0 r} \tag{5}$$

With the help of equations (1) and (2), the radius r and velocity v can be eliminated to give the result

$$B = - \frac{Z^2 \, me^4}{8 \, h^2 \, \epsilon_0^2 \, n^2} \tag{6}$$

Where $n = 1, 2, 3$ for the energy orbits in which the electron can possibly have total energy En,

where $\quad En = \dfrac{E_p}{n^2} \tag{7}$

The potential energy is twice the kinetic energy.

These equations can be used to calculate the energy of the electron in the different orbits of the hydrogen atom

$\therefore \quad E_1 = $ Energy in the ground orbit ($n = 1$)

$$E_1 = -\frac{Z^2 \, me^4}{8\epsilon_0^2 h^2} \times \frac{1}{(1^2)} \tag{8}$$

If we substitute in this equation, the value of the constants e, m, h, ϵ_0
i.e., $\quad e = -1.60 \times 10^{-19}$ coulomb

$m = 9.10 \times 10^{-31}$ kg

$h = 6.62 \times 10^{-34}$ joules-s (Planck's constant)

$\epsilon_0 = 8.854 \times 10^{-12}$ Fm^{-1}

$E_1 = $ the energy in the ground orbit $= 21.76 \times 10^{-19}$ J

Similarly, in the next higher orbit,

$$E_2 = \frac{E_1}{(2)^2}$$

The same method can be applied to other orbits to get the energy values.

Now, the frequency of any circular orbit

$$f = \frac{v}{2r} = \epsilon_0 \frac{2\pi \, e^2}{nh/2\pi \, r} \tag{9}$$

The value of the frequency can also be determined by applying Bohr's second postulate,

$$hf = E_{n2} - E_{n1}$$

i.e., $\quad f = \dfrac{E_{n2} - E_{n1}}{h}$

$\therefore \quad f = \dfrac{me^4 \, z^2}{8\epsilon_0^2 \, h^3} \left(\dfrac{1}{n_1^2} - \dfrac{1}{n_2^2} \right) \tag{10}$

In terms of the wave number,

$$\frac{1}{\lambda} = \frac{bu}{c} = \frac{me^4 \, z^2}{8 \, \epsilon_0^2 \, h^3 \, c} \left(\frac{1}{n_1^2} - \frac{1}{n_2^2} \right)$$

$$= R \left(\frac{1}{n_1^2} - \frac{1}{n_2^2} \right) \tag{11}$$

Where $R = $ the Rydberg constant of the Bohr formula

$$R = \frac{me^4}{8 \, \epsilon_0^2 \, h^3 \, c} = 1.097 \times 10^7 \text{ m}^{-1}$$

$c = $ the speed of light in vacuum

$z = 1$ for hydrogen

2.13.7 Spectral Series of Hydrogen Atom

Light which is discontinuous in frequency distribution forms a discrete set of light images that are called spectral lines. These lines can be analyzed by the distribution of their frequency or colour. The obvious place to start the study of spectra in the context of our present topic is with the spectrum of hydrogen. It is not surprising that this lightest element has the simplest spectrum and probably the simplest structure. The hydrogen spectrum is shown in Fig. 2.5. The regularity of the spectrum lines shows some interrelationship. In 1884, Balmer set out a formula on the wave numbers of lines in the visible region of the hydrogen spectrum. The Balmer formula is

$$\lambda \text{ (Angstrom)} = \frac{3645.6 n^2}{n^2 - 4}$$

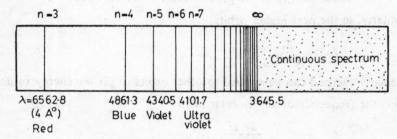

Fig. 2.5 *Balmer series of hydrogen atom*

Each different wavelength is obtained by substituting in the formula the different values of the running integers n, which are $n = 3, n = 4, n = 5$.

The success of Balmer led Rydberg to attempt a formulation which would apply to heavier elements. He proposed an equation of the form:

Wave number $\dfrac{1}{\lambda} = A - \dfrac{R}{(n + \alpha)^2}$

where n = running integer,

A and α are constants which depend on the element and part of the spectrum or spectral series to which the formula is applied. This could be used in many spectral series. The value of R is constant (1.09737×10^7 m^{-1}) when applied to different elements. It was further modified by Ritz to

$$\text{Wave number} = \frac{R}{(m + \beta)^2} - \frac{R}{(n + \alpha)^2}$$

Where α and β are adjustment constants which depend on the element. For different spectral series of a given element, m takes on different integral values. The different lines within a series are computed by changing the running integer $n = 3, 4, 5, 6$. It is easily shown that $\alpha = \beta = 0$ and $m = 2$ reduces to the Balmer formula for hydrogen. Electron jumping in this series is from any outer orbit to the next smaller orbit.

By substituting $m = 1$ and $n = 2, 3, 4$, etc., we obtain a series of spectrum lines in the ultraviolet region. These lines were first photographed and checked by Lyman. This series is now called the Lyman series. Bohr's orbital model of the hydrogen atom for various series is shown in Fig. 2.6.

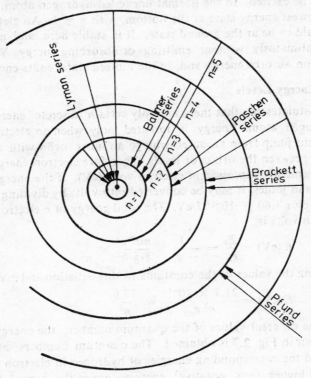

Fig. 2.6 *Spectral series of hydrogen atom*

On the orbital picture, the Lyman series of lines arises from electron jumps from any outer orbit directly to the innermost orbit, the normal state.

If $m = 3$ and $n = 4, 5, 6$, etc. the calculated frequencies predict spectrum lines in the infrared spectrum. These lines were first looked for and observed by F Paschen. The series is now known by his name. Another series of lines arising from electron jumps ending on orbit $m = 4$ was predicted and observed in the far infrared by Brackett. His series is called the Brackett series. In 1924, Pfund located another series in the same region.

2.13.8 Normal and Excited Atoms

When the single electron of a hydrogen atom is in the innermost orbit, $n = 1$, the atom is said to be in its normal state. If an electrical discharge is sent through a vessel containing hydrogen gas, cathode rays (electrons) moving at a high speed make frequent collisions with electrons, knocking some of them out of the atom completely and some into one of the outer allowed orbits $n = 2, 3, 4, 5, \ldots$

When the electron is completely removed from the atom, the atom is said to be ionized, whereas when it is forced into an outer orbit the atom is said to be excited. In the normal unexcited hydrogen atom, the electron is in its lowest energy state at the bottom, with $n = 1$. An eletcron in this state is said to be at the ground state. It is stable here and moves in this orbit continuously without emitting or absorbing energy. When excited the electron absorbs energy and, in the ionized state, emits energy.

2.13.9 Energy Levels

Bohr's postulates say that there are only certain discrete energy levels in the hydrogen atom. Energy is radiated only when an electron makes a spontaneous jump from an outer orbit to an inner orbit with a difference in energy between the orbits or levels. Therefore electron energy levels are most conveniently expressed in electron volts (eV). If the energy of n electrons E_n is in joules, it may be converted into volts by dividing by the conversion factor 1.60×10^{-19} J/eV. The total energy of n electrons in terms of electron volts is,

$$E(\text{eV}) = \frac{E_n}{e_c} = \frac{1}{-e_c} \times \frac{mz^2 e^4}{8\epsilon_0^2 h^2} \times \frac{1}{n^2}$$

Substituting the values of the constants in this equation and making $z = 1$,

$$E(\text{eV}) = \frac{21.7 \times 10^{-19}}{n^2 e_c} = \frac{13.6}{n^2}$$

Putting the different values of the quantum number, the energy level diagram shown in Fig. 2.7 is obtained. The quantum numbers are shown at the left and the corresponding energies of hydrogen in electron volts at the right. The higher (less negative) energies are at the top while the lower (more negative) are towards the bottom.

As stated above, the lowest energy state is at the bottom (ground state), when $n = 1$. The quantum number n is called the principal quantum number to distinguish it from others that we will discuss later in this chapter. In the commonly used notation, an electron in a level with $n = 1$ is said to be in the K shell. Correspondingly, if $n = 2$, $n = 3$, or $n = 4$, the electron is said to be in the L, M or N shells respectively.

The electron volt energies for the above shells are,

For ground state, $n = 1$ (K-shell) $E_1 = -13.6$ eV

$\qquad\qquad\qquad\quad n = 2$ (L-shell) $E_2 = E_1/2^2 = -3.4$ eV

$\qquad\qquad\qquad\quad n = 3$ (M-shell) $E_3 = E_1/3^2 = -1.51$ eV

$\qquad\qquad\qquad\quad n = 4$ (N-shell) $E_4 = E_1/4^2 = -0.86$ eV

$\qquad\qquad\qquad\quad n = 5$ (O-shell) $E_5 = E_1/5^2 = -0.54$ eV

$\qquad\qquad\qquad\quad n = \infty$ $E_\infty = 0$

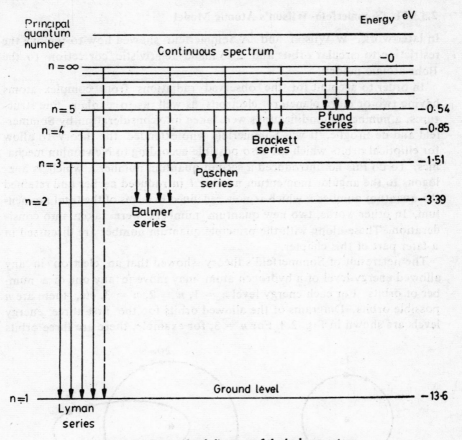

Fig. 2.7 *Energy level diagram of the hydrogen atom*

2.13.10 Limitations of Bohr's Model

Although the planetary model of the hydrogen atom is rather successful and the idea of quantized atomic energy is correct, the model is unsatisfactory in many respects, summarised below:

(i) The assumptions made that only circular orbits are allowed are inexplicable and arbitrary.
(ii) It cannot be generalised to deal with systems of two or more electrons.
(iii) No explanation is provided as to why certain spectral lines are more intense than others and there is no method to calculate the rate of transition between different energy levels.
(iv) It is unable to handle unbound systems.
(v) It deals with the planetary model which introduces only one quantum number (principle quantum number n).

2.13.11 Sommerfeld–Wilson's Atomic Model

In later work, W Wilson and A Sommerfeld showed how to remove the restriction to circular orbits and also made relativistic correction to the Bohr atomic model.

In order to account for the observed radiations from complex atoms having two or more planetary electrons as well as to explain fine structures, a number of modifications were taken into consideration by Sommerfeld and de Broglie. It was Sommerfeld who extended the theory to allow for elliptical orbits which are also possible according to Newtoniun mechanics. To do this he introduced a second quantum number l, which is analogous to the angular momentum number l introduced earlier and retained the quantum number n which is now redefined in terms of the total momentum. In other words, two new quantum numbers were taken into consideration. These along with the principle quantum number are discussed in a later part of this chapter.

The net result of Sommerfeld's theory showed that an electron in any allowed energy level of a hydrogen atom may move in any one of a number of orbits. For each energy level $n = 1$, $n = 2$, $n = 3$, etc, there are n possible orbits. Diagrams of the allowed orbits for the first three energy levels are shown in Fig. 2.8. For $n = 3$, for example, there are three orbits

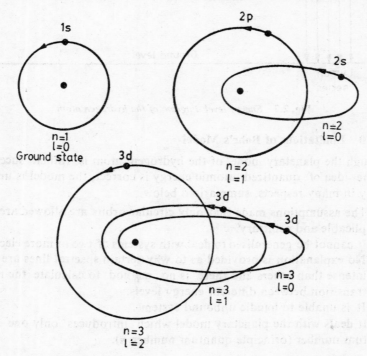

Fig. 2.8 *Electron orbitals for the hydrogen atom according to the Bohr-Sommerfeld theory*

with designations $l = 2$, $l = 1$ and $l = 0$. The diameter of the circular orbits is given by Bohr's theory and is just equal to the major axes of the two elliptical orbits. The minor axes are 2/3 and 1/3 of the major axis.

It is common practice to designate letters to the l-values, as follows:

$l = 0$	$l = 1$	$l = 2$	$l = 3$	$l = 4$
s	p	d	f	g

According to this system, the circular orbit with $n = 3$ and $l = 2$ is designated $3d$, while the elliptical orbit $n = 2$, $l = 0$ is designated $2s$, etc. n is the principal quantum number and l is the orbital quantum number. All orbits with the same value of n have the same total energy, the energy given by Bohr's equation of circular orbits. Each of the allowed orbits of the Bohr-Sommerfeld model of the hydrogen atom becomes a subshell into which electrons are added to build up the elements of the periodic table. The maximum number of electrons allowed in any one subshell is given by the relation $2(2l + 1)$.

2.14 VECTOR MODEL

The atomic model of today is the vector model. It is based on quantum mechanics and is also known as the quantum model. In this model all the principal quantised terms are represented by vectors. Attempts to calculate the fine structure separations of various energy levels arising in any given atom have been made by many investigators. Sommerfeld's atomic model does not explain this phenomenon and the effects of magnetic and electric fields on spectral lines. The vector model, with new explanation on electron movements, was introduced to overcome the deficiencies of Sommerfeld's atomic model. It takes into account electron spin while retaining the Sommerfeld feature of planetary movement of electrons in different orbits (circular or elliptical) and electron movement in different planes.

Using Bohr's atomic model of the hydrogen atom, it was observed that the angular momentum of the rotating electron was an integer of $h/2\pi$. However, after taking into consideration phenomena like the Zeeman and Stark effects, it was noted that there was spinning of electrons due to orbital angular momentum and magnetic momentum under the influence of magnetic and electric fields. Thus, many superfine structures were created which caused the electrons to move in different planes (horizontal or vertical) in order to give spatial motion. These energy levels are different in different planes. With the introduction of the vector model, two new quantum numbers, the magnetic spin quantum number (m_s) and the magnetic orbital quantum number (m_l), along with the principal quantum number (n) and the orbital or azimuthal quantum number (l), can be used to specify the state of electrons.

2.15 QUANTUM NUMBERS

We discussed the spectrum of hydrogen in terms of one quantum number n called the principal quantum number. Measurements of extreme precision on hydrogen and of less precision on other elements show that the energy levels have a fine structure. If the substance under study is placed in a magnetic or an electric field, the energy levels are further sub-divided, i.e., by the Zeeman and Stark effects. Additional quantum numbers are therefore necessary to specify in detail the energy levels of atomic electrons. It is found that four quantum numbers are required.

According to Bohr's theory, the angular momentum of an orbital electron is an integral multiple of $h/2\pi$. Although this integer n is associated only with angular momentum in the case of circular orbits, this association must be modified when elliptical orbits are considered, as the circular path is one extreme of an elliptical orbit and a straight line another. A linear orbit can be considered to have an energy associated with it although it has no angular momentum, while circular orbits have the maximum angular momentum.

Sommerfeld retained the principal or total quantum number, and another quantum number to characterize the angular momentum was introduced.

2.15.1 Orbital Quantum Number

The orbital quautum number l specifies the number of units of angular momentum $(h/2\pi)$ associated with an electron in a given orbit. This quantity can be represented vectorally by a straight line parallel to the axis of rotation whose positive sense is given by the direction of advance of a right hand screw rotated with the motion. It was found that the integer l could take on only positive values from zero to $n-1$. Thus, the electron in the smallest orbit $n = 1$ will have no angular momentum as l must equal zero. For a larger orbit with $n = 3$ there are three possible orbital shapes or eccentricities. First a straight line through the nucleus without angular momentum for $l = 0$, second, an ellipse with an angular momentum for $l = 1$, and third, a rounder ellipse with more angular momentum for $l = 2$.

By the nomenclature used in spectroscopy, the subshells in the main shell are designated by s, p, d, f, g and h with quantum number $l = 0, 1, 2, 3, 4$ and 5, respectively. Like hydrogen, many elements display a series of spectral lines. These series were given the descriptive names sharp (s), principal (p), diffuse (d) and fundamental (f) and sub-shells were designated alphabetically in continuing states.

This notation is extended by writing the value of n for a state before the letter and, if an atom has more than one electron with particular values of n and l, this number is written as a superscript of the letter. Thus, if an atom has six electrons for which $n = 3$ and $l = 2$ the atom is said to have $3p^6$ electrons.

The orbital quantum number (l) is also known as the azimuthal quantum number.

2.15.2 Magnetic Orbital Quantum Number

The third quantum number called the magnetic orbital quantum number, m or m_l, determines the possible quantised space orientation of the electron's elliptical orbit without any effect on energy levels.

When the atom is placed in an external magnetic field, each electronic orbit will be subjected to torque which tends to make the l-vector parallel to the field. Because of the righting torque of the field on the revolving system, the vector l will precess about the field, with the result that the spin and angular momentum vectors take up definite positions in space. These extra subshells with possible integral values will introduce another quantum number called the magnetic quantum number m or m_l.

2.15.3 The Magnetic Spin Quantum Number (m_s)

The projection of angular momentum along the direction of the magnetic field gives a new quantum number called the magnetic spin or electron spin quantum number (m_s).

The spin of the electron about its own axis as it revolves about the nucleus is analogous to the rotation of the earth as it moves along its orbit around the sun. The spin momentum has a numerical value $1/2$ ($h/2\pi$). The rotating electron has a magnetic moment and the spin vector is capable of orientation in only two ways, parallel or unparallel to the surrounding magnetic field. Therefore, the spin quantum number (m_s) has only two values, $+ 1/2$ and $- 1/2$. In the absence of a magnetic field there is no unique assignment to the direction of m_l or m_s.

With the help of quantum numbers, the state of an electron is clearly understood in terms of its position in the principal energy level or shell (n), the sub-level or subshell (l) or sub-subshell (m_l) and the direction of the spin located by m_s.

2.15.4 The Principal Quantum Number

The principal quantum number of any electron in an atom determines the main energy level or shell to which an electron belongs. It has integral values $1, 2, 3, 4, \ldots \infty$, and is the quantum number used in the Bohr and Sommerfeld atomic model. All the electrons that have the same value of this number are at nearly the same distance from the nucleus and have the same energy states. These electrons occupy the same atomic shell/level. Each of these shells is assigned a letter according to the scheme:

n	:	1	2	3	4	5	...
letter	:	K	L	M	N	O	...

With increase in distance of the shell from the nucleus, the energy in the level or shell increases. Therefore, different shells possess different values of energy.

2.16 PAULI'S EXCLUSION PRINCIPLE

Wave mechanics is not useful in assigning quantum numbers to each electron in the atom. It is found that each electron has its own set of quantum numbers which is different from that of every other electron in the atom. Pauli's Exclusion Principle assigns quantum numbers to electrons. It states that *no two electrons in an atom can exist in the same state*. The state of any electron can be completely specified by a group of four quantum numbers, n, l, m_l, m_s, or n, l, j, m_j. According to this principle, the group of values assigned to the four quantum numbers must be different for different electrons.

The origin of this principle may be partly understood from the following argument. The three quantum numbers n, l, and m_l, completely determine the shape of the wave function and hence also the charge distribution. If two electrons have the same value for these quantum numbers their charge distributions are directly superimposed on one another or, in other words, the electrons are in the same place. Now, this is a very unfavourable situation since the two electrons must repel one another very strongly. In practice, it occurs only if the electron spins are opposed; their quantum numbers m_s are then different. It seems that two electrons with counter rotating spins do not repel one another very strongly.

On the other hand, if the spins are parallel, the repulsion is so strong that they cannot occupy the same position and, therefore, must have a different set of values for n, l, and m_l.

Electrons with the same value of principal quantum number n form a definite group, shell or energy level. They are further subdivided according to the value of the orbital angular moment l. Differences in l values for the same value of n denote comparatively smaller energy differences than equal values of l and different values of n. Therefore, electrons that possess the same value of l for a given n are said to be in the same sub-shell or sub-group.

A combination of the Pauli's Exclusion Principle with the restrictions on the values of l, m and m_s leads to the following conclusions:

(i) The maximum numbers of electrons that can be contained in states characterized by principal quantum numbers 1, 2, 3, ..., n are equal to 2, 8, 18, ..., $2n^2$ respectively.

(ii) For a given value of n, the maximum number of electrons that can be contained in s, p, d and f states are 2, 6, 10 and 14, respectively, i.e. $2(2l + 1)$

The Pauli Exclusion Principle is of importance because, by placing a limit on the number of electrons in a given state, it leads to the concept of completed groups and sub-groups of electrons, and this explains a periodic repetition.

The Pauli's Exclusion Principle applies to molecules and other assemblies of atoms in which electrons occupy quantized energy states. In all such cases no states can contain more than one electron as defined by all the quantum numbers including the spin quantum number.

2.17 WAVE MACHANICS

In the early 19th century, it was experimentally observed that electrons and light, which were previously considered to be particle in nature, behave like waves. The fundamental principle of wave mechanics, first purposed by de Broglie, states that a moving electron or particle has associated with it a wave motion. The wave motion has a definite wavelength (λ) given by $\lambda = h/mv = h/p$, where h is the Planck's constant, m the mass v the velocity and p the momentum of the particle.

Since the mechanics of the particle is described in terms of a wave, we sometimes call this new type of mechanics wave mechanics. In atomic phenomena, wave mechanics replaces the orbital theory in order to explain many details of spectra, nuclear properties, electron diffraction and reflection, etc.

The method of solving problems on an atomic scale through the wave aspects of particles was achieved by Schrodinger with his wave equation which, in time-dependent form, is

$$\frac{h^2}{8\pi^2 m}\left(\frac{d^2\psi}{dx^2} + \frac{d^2\psi}{dy^2} + \frac{d^2\psi}{dz^2}\right) = \frac{h}{2\pi i}\frac{d\psi}{dt} + V\psi$$

where ψ is the wave function or the probability amplitude with its coordinates in space and time x, y, z and t and V is the potential energy of the particle. The use of the schrodinger wave equation is illustrated in cases where the electron is subjected to forces which hold it in a certain region of space, as in an atom or a metal. This is discussed in detail in the chapter on *Free-Electron Theory*.

2.18 THE PERIODIC TABLE

Mandeleev, a Russian chemist, devised a way of tabulating the elements on the basis of the periodicity of their chemical properties. The table is now known as the periodic table of the elements. It is very useful in the study of elements, because those from the same group have similar chemical properties. They are arranged in ascending order of atomic number.

To construct the periodic table the electronic structure (electron-shell) concept is used, assuming that the energy levels are filled sequentially with rising values of quantum numbers n and l. In the vertical columns of the table are placed the elements with similar properties, to form a periodic group. The horizontal rows in the table indicate the periods. There are seven rows in the periodic table. (See. Appendix)

The elements with atomic numbers 2, 10, 18, etc., are the *rare gases* which are exceedingly inert as their electronic shells are completely filled.

The first element in the table is hydrogen with one electron which occupies one of the two $1s$ states (lowest energy state). The second element helium has two electrons which fill the first shell. Lithium has an atomic number of three. The first shell is filled with two electrons and the third electron goes into the lowest subshell of the $2s$ level. A new row is begun whenever a p-shell has just been filled with electrons. For example, the rare gases like neon with its $2p$ shell or argon with its $3p$ shell finishes each row in the table.

We observe similarities between the elements in the first column, i.e., lithium, sodium, potassium, etc. each with one electron outside the filled p-shells. This is known as the first group or first column of the periodic table. These elements have the same chemical properties—all are very reactive or alkali metals. The second column or group consists of beryllium, magnesium, calcium, etc. with valencies of two. They form the *alkaline earths*. The first and second groups form light metals in periodic table.

We will now discuss that portion of the periodic table where the rule concerning the order of filling of the various shells or levels has been broken and, instead of the outermost electrons of elements like potassium and calcium entering $3d$ shells go into $4s$ shells, thus leaving an incomplete $3d$ shell. Elements which have incomplete d shells are known as *transition elements*. They occupy three horizontal rows, each forming first, second and third transition series in the fourth, fifth and sixth rows of the table.

The transition elements occupy the central position in the periodic table. Most engineering metals come from this portion of the periodic table. Transition elements have higher melting points and higher densities as compared to light metals.

The fifth and sixth rows of the periodic table form another transition series known as the *rare earth metals* or the *Lanthanide series*. The $5s$ states are occupied before the $6p$ shell begins to fill. Like rare earth metals, another group of elements occupy the seventh row of the periodic table. This forms the last series of transition elements. These elemdnts do not occur in nature. They are prepared artificially and are called *Actinide elements*.

REVIEW QUESTIONS

1. Define the following:
   (a) Atomic structure   (AMIE Summer 1983)
   (b) Electron
   (c) Protons
   (d) Nucleus
   (e) Atomic number   (AMIE Summer 1982)
   (f) Molecule   (AMIE Summer 1982)

2. (a) Explain the simplified concept of the atom and the significance of the same in relation to properties of bulk materials.
   (b) Explain the difference between atomic weight and atomic number and their importance in the periodic table.   (AMIE Winter 1982)
3. Write short note on the electron theory of metals.   (AMIE Summer 1984)
4. Describe the atomic models briefly.
5. (a) Specify the different quantum numbers describing the state of an electron in an atom. State Pauli's Exclusion Principle.
   (b) Explain how atomic shells and sub-shells are formed.   (AMIE 1979)
6. Write short notes on:
   (a) Non-acceptance of Thomson's model for an atom
   (b) Pauli's Exclusion Principle   (AMIE 1980)
7. Write short notes on:
   (a) Thomson's atom model
   (b) Rutherford's nuclear atom model
   (c) The spectrum of hydrogen   (AMIE 1975)
   (d) Sommerfeld-Wilson's atomic model
8. Justify the Mosley law on the basis of Bohr's theory of hydrogen atom.
    (AMIE 1976)
9. Write briefly on the modern concept of the structure of the atom.
10. Explain how the experiment on α-particle scattering led to the concept of the nuclear model of atom.   (AMIE 1974)
11. State Bohr's postulates.   (AMIE 1974 and 1976)

3
CRYSTAL GEOMETRY AND STRUCTURE

3.1 INTRODUCTION

Solids can be either crystalline or non-crystalline (amorphous). A crystal is a solid whose constituent molecules or atoms are arranged in a systematic pattern. Crystalline solids are usually built up of a number of crystals which may be similar or of widely varying sizes and metallic or non-metallic. In materials (such as glass) which are non-crystalline, called amorphous, the internal structure is not based on a regular repetition pattern (Fig. 3.1). In crystallography the structure implies the arrangement and disposition of atoms within a crystal. It is the basis of understanding the properties of materials.

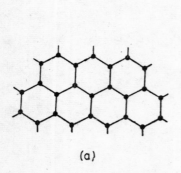 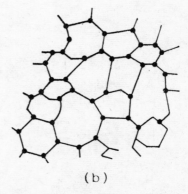

(a) *Crystalline structure (highly ordered state)*   (b) *Amourphous structure (disordered state)*

Fig. 3.1 *Structure of a solid*

3.2 CRYSTAL STRUCTURE

When a metal freezes from a state of fusion it crystallises. During solidification, the atoms of the liquid metal arrange themselves in a systematic pattern. The repeating three dimensional pattern in crystals is due to atomic co-ordination within the material. In addition, the pattern sometimes controls the external shape of the crystal. The planar surfaces of gems and quartz and the six-pointed outline of snow flakes are probably the most familiar examples of this.

The internal atomic structure of metals can be studied by the X-ray diffraction method. A crystalline solid can be either a single crystal, where the solid consists of only one crystal, or an aggregate of many crystals separated by well-defined boundaries. In the latter form, the solid is said to be polycrystalline. Metals, some ceramics and many ionic and covalent solids are found in this form. In Fig. 3.2 a polycrystalline solid is shown. The boundary between crystallites is called a grain boundary. Grain boundaries can be shown up by etching a metal with a suitable acid. The metal dissolves more readily at the boundary.

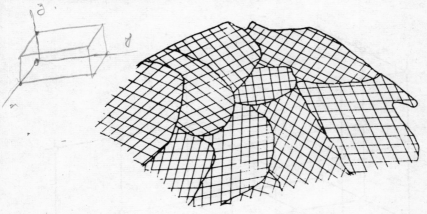

Fig. 3.2 *A polycrystalline solid composed of many grains oriented at random and separated by grain boundaries*

3.3 SPACE LATTICE

A space lattice is, by definition, an infinite array of points in three dimensions in which every point has surroundings identical to those of every other point in the array. If the centres of the atoms are considered to be connected together by straight lines, then a system will be obtained comprising a great number of equal parallelopipeds. This system is known as a space or crystal lattice. The space lattice of a crystal is described by means of a three-dimensional coordinate system in which the co-ordinate axes coincide with any three edges of the crystal that intersect at one point and do not lie in a single plane. It is very useful as a reference in corelating the symmetry of actual crystals

The complex lattice most frequently encountered in metals may be considered to comprise of several primitive translation lattices displaced in

30 *Materials Science*

relation to each other. Crystals of most metals have highly symmertrical structures with close-packed atoms. The most common types of space lattices are (BCC) body-centred cubic lattices, (FCC) face centred cubic lattices and close-packed hexagonal space lattices. See Fig. 3.3 (a) for a simple cubic space lattice.

3.4 UNIT CELL

The unit cell is the smallest component of the space lattice. Space lattices of various substances differ in the size and shape of their unit cells. The distance from one atom to another atom measured along one of the axes is called the space constant. In the cubical cell, the lattice constant has the same value in all the three dimensions. To describe one unit cell is to describe the whole crystal.

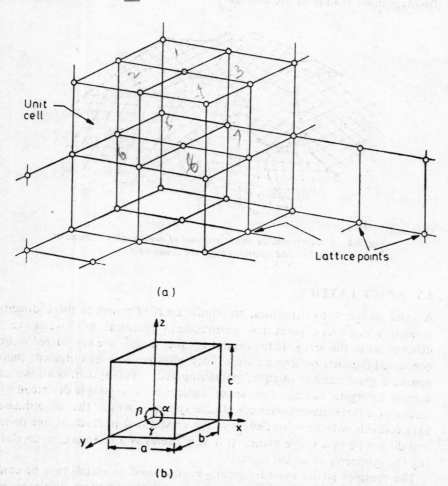

Fig. 3.3 *(a) Simple cubic space lattice (b) Unit cell*

Figure 3.3 (b) shows the unit cell in the shape of a parallelopiped having equality or inequality of lengths of the unit cell edges (a, b, c) according to whether or not the angles (α, β, γ) between the edges are or are not right angles. The lengths a, b, c and angles α, β, γ are called the lattice parameters. Knowing the lattice parameters, the form and actual size of the unit cell can be known.

3.5 CRYSTAL SYSTEMS

By pure symmetry considerations, there are only fourteen independent ways of arranging points in three-dimensional space, such that each arrangement conforms to the definition of a space lattice. These 14 space lattices are called Bravais lattices after their orginator (Fig. 3.4). Each space lat-

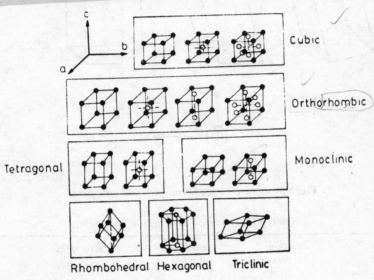

Fig. 3.4 *The fourteen Bravais lattices*

tice can be defined by reference to a unit cell which, when repeated in space an infinite number of times, will generate the entire space lattice. To describe basic crystal structures, seven different co-ordinate systems of reference axes are required, as shown in Fig. 3.5. The characteristics of the seven systems are summarized in Table 3.1.

3.6 ATOMIC PACKING

The complex crystal lattice most frequently encountered in metals may be considered to comprise of several primitive or basic translation lattices displaced in relation to each other. Crystals of most metals have highly symmetrical structures with close-packed atoms.

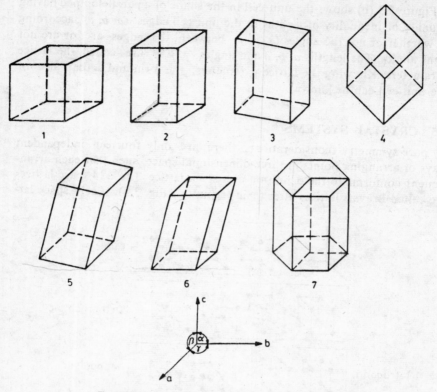

Fig. 3.5 *Basic crystal systems with reference axes*

Table 3.1 *The seven basic crystal systems*

Crystal System	Relation between Primitives	Interface Angles	Examples
Cubic	$a = b = c$	$\alpha = \beta = \gamma = 90°$	CaF_2, NaCl, Au
Monoclinic	$a \neq b \neq c$	$\alpha = \beta = 90° \neq \gamma$	$FeSO_4$, $CaSO_4$, $2H_2O$, $NaSO_4$
Triclinic	$a \neq b \neq c$	$\alpha \neq \beta \neq \gamma \neq 90°$	$CuSO_4$, $K_2Cr_2O_7$
Tetragonal	$a = b \neq c$	$\alpha = \beta = \gamma = 90°$	SnO_2, $NiSO_4$, Sn, TiO_3
Orthogonal	$a \neq b \neq c$	$\alpha = \beta = \gamma = 90°$	$BaSO_4$, $MgSO_4$, KNO_3
Rhombohedral	$a = b = c$	$\alpha = \beta = \gamma \neq 90°$	$CaSO_1$, SiO_2, $CaCO_3$
Hexagonal	$a = b \neq c$	$\alpha = \beta = 90°$ $\gamma = 120°$	SiO_2, AgCl, Zn, Graphite

The role of interatomic bonding forces in the structure of materials will now be considered. Atoms in a solid state may be regarded, for most purposes, as rigid spheres with all the atoms of one element being identical in size. By packing identical spheres together as closely as possible, a number of structures can be built up. As shown in Figs 3.6 (a) and (b), a single layer of spheres pushed together arrange themselves in a hexagonal close-packed network with the six neighbours. This layer offers recesses into which additional spheres can rest to form a second layer. This layer will occupy all the X-sites or all the Y-sites. A third layer is again presented with alternatives either over the first layer or over the unoccupied Y-sites. If the spheres in the third layer are positioned vertically above those in the first layer, a hexagonal lattice is formed which is known as the hexagonal close-packed structure (HCP) with a stacking arrangement $ABABAB\ldots$ If the third layer of spheres is placed in the Y-sites, the fourth layer will be directly over the first layer. Thus the stacking arrangement is now $ABC\ ABC\ ABC\ldots$ etc., and a face-centred cubic structure is built up.

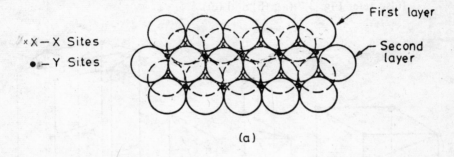

(a)

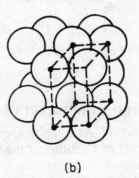

(b)

Fig. 3.6 *Atomic packing (a) Close packing of identical spheres (b) Close-packed hexagonal structure*

Although these are the only two ways in which identical spheres can be closely packed together, a very important structure where the packing is slightly loose is the body-centred cubic lattice structure.

3.7 CO-ORDINATION NUMBER

This is defined as the number of nearest atoms directly surrounding a given atom. Its value is six for simple cubic, eight for BCC and twelve for FCC structures.

3.8 CRYSTAL STRUCTURES FOR METALLIC ELEMENTS

The most common types of space lattice or unit cells, with which metallic elements crystallise, are

(i) Body-centred cubic structure (BCC)
(ii) Face-centred cubic structure (FCC)
(iii) Hexagonal close-packed structure (HCP)

(also refer Fig. 3.7 (a), (b) and (c))

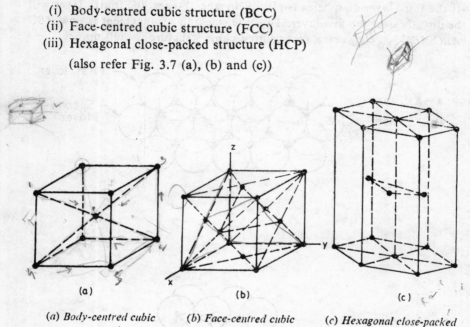

(a) Body-centred cubic   (b) Face-centred cubic   (c) Hexagonal close-packed

Fig. 3.7 Common unit cells

3.9 BODY-CENTRED CUBIC STRUCTURE (BCC)

Iron has a cubic structure. At room temperature the unit cell of iron has an atom at each corner and another at the body centre of the cube. Such crystals have been named body-centred cubic. Each iron atom in a BCC structure is surrounded by eight adjacent iron atoms, whether located at a corner or at the centre of the unit cell. Therefore the geometric environment of each atom is the same. The lattice constant is equal to 9 and the atomic packing factor is 0.68. Other metals possessing this structure are Mo, V, Mn, Ta, α—Cr, and Nb.

The unit cell of a cubic cell contains eight atoms at the corners which are shared by the adjoining eight cubes. Hence the share of each cube = 1/8th of each corner atom.

∴ Total no. of atoms = 1/8 × 8 = 1 atom
BCC crystal has one atom at centre = 1 atom

∴ Total atoms in BCC = 2 atoms

3.10 FACE-CENTRED CUBIC STRUCTURE

In this type of lattice structure, atoms are located at the corners of the cube and at the centre of each face. This type is typical of the metals—Cu, Al, Pb, Ag, Au, α-Fe, Ca, β-Co, etc.

A metal with FCC structure has four times as many atoms as it has unit cells. This shows that the FCC structure is more densely packed than BCC structure. The packing factor is 0.74. No other structure possesses such a large number of close-packed planes and directions. For this reason, metals with an FCC structure can be deformed critically. The co-ordinate number in an FCC metal is 12 which accounts for the higher packing factor.

An FCC cube has one atom at each corner and, in addition, one atom at the intersection of the diagonals of each of the six faces of the cube. In this case there are 8 atoms, one at each corner of the cube plus 6 face-centred atoms at the 6 planes of the cube.

∴ Total atoms in FCC unit cell = 1/8 × 8 + 1/2 × 6
= 1 + 3 = 4 atoms

3.11 CLOSE-PACKED HEXAGONAL STRUCTURES

A lattice structure of this type has an atom at each corner of the hexagon, one atom each at the centres of the two hexagonal faces and one atom at the centre of the line connecting the perpendiculars in the three rhombuses which combine and form the hexagonal close-packed structure. The atomic packing factor for an HCP metal is found to equal 0.74. This is identical to the packing factor of an FCC metal because each has a co-ordinate number of 12. HCP is found in such metals as Be, Mg, Ca, Zn, Cd, Ti and others.

3.12 CRYSTAL SYMMETRY

The symmetry of a crystal indicates that in a lattice if the parts of an ideal crystal are inter-changed, the various directions are geometrically equivalent. The unit atomic groups also have similar properties just like the original crystal. Symmetry of a crystal results when it possesses an ordered arrangement of atoms, ions or molecules in the internal structure. A cubic cell is symmetrical about it diagonal. The symmetry of a crystal form is determined by the positions of similar faces, edges, diagonals, etc.

When crystals possess identical square (cubic) or triangular (octahedral) faces only, they are said to have simple symmetry, whereas crystals possessing a combination of square and triangular faces are called combination systems. In the latter, the sets of faces are combinations of two or more simple forms.

The fundamental elements in determining the symmetry are shown in Fig. 3.8 (a), (b) and (c). If we cut the crystal along a plane which divides it into two similar halves so that one half is the reflection of the other half, such a plane is called a *plane of symmetry*. In a cube there are three such planes parallel to the faces and six diagonal planes. Thus a cube has nine planes of symmetry.

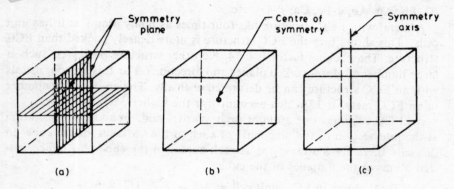

Fig. 3.8 *Crystal symmetry*

The next element of symmetry is the centre of symmetry. It is a point in a crystal which is equidistant from all the faces of the crystal. If a line is drawn from this, it represents a reflection through a point instead of a reflection in a plane. A crystal may have one or more axis and plane of symmetry but the centre of symmetry is only one.

The axis of symmetry is the imaginary line, passing through the centre of the crystal, about which the crystal may be rotated so that it presents an identical appearance more than once in the course of its rotation. If rotation around its axis presents the same appearance once, it is called *onefold symmetry*. The folds of symmetry can be determined by rotating the crystal through 360°, 180°, 90°, 60° and so on to get 1, 2, 3, 4, . . . fold symmetry. The limit for crystalline solids is six-fold symmetry.

3.13 ATOMIC RADIUS

Atomic radius is defined as half the distance between the nearest neighbours in the crystal structure of a pure element. It is expressed in terms of the cube edge element a and is denoted by r. It is possible to calculate the atomic radius by assuming that atoms are spheres in contact in a crystal. Atomic radius can be calculated in various crystal structures as shown in Fig. 3.9.

3.13.1 Simple Cubic Structure

In this structure, atoms touch each other along the lattice, as shown in Fig. 3.9 (a)

$$a = 2r$$

or $$r = a/2$$

3.13.2 BCC Structure

In this Fig. 3.9 (b) the atoms touch each other along the diagonal of the cube. Therefore the diagonal in this case is $4r$

Also, $$AC^2 = AB^2 + BC^2$$

But $$AB^2 = a^2 + a^2 = 2a^2$$

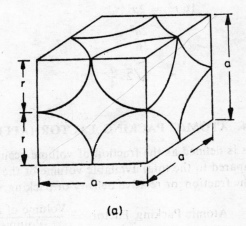

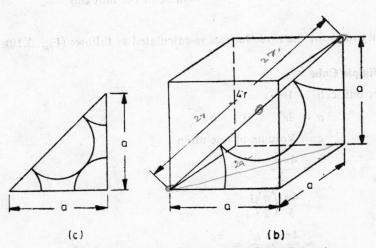

Fig. 3.9 *Atomic radii for three unit cells (a) Simple cubic (b) Body-centred cubic (c) Face-centred cubic*

and
$$AC = 4r = \sqrt{2a^2 + a^2} = \sqrt{3a^2}$$
$$= a\sqrt{3}$$
or
$$r = \sqrt{\frac{3}{4}}\, a$$

3.13.3 FCC Structure

As shown in Fig. 3.9 (c), atoms touch along the diagonal of any face of the cube. The diagonal of the face has a length of $4r$.

$$AC^2 = AB^2 + BC^2$$
$$(4r)^2 = a^2 + a^2$$
$$16\, r^2 = 2a^2$$

or
$$r^2 = \frac{2a^2}{16}$$

$\therefore$
$$r = \sqrt{2}\,\frac{a}{4}$$

3.14 ATOMIC PACKING FACTOR (APF)

This is defined as the fraction of volume occupied by spherical atoms as compared to the total available volume of the structure. It is also known as the fraction or relative density of packing. It is given by the relation,

$$\text{Atomic Packing Factor} = \frac{\text{Volume of atoms in a unit cell}}{\text{Volume of the unit cell}}$$
$$= v/V$$

The density ratio for the three lattices is calculated as follows (Fig. 3.10).

3.14.1 Simple Cube

Atoms per unit cell $= 1$

$$a = 2r$$

$v =$ Volume of one atom

$$= \frac{4\pi\, r^3}{3}$$
$$= \frac{4\pi}{3}\left(\frac{a}{2}\right)^3$$
$$= \frac{4\pi}{3} \times \frac{a^3}{8} = \frac{\pi a^3}{6}$$

∴ Atomic packing factor

$$= \frac{v}{V} = \frac{a^3\,\pi}{6} \Big/ a^3$$

$$= \frac{\pi}{6} = 0.52$$

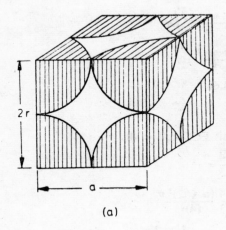
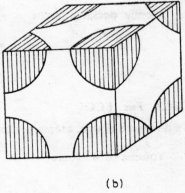
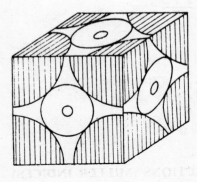

Fig. 3.10 *Atomic packing factor (a) Simple cubic (b) Body-centred cubic (c) Face-centred cubic*

3.14.2 BCC Crystal

Atoms per unit cell $= 2$

Volume $= 2 \times \dfrac{4}{3} \pi r^3$

$= 2 \times \dfrac{4}{3} \pi \left(\dfrac{a\sqrt{3}}{4}\right)^3$

$$\because \quad r = \frac{a\sqrt{3}}{4}$$

$$= 2 \times \frac{4}{3} \pi a^3 \times \frac{3\sqrt{3}}{64} = \frac{\pi a^3 \sqrt{3}}{8}$$

Volume of unit cell $= a^3$

$$\therefore \text{ Atomic packing factor} = \frac{\pi a^3 \sqrt{3}}{8} \Big/ a^3$$

$$= \frac{\pi \sqrt{3}}{8}$$

$$= 0.68$$

3.14.3 For FCC

As discussed above, atoms per unit FCC cell $= 4$

$$\therefore \text{ Volume of 4 atoms} = 4 \times \frac{4}{3} \pi r^3$$

$$= 4 \times \frac{4}{3} \frac{(a\sqrt{2})^3 \pi}{(4)^3}$$

$$= \frac{16}{3} \pi a^3 \times \frac{2\sqrt{2}}{64}$$

$$= \frac{\pi}{6} \sqrt{2} \, a^3$$

$\therefore$ Atomic packing factor

$$= \frac{\pi}{6} \sqrt{2} \, a^3/a^3$$

$$= \frac{\pi}{6} \times \sqrt{2} = 0.74$$

3.15 LATTICE PLANES AND DIRECTIONS (MILLER INDICES)

A complete description of the crystal structure can be noted from the atomic positions in a unit cell. In order to describe a plane, such as the cube faces in the FCC structure or the sheets of atoms in graphite, a shorthand notation has been devised comprising a set of three numbers called Miller indices which identify a given group of parallel planes.

Since the primary interest in lattice planes is confined to the major planes in the unit cell of a crystal, they are defined in terms of their intercepts on the three translation vectors (Fig. 3.11). Any plane $A'B'C'$ can be defined by its intercepts OA', OB', OC' on the three principal axes of the unit cell OA, OB and OC.

The usual notation is given by taking the reciprocals of the ratios of these intercepts to the unit cell parameters. Thus the plane $A' B' C'$ is denoted by $OA/OA'\ OB/OB'\ OC/OC'$ which is generally written as (hkl). If $OA' = a/2$, $OB' = b/4$ and $OC' = c/2$, then the plane $A'B'C'$ is the $(1/\tfrac{1}{2}\ 1/\tfrac{1}{4}\ 1\tfrac{1}{2})$ or the (242) plane.

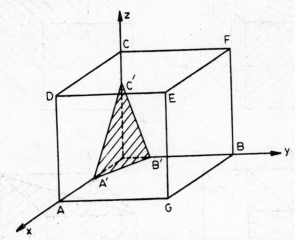

Fig. 3.11 *Designation of a crystal plane by Miller indices*

Similarly, a plane ABC with intercepts OA, OB and OC is the $(OA/OA\ OB/OB,\ OC/OC)$ or (111) plane and the plane $DFBA$ is the OA/OA, OB/OB $OC/\infty)$ or (110) plane, i.e. a plane parallel to any axis is taken to intercept that axis at infinity.

If the intercept is negative then the corresponding Miller index is also negative and is denoted by placing a bar over the numerical value, e.g. $\bar{2}$. Figure 3.12 shows the planes in a cubic structure. The Miller indices of the sides of a unit cell of a cubic lattice are (100), ($\bar{1}$00), (010), (0$\bar{1}$0), (001) and (00$\bar{1}$). These are equivalent planes of the same form and are collectively represented by {100}, called families of planes.

The crystallographic or lattice direction can be defined as a line joining any two points of the lattice. The direction of a line in a lattice with respect to the unit cell vectors can be described using a similar notation. Miller indices of a direction are simply the vector components of the direction resolved along each of the co-ordinate axes, expressed as multiples of the unit cell parameters and reduced to their simplest form. They are denoted by [hkl] (note square brackets to distinguish it from the (hkl) plane).

Just like the principal planes of importance, the directions with which we shall be mainly concerned are [110], [100] and [111]. These are, respectively, a cube face diagonal, a cube edge and a body diagonal. Families of directions are labelled by special brackets as are families of planes. Thus ⟨100⟩ denotes the family of directions which includes [100], [010], [001], [$\bar{1}$00], [0$\bar{1}$0], and [00$\bar{1}$]. Figure 3.13 shows the Miller indices for directions.

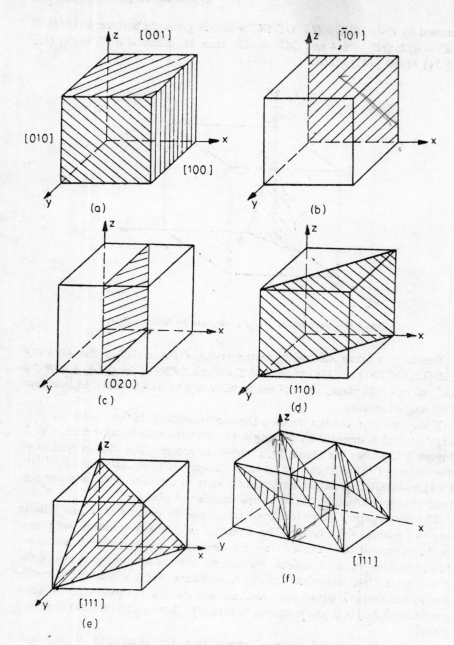

Fig. 3.12 *Lattice planes in the cubic system. Negative intercepts are drawn on negative co-ordinates*

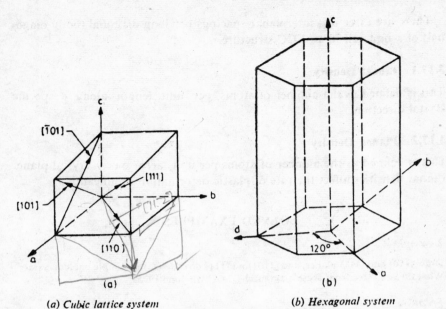

(a) *Cubic lattice system* (b) *Hexagonal system*
Fig. 3.13 *Miller indices for directions*

3.16 IMPORTANT FEATURES OF MILLER INDICES
A few important features of Miller indices are listed below:
  (i) Miller indices do not only define a particular plane but a set of parallel planes.
  (ii) A plane which is parallel to any one of the principal axes has an intercept of infinity (∞) and the Miller index is therefore zero.
  (iii) All equally spaced parallel planes with a particular orientation have the same index number [*hkl*].
  (iv) Only the ratio of indices is important.
  (v) The directions in space are represented by square brackets [ ] and the letters *xyz*, whereas Miller indices of planes are denoted by ( ) and the letters *hkl*.
  (vi) The common inside brackets are used separately and not combined. Thus (111) is read as one-one-one and not "one hundred eleven".
  (vii) Negative indices are represented by putting a bar over the digit $(0\bar{1}0)$.

3.17 INTERPLANAR SPACINGS
Interplanar spacings are the distances between planes and are represented by a number of parts of the body diagonal of a unit cell. In the cubic system, the interplanar spacings can be determined from the following relation:

$$d_{hkl} = \frac{a}{\sqrt{h^2 + k^2 + l^2}}$$

where *a* is the lattice constant and *h, k, l* are the indices of the planes.

There are three d_{111} interplanar spacings per long diagonal (body diagonal) of a unit cell in an FCC structure.

3.17.1 Linear Density

This is defined as the number of atoms per unit length along a specific crystal direction.

3.17.2 Planar Density

This is defined as the number of atoms per unit area on a crystal plane. Planar densities affect the rate of plastic deformation significantly.

SOLVED EXAMPLES

Example

Draw (110) and (111) planes, and [110] and [111] directions in a simple cubic crystal. What do you infer from these diagrams?   (AMIE June 1975)

Solution

The required planes are shown in Fig. 3.14.

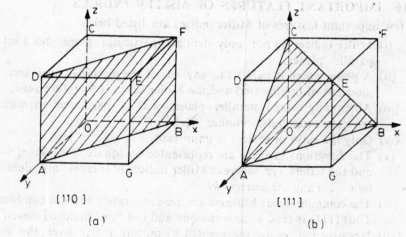

Fig. 3.14 *Miller indices*

The directions of [110] are denoted by OG and [111] by OE (The directions are described by giving the co-ordinates of the first whole-number point xyz through which they pass.

Example

Obtain the Miller indices of a plane which intercepts at a, $b/2$, $3c$ in a simple cubic unit cell.   (AMIE Summer 1978).

Solution

The given parameters are a, $b/2$, $3c$. Its numerical parameters are 1, ½, 3. Therefore the Miller indices are:

$$\left(\frac{1}{1}, \frac{1}{1/2}, \frac{1}{3}\right) \text{ (Reciprocals)}$$

or   (1, 2, 1/3)

or   (3 6 1) Ans

Example

Draw the (112) and (111) planes in a simple cubic cell.

or

Draw the different (111) planes in the unit cell of a simple cubic structure.

(AMIE Dec. 1977)

Solution

For the planes (112), intercepts on the principal axes are

$$\left(\frac{1}{1} \quad \frac{1}{1} \quad \frac{1}{2}\right) \text{ (Fig. 3.15 (a))}$$

Similarly, for the plane (111), the intercepts on the principal axes are

$$\left(\frac{1}{1} \quad \frac{1}{1} \quad \frac{1}{-1}\right)$$

The four (111) triangle planes (Fig. 3.14 (b)) are

ABC, ABE, BCE and ACE.

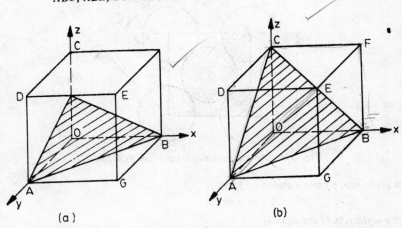

Fig. 3.15 *Miller indices*

Example

The ratio of the intercepts of an orthorhombic crystal are $a : b : c :: 0.429 : 1 :. 379$. What are the Miller indices of the faces with the following intercepts?

    0.214 : 1 : 0.188
    0.858 : 1 : 0.754
    0.429 : 00 : 0.126

Solution

The intercepts in unit axial dimensions are:

$\frac{1}{2} : 1 : \frac{1}{2}$

$2 : 1 : 2$

$1 : \infty : 1/3$

Their reciprocals are

$2 : 1 : 2$

$\frac{1}{2} : 1 : \frac{1}{2}$

$1 : 0 : 3$

Reducing them to the smallest whole number and enclosing them in brackets, we get the Miller indices

(212), (121), (103)

Example

How many atoms per square millimeter are there on the (100) plane of lead. Assume the interatomic distance to be 3.499 Å. (AMIE Winter 76)

Solution

As we know, the structure of lead is FCC (Fig. 3.16)

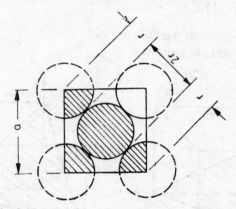

Fig. 3.16 *Atomic concentration in an FCC lattice.*

The given interatomic distance = 3.499 Å

∴ $2r = 3.499$ Å

If a is the side of the square,

$\sqrt{2a^2} = 4r$ (length of the diagonal)

∴ $a = \dfrac{4r}{\sqrt{2}} = \dfrac{2 \times 3.499}{\sqrt{2}} = 4.95$ Å

Area of (100) plane = $(4.95 \times 10^{-7})^2$ mm²

No. of atoms in the plane = 2

$$\text{Atoms/mm}^2 = \frac{2}{(4.95 \times 10^{-7})^2}$$

$$= 8.2 \times 10^{12} \text{ atoms/mm}^2$$

REVIEW QUESTIONS

1. (a) What do you understand by the term 'crystal lattice' and how many types of this are found in metals? What is a unit cell?
   (b) What characteristic of the atomic structure of metals accounts for their relatively high thermal and electric conductivity?   (AMIE, Summer 1984)
2. Describe with examples:
   (a) Unit cell
   (b) Atomic packing factor
   (c) Miller indices   (AMIE, Summer 1983)
3. Distinguish between amorphous and crystalline substances. (AMIE, Winter 1982)
4. Describe the crystal structures of metallic elements   (AMIE, Winter 1976, 77)
5. Write short notes on the lattice parameters of a unit cell. (AMIE, Winter 1977)
6. Describe briefly the following:
   (a) Atomic radius   (AMIE, Winter 1976)
   (b) Crystal symmetry   (AMIE, Winter 1976)
   (c) Co-ordination number   (AMIE, Winter 1976)
   (d) Important features of Miller indices   (AMIE, Summer 1975)
7. How are crystal planes identified by means of Miller indices?
8. What is a crystal. In what patterns do formations of crystals take place?
9. Explain the system of Miller indices for planes and directions of crystal lattices.
10. Write short notes on:
    (a) Bravis space lattice
    (b) Anistropy
    (c) Polytropy
    (d) BCC, FCC and HCP
11. Explain 'Miller' indices for denoting crystal planes. Draw the following planes and directions in an FCC structure:
    (i) [321], (ii) [102], (iii) [201], (iv) [111].   (AMIE, Summer 1986)
12. (a) What do you understand by crystallographic notation of atomic planes? Explain with the help of examples.   (AMIE, Winter 1986, 87)
    (b) Draw the following planes and directions in an FCC structure:
    (i) (010), (ii) (111), (iii) (011) (iv) (001)   (AMIE, Winter 1986)

4
BONDS IN SOLIDS

4.1 INTRODUCTION

The arrangement of atoms in a solid is mainly influenced by the nature of the chemical bond that holds them together. The atoms and molecules in a solid state are more closely packed than in the gaseous and liquid states and are held together by strong mutual forces of attraction and repulsion. Atomic arrangements in elements and compounds can be described on the basis of this.

Bonds hold atoms at different distances such that the inter-atomic forces just balance. The process of holding is known as bonding. The attractive forces between atoms are basically electrostatic in origin and the classification of different types of bonding depends on the electronic structure of the atoms concerned and are hence directly related to the periodic table. The type of bonding determines the electrical, chemical and physical properties of the material concerned.

4.2 TYPES OF BONDS

There are basically two groups which classify common bonds on the basis of strength, directionality of bonding forces, cohesive forces (chemical bonds) and the character of any solid material. There are primary bonds and secondary bonds.

Primary bonds are the strongest bonds between atoms by virtue of their interatomic nature. These bonds are also known as attractive bonds. The attractive forces are directly associated with the valence electrons in the respective orbitals. Primary bonds are further classified on the positions taken by bond electrons during the formation of the bond. The principal types are (Fig. 4.1):

    (i) Ionic or electrostatic bonds
    (ii) Covalent, atomic or homopolar bonds
    (iii) Metallic bonds

Secondary bonds are weaker than primary bonds. The attractive forces exist between atoms or molecules. These are also known as intermolecular bonds as they result from intermolecular or dipole attractions. Vander Waals bonds and hydrogen bonds are common examples of secondary bonds.

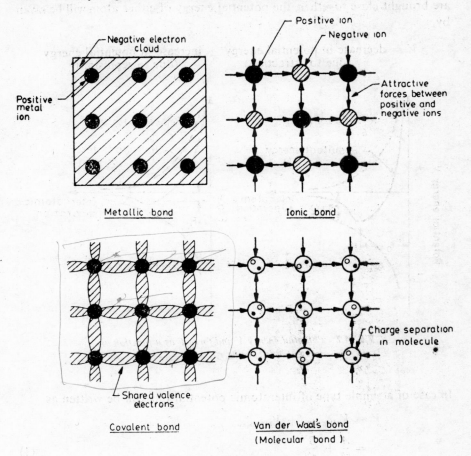

Fig. 4.1 *Types of bonding*

Few materials have pure bonds of one type or the other. Mixed bonds are observed in the common materials.

4.3 MECHANISM OF BOND FORMATION

The attraction between atoms brings them closer together until the individual electron clouds begin to overlap and a strong repulsive force arises to comply with Pauli's Exclusion Principle. When the attractive force and the repulsive force between any two atoms are equal, the two atoms should be in a stable situation with a minimum potential energy. The forces between two atoms or ions as a function of their distance of separation r are schematically shown in Fig. 4.2.

Consider two atoms in their ground states with infinite separation and, consequently, an interaction potential energy equal to zero. As the atoms are brought close together, the potential energy of either atom will be given by,

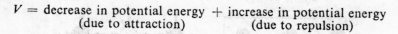

$$V = \text{decrease in potential energy} + \text{increase in potential energy}$$
$$\text{(due to attraction)} \qquad \text{(due to repulsion)}$$

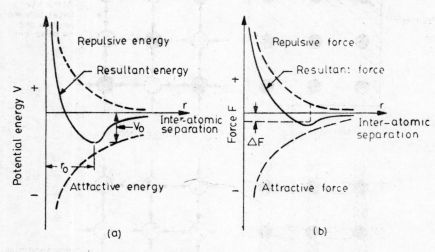

Fig. 4.2 *Potential energy V and force F as a function of inter-atomic spacing r*

In case of a simple type of interatomic potential this can be written as

$$V = V_{\text{attractive}} + V_{\text{repulsive}}$$
$$= \frac{A}{r^n} + \frac{B}{r^m} \qquad (1)$$

where A is the proportionality constant for attraction, and B for repulsion. Figure 4.2 (a) shows the potential energy curve. In order to give a minimum potential energy condition at the equilibrium separation r_0, m must be greater than n so that d^2v/dr^2 is positive. The resultant force of interaction F is given by

$$F = -dv/dr \qquad (2)$$
$$= \frac{nA}{r^{n+1}} + \frac{mB}{r^{m+1}} \qquad (3)$$

The net force is indicated by the solid curve in Fig. 4.2 (b).

In Fig. 4.2, the potential energy due to attraction is negative, as the atoms do the work of attraction. Similarly, the repulsion energy is positive, since external work must be done to bring the atoms together.

At the equilibrium separation the net force is zero, i.e., $nA/r_0^{n+1} = mB/r_0^{m+1}$. When the atoms are separated by only a few atomic diameters, the repulsive forces between like charges start to assert themselves. At the equilibrium separation, the forces of attraction is just equal to the forces of repulsion and the potential energy is at a minimum (V_0). This is the bonding energy of the system and is approximately equal to the heat of dissociation of the molecule.

The strength of a bond is best measured by the energy required to break it. It is the amount of heat which must be supplied to vaporize the solid, and hence separate the constituent atoms. The energy required for the formation of one mole of a substance from its ions or atoms is called the bond energy. The melting points of the elements are dependent on the strength of the bond, the stronger the bond the higher are the melting points.

4.4 IONIC BOND

An ionic bond can only be formed between two different atoms, one electropositive and the other electronegative. These positive and negative ions, when brought into close proximity, result in an attractive force which forms the ionic bond. These ions, of course, are formed when the atoms involved, lose or gain electrons to stabilise their outer-shell electron configuration. Elements which tend to lose electrons in order to achieve this stable outer-shell configuration are called electropositive and those which gain electrons are called electronegative. Generally, metals are electropositive and non-metals electronegative due to their abilities to form positive and negative ions, respectively, under favourable conditions during chemical reaction. Let us consider the reaction which takes place between sodium (metal) and chlorine (non-metal) to form sodium chloride. Sodium has a single electron in its outer shell and this transfers to join the seven electrons in the outer shell of the chlorine atom (Fig. 4.3). This type of atomic interaction involving the outright transfer of an electron from one atom to another, leads to the formation of ions which are held together by electrostatic attraction. Because of the electrostatic nature of the binding force, the bond is said to be ionic or electrovalent.

Ionic bonds are formed mainly in inorganic compounds like sodium chloride (common salt NaCl), MgO, CuO, CrO_2 and MoF_2. In MgO the ions are doubly ionized leading to a stronger interatomic bond and hence a higher melting point, 2800 °C, compared with 800 °C for NaCl. Formation of ionic bonds, as seen in cupric oxide, chromous oxide and molybdenum fluoride, shows that the metallic element need not be from Group I or II but that any metal may get ionized by losing its valence electrons.

52 *Materials Science*

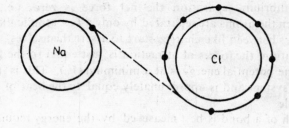

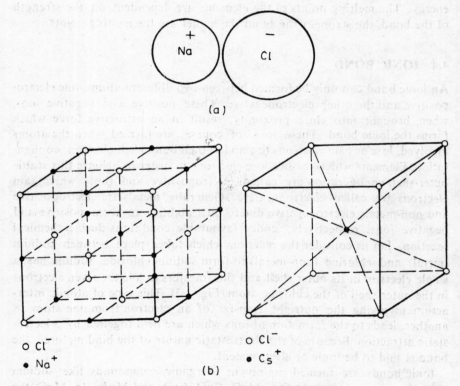

Fig. 4.3(a) *Ionic Bond* **(b)** *Ionic crystal structures*

4.4.1 Ionic Solid

Ionic solids contain atoms of different sizes which are held together by non-directional bonds. Such solids cannot form close-packed structures like FCC and HCP, because the atoms are not of the same size. The charge on the ions requires an alternating arrangement of anions and cations. The type of crystal structure is mainly determined by the packing of the larger anions which are normally the negative ions. The structure framework is made by anions within which the small cations (positive ions) fit interstitially. Figure 4.3 shows NaCl and CsCl structures.

Three common ionic crystal structures are the sodium chloride, caesium chloride and zinc blende structures. The main factor determining the structure adopted is the ratio of the ionic radii r_c/r_a, since like ions must not touch because of the large electrostatic repulsion.

The NaCl structure can be regarded as two interpenetrating FCC lattices. Each has six nearest neighbours of opposite charge, i.e., the co-ordination number is 6 (which is the number of nearest neighbours of an atom in a crystal). The ionic radius is the radius of an ion. It should be noted that ionic radii can be larger or smaller than the atomic radii depending on whether an anion gains or loses electron in the ionization process. There is only one value of the ratio of cation to anion radius at which this configuration can exist. The ionic radii ratio $r_c/r_a = 0.5$, where r_c is the ionic radius of a cation and r_a of an anion.

The CsCl structure is another ionic structure showing two interpenetrating simple cubic lattices, one for each kind of ion. It has a coordination number of 8 and r_c/r_a is about 0.92. In this arrangement the cation is surrounded by eight anions.

Zinc blende (ZnS) is also an ionic solid and can be shown as two interpenetrating FCC lattices with four nearest neighbours (co-ordination number $= 4$). The ionic radii ratio in this structure is 0.48.

In ionic solids, the force of attraction between ions is the coulomb's force $-e^2/4\pi \epsilon_0 r_0^2$, where r_0 is the equilibrium separation. Let us use this force to calculate the total energy of a single ion in the presence of others. In the NaCl structure, the central Na ion is surrounded by six nearest neighbours of opposite electric charge at a distance r_0, 12 next neighbours of like charge at a distance $\sqrt{2}\, r_0$, 8 second nearest neighbours of unlike charge at a distance $\sqrt{3}\, r_0$, etc.

The potential energy for a single molecule is given by the relation:

$$V_0 = -\left(\frac{e^2}{4\pi \epsilon_0 r_0^2}\right) + \frac{B}{r_0^m}$$

Now, the total energy of the central Na ion will be

$$V_0 = -\frac{6e^2}{4\pi \epsilon_0 r_0} + \frac{12e^2}{4\pi \epsilon_0 \sqrt{2}\, r_0} - \frac{8e^2}{4\pi \epsilon_0 \sqrt{3}\, r_0} + \cdots + \frac{C}{r_0^m}$$

Where C is another constant representing the net repulsion due to all the overlapping electron clouds.

The potential energy can also be written as

$$V_0 = \left(\frac{Me^2}{4\pi \epsilon_0 r_0}\right) + \frac{C}{r_0^m}$$

where M is a constant known as the Madlung constant, a geometric term. It is generally used to obtain the energy of a pair of ions in a crystal by multiplying the coulomb energy term for a pair of isolated ions by this constant.

The Madlung constant M for the NaCl structure is

$$\frac{6}{\pi} - \frac{12}{\sqrt{2}} + \frac{8}{\sqrt{3}} - \frac{6}{\sqrt{4}} + \frac{24}{\sqrt{5}} - \ldots$$

Its value is equal to 1.7475.

For the CsCl structure, $M = 1.7626$.

4.4.2 Characteristics of Ionic Compounds

(i) In ionic bonding, a metallic element loses from the outer electron shell of its atom a number of electrons equal to its numerical valency. Thus, an electrostatic attraction between positive and negative ions occurs.

(ii) Ionic compounds are generally crystalline in nature and rigid.

(iii) They have high melting and boiling points due to the strong electrostatic forces binding them.

(iv) Since each positive ion attracts all neighbouring negative ions and vice versa, the bond itself is non-directional.

(v) They are generally non-conductors of electricty.

(vi) They are insoluble in organic solvents but highly soluble in water.

4.5 COVALENT BOND

Elements from the central groups of the Periodic Table, notably Group IV, are not readily reduced to a closed-shell electronic configuration, because the energy required to remove all the valence electrons is too large and so ionic bonding is unlikely. However, it is still possible for each atom to effectively complete its outer electron shell by sharing electrons with its neighbours. This sharing of electrons gives rise to covalent bonding.

The covalent bond is formed by sharing of electrons between atoms rather than by transfer of electrons. Only a few solids are held together by covalent bonds. Covalent bonding alone is not sufficient to build three-dimensional solids. The majority of solids incorporating covalent bond are also bound by either ionic or Vander Waals bonds.

An excellent example of covalent bonds is seen in the chlorine molecule. The outer shell of each atom possesses seven electrons. Each chlorine atom would like to gain an electron and thus form a stable octet. This can be done by sharing of two electrons between pairs of chlorine atoms, thereby producing stable diatomic molecules. In other words, each atom contributes one electron for the sharing process. The nature of sharing of the covalent bonds in molecules of chlorine, hydrogen and hydrogen fluoride are illustrated in Fig. 4.4.

In another example, consider germanium which has four electrons in the outer shell. Four more electrons are required to fill this shell and this can be done by sharing between opposite spin pairs so there is no violation of Pauli's Exclusion Principle. Attractive forces arise from the interaction of these anti spin electrons. The electrostatic repulsion between the four electron clouds concentrates them in space as far as possible from each other, the four bonds arranging themselves pointing towards the corners of an imaginary tetrahedron.

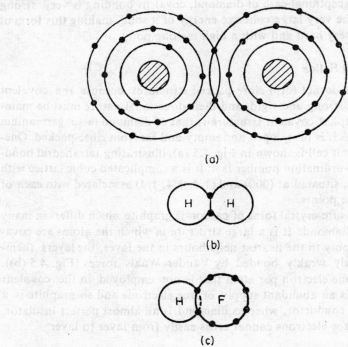

Fig. 4.4 *Nature of covalent bond* **(a)** *a molecule of chorine* **(b)** *a molecule of hydrogen* **(c)** *a molecule of hydrogen fluoride*

This directional nature of the bonds plays an important role in the resulting crystal structure, as will be seen later. This rearrangement of orbitals is called hybridization and four identical hybrid orbitals result. This is an important feature of the molecular orbital theory. The molecular orbitals of methane, diamond and graphite, carbon compounds, illustrates hybridization and directional bonding.

4.5.1 Characteristics of Covalent Compounds

(i) Covalent compounds are directional in nature.
(ii) They can exist in all states of matter.
(iii) They are generally electric insulators.

(iv) They are homopolar, i.e., the valence electrons are bound to individual or pairs of atoms and electrons cannot move freely through the material as in the case of metallic bonds.
(v) They are insoluble in water, but soluble in non-polar solvents such as benzene, alcohol, chloroform and paraffins.
(vi) They are soft, rubbery elastomers, and form a variety of structural materials commonly known as plastics. Their melting and boiling points are low.
(vii) In the exceptional case of diamond, covalent bonding is very strong due to the very large cohesive energy of a solid, making this form of carbon very hard and with a high melting point.

4.5.2 Covalent Solids

Covalent solids do not form close-packed structures because the covalent bonds are very strong and rigid, and their directional nature must be maintained. The simplest covalent structure is that of diamond (also, germanium and silicon) which is fairly open and empty and far from close-packed. One-eighth of the unit cell is shown in Fig. 4.5 (a), illustrating tetrahedral bonding, i.e., the co-ordination number is 4. It is a complicated cubic lattice with two atoms, e.g., situated at (000) and (1/4, 1/4, 1/4) associated with each of the cubic lattice points.

The low pressure crystal form of carbon is graphite which differs in many respects from diamond. It is a large structure in which the atoms are covalently bonded only to the nearest neighbours in the layer, the layers themselves being only weakly bonded by Vander Waals forces (Fig. 4.5 (b)). Since there is one electron per atom that is not employed in the covalent bonding there is an abundant supply of free electrons and so graphite is a good electrical conductor, whereas diamond is an almost perfect insulator. However, the free electrons cannot cross easily from layer to layer.

4.6 METALLIC BOND

Metallic bonding is confined to metals and near-metals many of which are to be found in Groups I, II and III of the Periodic Table. Metallic bonding is the principal force holding together the atoms of a metal. A metallic bond results from the sharing of a variable number of electrons by a variable number of atoms. This type of bond is characteristic of atoms with a small number of loosely held valence electrons that can be easily released to the common pool. Bonding results when each atom of the metal contributes its valence electrons to the formation of an electron cloud (frequently referred to as an electron gas) that pervades the solid metal. Valence electrons are not bonded directly to individual atoms but move freely in the sphere of influence of other atoms, and are bound to different atoms at different times for a short period of each time. At the simplest level, in the metallic bond, metal ions are bound together in a solid metal by the interaction of all the valence electrons—the electron cloud or gas that may be considered to move freely.

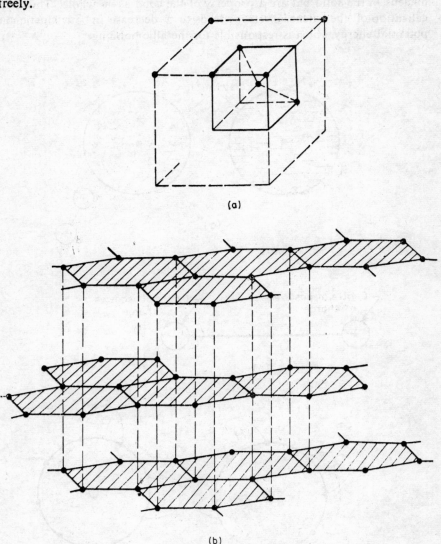

Fig. 4.5 *Covalent solids* **(a)** *Diamond structure* **(b)** *Sheet structure of graphite*

Figure 4.1 is a schematic picture of metal ions (positive) and electron clouds (negative). The bonding of a metallic crystal is due to the attraction of the positive nuclei and the electrons passing between them. A metallic bond thus conceived can exist only between a large aggregate of metallic atoms and must therefore be non-directional. The high electrical conductivity of metals is due to the free electrons moving freely in a electric field.

Theoretically, it can be shown that the wave functions of valence electrons are so spread out that these electrons can no longer belong to any one nucleus in the solid but are a property of the solid as a whole. This delocalization of the valence electrons leads to a decrease in both kinetic and potential energy which is responsible for metallic bonding.

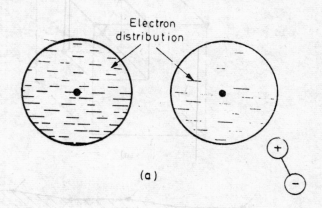

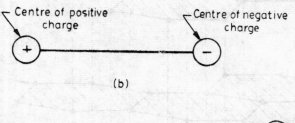

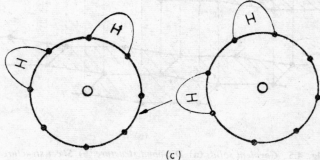

Fig. 4.6 *Mixed bonds* (a) *Dispersion bond* (b) *Electrical dipole* (c) *Hydrogen bond in* H_2O

Metallic bonding is also sometimes called unsaturated covalent bonding; thus the similarity to the covalent bond lies in a sharing of electrons. The difference is that these electrons belong to the metal as a whole rather than to any particular atom. As the number of valence electrons increases, the bonding tends to become partially covalent.

4.6.1 Characteristics of Metallic Compounds
(i) Metallic compounds are crystalline in nature
(ii) They are good conductors of heat and electricity
(iii) As the free electrons in a metal absorb light energy, metals are opaque to light.
(iv) Metals have good lustre and high reflectivity
(v) Metals are moderately strong and generally have high melting points.

4.7 COMPARISON OF BONDS
The various types of bonds are compared in this section.

4.7.1 Ionic and Covalent Bonds
Ionic compounds have crystalline structures. They are non-directional and rigid. They have high melting points. Ionic bonded materials are highly soluble in water but insoluble in organic solvents.

Covalent bonds can be observed in gases, liquids and many solids. With the exception of diamond and graphite, compounds of covalent bonded materials are usually soft, easily fusible and electrically insulators. Covalent bonds are directional.

4.7.2 Ionic and Metallic Bonds
In ionic bonding, a metallic element loses, from the outer electron shell of its atom, a number of electrons equal to its numerical valency. These lost electrons are transferred to the outer electron shells of non-metallic atoms with which the metal is combining. A complete octet of electrons is left behind in the metallic particle and completed in the non-metallic particle, resulting in both having stable outer shells.

Ionic bonds result from the mutual attraction of positive and negative ions. Ionic compounds have fairly high melting points, are stronger than metallic bonds, but have low electrical and thermal conductivity. The bond itself is non-directional.

A metallic bond results from the sharing of a variable number of valence electrons by a variable number of atoms of the same metal. The valence electrons lie much further from the nucleus than in non-metals. The valence electrons of metals are not bound to individual atoms or pairs of atoms but move freely through the metal.

Metallic bonds are weaker than ionic bonds but their compounds have higher thermal and electrical conductivity and moderate melting points.

4.7.3 Covalent and Metallic Bonds

In some respects, the metallic bond is similar to the covalent bond and yet quite different. In metallic bonding the valence electrons which hold the atoms together move freely through the metal. In the covalent bond, two atoms are paired or shared by a particular pair of electrons. Metallic bonding, due to unlimited sharing of electrons, is also called unsaturated covalent bonding. Common to both bonds is the sharing of electrons. Metallic bonds are not directional unlike covalent bond.

4.8 SECONDARY BONDS

Secondary bonds are called molecular bonds. They are formed in case of elements or compounds whose electron configuration is such that very little electron transfer takes place between atoms. These bonds are formed as a result of dipole attractions, the dipoles forming as a result of unbalanced distribution of electrons in asymmetrical molecules. These bonds are also formed as a result of weak Vander Waals forces of attraction which exist between atoms. These forces are due to the electrostatic attraction between the nucleus of one atom and the electrons of the other.

The rare gases, for example, argon and helium, condense to form solids at sufficiently low temperatures, forming secondary bonds which result neither from the transfer of electrons (metallic, ionic) nor from the sharing of electrons (covalent). These bonds are weaker than the primary bonds discussed earlier.

Secondary bonds are very important in determining the structure and properties of many non-metallic materials that include plastics, graphite and the paraffins. Common secondary bonds are (i) dispersion bonds, (ii) dipole bonds and (iii) hydrogen bonds.

Characteristics of secondary compounds:

(i) Molecular compounds are generally transparent to light and good insulators, with the exception of water (H_2O).

(ii) They have low melting points.

4.9 MIXED BONDS

In many substances bonding between atoms does not occur as one of the above-mentioned ideal types but, rather, as a mixture of these bonds. Mixed ionic covalent bonding occurs in the structure of quartz, glass or silica.

In the lattice of materials, many structural defects are produced by ionic and metallic bonding. We find that there is a continuous change in bonding character in a series of alloys of metals such as copper-nickel, copper-zinc and copper-tin. Bonds other than metallic are present but certain metallic properties are retained.

(a) We know that an atom with a closed electron shell consists essentially of a positive nucleus surrounded by a spherical cloud of negative charge. If the electron clouds of two such atoms were static and undeformable, no force would exist between them. Actually, however, the electron clouds result from the motion of various electrons around the nucleus. Thus, an atom can on an average have no electrical dipole moment but have a rapidly fluctuating dipole moment. At any moment the centre of the negative charge distribution does not coincide with the nucleus but fluctuates rapidly about it. This fluctuating charge on one molecule tends to interact with the fluctuating charge on a neighbouring molecule, resulting in a net attraction (Fig. 4.6 (a)). Molecules of inert gases which consist of single atoms are held together by dispersion forces when the gases are solidified. In many organic solids the most important bonding forces between the molecules are of this type.

(b) When two atoms approach each other, the rapidly fluctuating dipole moment of each affects the motion of electrons in the other, and a lower energy is produced if the fluctuations occur in asympathy with one another. This can be seen in Fig. 4.6 (b) where each atom is shown as a small electrical dipole. Consequently, the centre of positive charge and the centre of negative charge do not coincide and an electrical dipole is produced. This displays a marked attractive force providing a mechanism for molecular bonding. This is also due to the Vander Waals forces of attraction arising from electrical dipoles. Vander Waals' bonds are non-directional. These bonds are also formed between molecules that have no permanent dipoles. This type of binding is common in polar compounds like HCl and PVC.

(c) Hydrogen bonds are a special type of dipole bond which can be formed between atomic groups with no electrons to share. They are similar to Vander Waals bonds, but occur only when a hydrogen atom is present. The hydrogen bond occurs frequently in organic materials in which hydrogen often plays a major role. Hydrogen bonds are responsible for the unusual physical properties of ice and water (Fig. 4.6 (c)). Consider, for example, the water molecule, H_2O, where the electrons shared between the oxygen and hydrogen atoms tend to stay close to the oxygen atom than the hydrogen atoms because of the greater electronegativity of oxygen. As a result, oxygen acts as the negative end of the dipole and the hydrogen atoms act as the positive ends. The positive end can then attract the negative end of another water molecule and bond the molecules together. The hydrogen bond is also called a hydrogen bridge and is important in many biological molecules, such as DNA.

4.10 CHEMICAL BONDING AND PROPERTIES OF SOLID MATERIALS

As discussed above, we can classify solid materials according to the nature of their chemical bonds, i.e. ionic, covalent, metallic and molecular. Table 4.1 shows the properties of various chemical bonds.

Table 4.1 *Classification of Bonds and their Properties*

Bond type	Material	Heat of vaporization H(kcal/mole)	Melting point (K)	Electrical Resistivity Ω m	Thermal conductivity $W_m^{-1} k^{-1}$
Ionic	NaCl, LiF CaCl$_2$, MgO	121-242	700-3300	10^{12}-10^{20}	2-20
Covalent	Diamond, Si, Ge. SiC	85-405	500-3700	10-10^{20}	4-600
Metallic	Na, Fe, Cu	26-94	230-4150	10^{-3}-10^{-8}	40-400
Molecular	Ar, Ne, He, H$_2$, CH$_4$	0.02-2.4	<600	10^{12}-10^{20}	0.04-4

In order to know about their properties, it is of prime importance to confine our attention to the strength of a bond. It is best measured by the energy required to break the bond, i.e. the amount of heat which must be supplied to vaporize the solid, and hence separate the constituent atoms. As shown in Table 4.1, the weakest are the Vander Walls and hydrogen bonds. Next comes the metallic, followed by the ionic and covalent bonds whose strengths are nearly comparable.

Melting points are also indicative of bond strength. Melting occurs when the thermal vibration becomes so great that bonds are broken and atoms become mobile. The strength of ionic bonds increases with valency, and this results in the high melting point of magnesium oxide as compared to sodium chloride.

Bond strength is important in determining the strength of materials and their crystal structure.

Materials bonded ionically or covalently are non-conductors since their electrons are bound tightly in the bonds.

Because of the fluctuation of the electron cloud on application of a potential difference, metals are excellent conductors.

Metals possess ductility due to the shifting of electrons. As the impinging radiant energy is easily absorbed by the valence electrons, metallic bonded materials are opaque and lustrous.

4.11 CHEMICAL BONDING AND THE PERIODIC TABLE

The degree to which an atom can attract electrons to itself is called its electronegativity. In general, electronegativity increases steadily across the Periodic Table Group I to Group VII. It varies within a group, decreasing with increasing atomic number, except in the case of transition elements.

Two atoms of similar electronegativity form either a metallic or covalent bond, according to whether they can release or accept electrons.

When the electronegativities differ, the bond is partially ionic, the ionic character increasing with the difference in electronegativity.

In any one row of the periodic table, from right to left, the metallic character of the bond increases and covalent character decreases. In any one column, from top to bottom, the metallic character increases or the covalent character decreases. Li, Na, K, Rb, Ls and Fr are metals with increasing metallic character. At the other end, F, Cl, Br and I form covalent bonds of decreasing strength. Cs and Fr at the bottom left of the table are more metallic than others and have the lowest ionization potentials. F and Cl at the top right-hand corner are very non-metallic and have the greatest electron affinity.

There exists a weak secondary bond between atoms of inert elements, between molecules of groups VII, and VI and V, between molecules of unlike atoms.

Transition metals have a partial covalent character and hence high melting points. The melting points of common engineering metals from the transition series are Fe (1535 °C), Ni (1453 °C), Co (1495 °C), V (1900° C), Nb (2415 °C), Ta (2996 °C), Mo (2610 °C), Zr (1852 °C), Ti (1668 °C) and W (3410 °C).

REVIEW QUESTIONS

1. List the various types of bonds occurring in a crystal. Discuss any one of them in brief.   (AMIE Summer 1978)
2. List the different types of bonds occurring in a crystal. Describe the characteristics of metallic bonds.   (AMIE Summer 1979, 1987)
3. Distinguish between ionic and metallic bonds in solids.   (AMIE Dec 1973)
4. How are atoms held together in a metallic bond? Explain diagrammatically.
5. Describe the essential features of the followings with one example of each category:
   (a) Metallic bond
   (b) Ionic bond   (AMIE Summer 1986)
   (c) Covalent bond   (AMIE Summer 1986)
6. What are the various types of bonding in materials? Explain the different types of bonding and illustrate them with suitable examples.   (AMIE Winter 1982)
7. Explain briefly:
   (a) Covalent bonding   (AMIE Summer 1982)
   (b) Ionic bonding   (AMIE Summer 1983)

5
CRYSTAL IMPERFECTION

5.1 INTRODUCTION

The perfectly regular crystal structures that have been considered upto now are called ideals crystals in which atoms were arranged in a regular way. In actual practice, defects or lattice imperfections are found in most engineering alloys. These imperfections affect the properties of crystals, such as mechanical strength, chemical reactions, electrical properties, etc. to a great extent.

5.2 TYPES OF IMPERFECTIONS

All defects and imperfections in crystals can be classified in three main types, namely;
  (i) Point imperfections
  (ii) Line imperfections
  (iii) Surface and grain boundary imperfections

5.2.1 Point Imperfections or Defects

In some cases, when missing atoms, displaced atoms or extra atoms are involved, there may be point defects (Fig. 5.1) These defects are completely local in effect, e.g., a vacant lattice site. Point imperfections are always present in crystals and their presence results in a decrease in the free energy. The number of defects at equilibrium concentration at a certain temperature can be computed as,

$$n = Ne^{-Ed/kT}$$

where n = Number of imperfections

N = Number of atomic sites per mole

K = Boltzmann's constant

Ed = The free energy required to form the defect

T = Absolute temperature

VACANCIES The simplest point defect is a vacancy which simply involves a missing atom within a metal (Fig. 5.1 (a)). Such defects can be a result of imperfect packing during original crystallisation or may arise due to increased thermal energy causing individual atoms to jump out of their position of lowest energy. The thermal vibrations of atoms increase at higher temperatures. The vacancies may be single or two or more may condense into a di-vacancy or tri-vacancy. The atoms surrounding a vacancy tend to be closer together, thereby distorting the lattice planes. Vacancies exist in a certain proportion in a crystal at thermal equilibrium, leading to an increase in randomness of the structure.

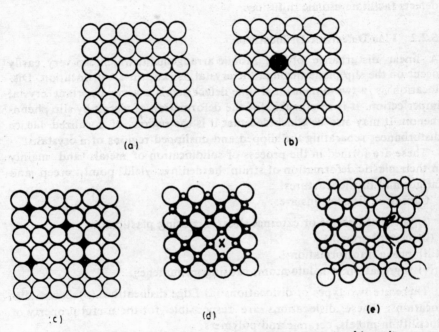

Fig. 5.1 Point imperfections, (a) Vacancies (b) Substitutional impurity
(c) Interstitial impurity (d) Schottky defect (e) Frankel defect

IMPURITIES Impurities may produce compositional defects in the crystal structure. When impurities in the form of foreign atoms (Fig. 5.1 (b)) occupy lattice sites where regular atoms are missing, they produce substitutional impurity. Impurities such as slag inclusions in metals having atoms of a smaller radius than the host atoms, will produce interstitial defects as shown in Fig. 5.1 (c). Such defects cause atomic diffusion and phase transformation and play an important role in changing the thermal and electrical conductivity of metals and alloys.

In brass, zinc is a substitutional atom in the copper lattice. The presence of carbon in iron is an example of carbon occupying void spaces (interstices) in the iron lattice without displacing any of the parent atoms of iron from their site.

FRANKEL DEFECT   As shown in Fig. 5.1 (d), a Frankel defect is closely related to interstices. An ion displaced from the lattice site into an interstitial site is called a Frankel defect. Closed-packed structures have fewer interstalices and Frankel defects because additional energy is required to force the atom into a new position.

SCHOTTKY DEFECT   This is closely related to vacancies and is obtained when an atom or ion is removed from a normal lattice site and replaced by an ion on the surface of the crystal. Both vacancies and Schottky defects facilitate atomic diffusion.

5.2.2 Line Defects or Dislocations

A linear disturbance of the atomic arrangement, which can very easily occur on the slip plane through the crystal, is known as dislocation. Dislocation is a two-dimensional line defect and is a very important crystal imperfection. It is responsible for the deformation of metals by slip phenomenon. It may also be concluded that it is the region of localized lattice disturbances separating the slipped and unslipped regions of a crystal.

These are formed in the process of solidification of metals and mainly in their plastic deformation of strain hardening, yield point, creep and fatigue and brittle fractures.

Causes of dislocations are:

(i) thermal stresses or external stresses causing plastic flow
(ii) crystal growth
(iii) phase transformation
(iv) segregation of solute atoms causing mismatches.

There are two types of dislocations (a) Edge dislocations (b) Screw dislocations. These dislocations are responsible for the useful property of ductility in metals, ceramic and polymers.

EDGE DISLOCATION   An edge dislocation is formed by adding an extra partial plane of atoms to the crystal. Figure 5.2 shows the introduction of an extra half plane in a perfect crystal, resulting in an edge dislocation. The position of the dislocation line is marked by the symbols $\perp$ and $\top$ indicating the involvement of extra planes from the top (positive sign) and bottom (negative sign) of the crystal, respectively. The vertical line of the symbol $\perp$ points in the direction of the dislocation line in the extra partial plane. Near the dislocation, the crystal is distorted due to the presence of zones of compression and tension in the crystal lattice. The dislocation line is a region of higher energy than the rest of the crystal. The lattice above the dislocation line is in a state of compression, whereas below this line, the lattice is in tension.

The magnitude and direction of the dislocation is measured by a vector called the Burger vector '*b*'. The Burger vector is found by traversing a path around the dislocation line. In Fig. 5.2 (a) starting point is selected and we see that, when there is a movement from one atom to another atom, the circuit closes itself, i.e., the end point meets the starting point in a perfect crystal.

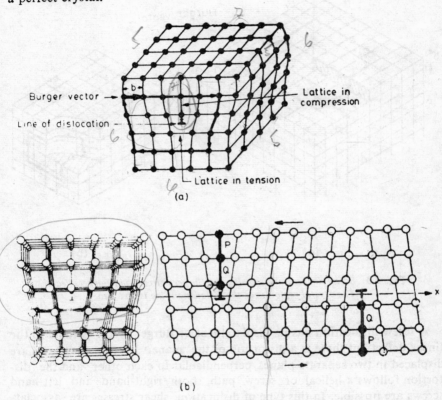

Fig. 5.2 *Edge dislocation caused by an extra partial plane of atoms in the crystal. (a) Positive dislocation. Burgers circuit shown by dark line (b) Positive and negative edge dislocation. Crystal symmetry is along the dislocation line.*

On the other hand, when the same step by step circuit is made around the dislocation in an imperfect crystal, the end point of the circuit does not meet the starting point. The vector '*b*' connecting the starting point with the end point is the Burger vector of the dislocation. In determining the Burger vector, the circuit movement along the atoms should be positive, i.e., like rotating a right-hand screw. In edge dislocation, the Burger vector lies at right angles to the line of dislocation.

SCREW DISLOCATION The formation of a screw dislocation is shown in Fig. 5.3. A perfect crystal and a plane cutting part way through it are also shown. The geometry of the screw dislocation has an interesting effect on the solidification process.

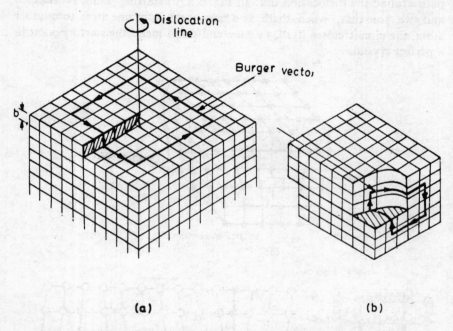

Fig. 5.3 *Screw dislocations showing the Burgers circuit and Burgers vector. The crystal region near the dislocation line is strained*

A screw dislocation has its displacement or Burger vector parallel to the linear defect but there is a distortion of the plane. In this the atoms are displaced in two separate planes perpendicular to each other and the distortion follows a helical or screw path, both right-hand and left-hand screws are possible. In this type of dislocation, shear stresses are associated with adjacent atoms and extra energy is involved along the dislocation. A screw dislocation does not exhibit climb motion.

Three effects of a screw dislocation are of great importance:
1. The force required to form and move a screw dislocation is probably somewhat greater than that required to initiate an edge dislocation.
2. Plastic deformation is possible under low stress, without breaking the continuity of the lattice.
3. Screw dislocation causes distortion of the lattice for a considerable distance from the centre of the line and takes the form of spiral distortion of the planes. Dislocations of both types (combinations of edge and screw) are closely associated with crystallization as well as deformation.

5.4.2 Dislocation Climb

It is possible for dislocations to change their slip planes and thus avoid obstacles in their path. This, of course, increases their mobility. This can be done by climb and cross slip. Climb applies to edge dislocations while cross slip is applicable to screw dislocations.

In edge dislocation, the Burger vector is perpendicular to the dislocation line and the slip plane contains the dislocation line. By the climb mechanism the slip plane can be changed without affecting the slip direction. Suppose there is an obstacle on the slip plane and the edge dislocation cannot slip past that. One method of climbing is the movement of vacancy towards the slip plane, i.e., one of the atoms in the half plane has moved to the site which contained the vacancy. If a similar vacancy migration occurs on every plane in the crystal, the dislocation line would be raised by an amount equal to the lattice spacing and a new slip plane would be established. Because it is not blocked by the obstacle, slip is unobstructed. The process is called dislocation climb and it increases mobility. It requires the presence of moving vacancies which can be increased with increasing temperature. The climb process is much more prevalent at high temperatures than at low ones. It is one of the important mechanisms that take place in the annealing process. Figure 5.9 shows the dislocation climb.

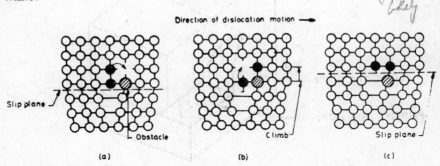

Fig. 5.9 *Dislocation climb. (a) Obstacle on the slip plane (b) Vacancy movement starts towards the slip plane (c) Vacancy movement causing the slip plane to climb*

5.4.3 Cross Slip

A screw dislocation can change its slip orientation more readily than an edge dislocation. In a screw dislocation the Burger vector is parallel to the line of dislocation, therefore the possible slip planes of screw dislocation need not be parallel as is required for edge dislocation. Screw dislocations are capable of avoiding obstacles by a process known as cross-slip where the dislocation simply changes slip planes. A screw dislocation can cross-slip from one easy slip-plane to another, provided the planes of easy slip intersect each other. The resolved shear stress is not the same on all easy slip-planes; it is of lower value on the plane of cross-slip than that on the slip-plane containing the obstacle. As a result, cross-slip occurs at higher stresses than normal slip.

Cross-slip occurs most readily in crystals with a large number of slip systems. The FCC and BCC structures have four and six planes of easy slip, respectively, and these planes intersect. Thus they meet the cross-slip condition. The work-hardening characteristics of materials depend on the presence of cross-slip or occurrance of higher stress values.

5.4.4 Jogs in Dislocation

A jog in a dislocation may be regulated as a short length of dislocation not lying in the same plane as the main plane but having the same Burger vector. Where the vector dislocation jumps from one plane to

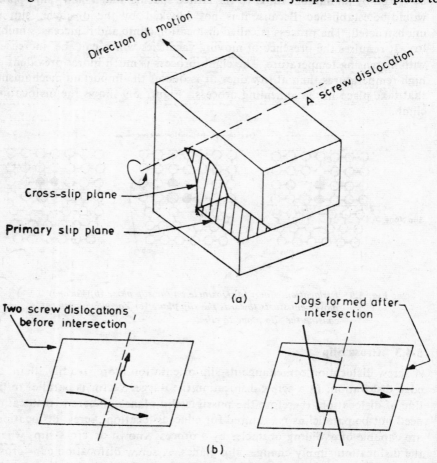

Fig. 5.10 (a) *Cross slip* (b) *Jogs in a crystal*

another, it is known as jog. This is due to the fact that dislocation may not be confined to a single plane but may move from one slip plane to other. Jogs may be formed at the utilisation of two edge or screw dislocations. Figure 5.10 shows jogs in dislocation.

5.4.5 Perfect and Partial Dislocations

Dislocations in real crystals are of two types
  (i) imperfect or partial dislocations
  (ii) perfect or full dislocations.

Partial dislocations have a Burger vector that is only a fraction of the lattice translation vector. This dislocation is always associated with a surface imperfection or fault in the stacking arrangement of planes in the crystal. Perfect dislocations are surrounded by good regions of the crystal, i.e., on both sides of the dislocation, the vertical planes match across the slip plane. This is possible because the Burger vector of full dislocation is an integer number of the lattice translation vector.

REVIEW QUESTIONS

1. What are the defects and imperfections in a crystal? Describe them with neat sketches.   (AMIE Summer 1986, 87)
2. Differentiate between edge dislocation and screw dislocation.
3. How is displacement expressed by means of the Burger vector?
4. What are the point, line and surface imperfections found in solid materials? Illustrate these imperfections with suitable sketches.   (AMIE Winter 1982)
5. (a) What is meant by crystal imperfections? State the effect of their presence in materials.
   (b) Explain with neat sketches the mechanism for dislocations.
6. Explain the Frank-Read source with suitable sketches.
7. Explain briefly the following:
   (a) Stacking fault
   (b) Tilt boundary
   (c) Twin boundaries
   (d) Surface and grain boundary defects
   (e) Frankel defect
   (f) Schottky defect
   (g) Vacancies
8. What is meant by crystal imperfection? Classify them in order of their geometry. What is a Burgers vector? Where and why this is used?
      (AMIE Winter 1987)

6
X-RAYS

6.1 INTRODUCTION

Of all the discoveries made by man, there is probably none that attracted more public attention more quickly than the discovery of X-rays. The fact that the rays permitted one to see through opaque objects was sensational and there was wide appreciation of the value of X-rays when quickly put to many uses.

Although optical microscopy was an invaluable tool in the early days of crystallography, the information that it revealed was rather limited as the wavelength of visible radiation is too large to resolve the detail on an atomic scale. X-rays are electromagnetic waves like visible light but with much smaller wavelength (10^{-10} m), of the same order of magnitude as the interatomic spacings in crystals. In engineering, the use of X-rays has provided valuable information on many aspects of structural studies including crystal imperfections (such as grain boundaries and dislocation) preferred orientation in polycrystalline sheets and the network of atoms in solids.

6.2 DISCOVERY OF X-RAYS

In 1895 Roentgen, a German physicist, discovered that X-rays are produced when a beam of cathode rays strikes a solid target. He found that a fluorescent screen coated with barium-platino cyanide placed at some distance from the discharge tube becomes luminous. He further noticed that the radiations affected photographic plates and produced ionisation in any gas through which they pass. Since their nature was not known they were called X-rays. The most important characteristic of these rays is that they could pass through opaque bodies.

6.3 PRODUCTION OF X-RAYS

Figure 6.1 shows the essential features of a modern Coolidge X-ray vacuum tube. The X-ray tube produces X-rays by causing cathode rays to strike a solid target. In a modern Coolidge tube, the source of cathode-ray electrons is a heated filament. There is no necessity for a residual gas to be

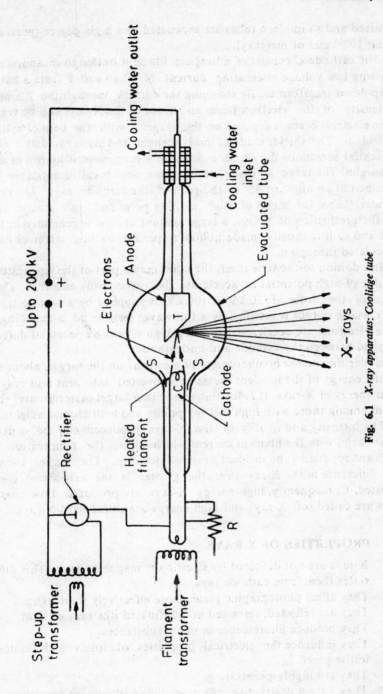

Fig. 6.1 *X-ray apparatus: Coolidge tube*

ionized and so modern tubes are evacuated to a high degree (pressure less than 10^{-10} mm of mercury).

The cathode C consist of a tungsten filament heated to incandescence by passing low voltage alternating current (4 to 10 volts) from a battery or step-down transformer. By changing the current through the filament, the intensity of the electron beam and hence of the X-rays can be regulated. The electron beam is focused on the target T with the help of cylindrical shields S. The shield, made of molybdenum and maintained at a negative potential surrounds the cathode. Such an arrangement is known as an electron gun. The target is made of tungsten and is called anticathode. It is inclined at an angle of 45° to the path of the cathode rays. The target of modern tubes is a metal of high melting point and high atomic number. In the production of X-rays, a large amount of heat is generated in the target and so it is usually made hollow to permit cooling water or oil to be circulated through it.

In addition to the tube itself, the other major part of the apparatus is a source of high potential to accelerate the cathode-ray electrons. The voltage, of the order of 50 kV to 100 kV, is supplied by a step-up transformer whose output is rectified by a full-wave rectifier to a matching filter. The electrons are accelerated to very high speeds by potential differences applied between the cathode and anticathode.

Under the intense bombardment of electrons on the target, about 99.8% of the energy of the incident beam is converted into heat and only 0.2% into energy of X-rays. It is for this reason that target materials are selected from among those with high melting points and high atomic weights.

The intensity and quality of the X-rays produced can be controlled. This varies with the filament current which adjusts the temperature of the filament producing the incident beam of electrons. The higher the potential difference in the X-ray tube, the greater is the velocity of electrons emitted. Consequently, high-energy X-rays are produced. Low energy X-rays are called soft X-rays and high energy are called hard X-rays.

6.4 PROPERTIES OF X-RAYS

1. X-rays are not deflected by electric or magnetic fields. This differentiates them from cathode rays.
2. They affect photographic plates more effectively than light.
3. They are reflected, refracted and diffracted like rays of light.
4. They produce fluorescence in many substances.
5. They influence the electrical properties of solids and liquids and ionise gases.
6. They are highly penetrating.
7. They have a destructive effect on living tissues on excessive exposure. X-rays damage and kill living cells.

8. They produce a photo-electric effect. This is observed from the knock out or impact of a beam of X-rays on certain metal surfaces, producing high-velocity electrons.
9. They are propagated in straight lines with the velocity of light.
10. When X-rays fall upon a plate of some chosen material, they give rise to other rays of similar character because of a complex phenomenon of secondary radiations which are of the following types.

6.5 SCATTERED X-RAYS

In the case of substances of low atomic number and gases, the secondary X-rays are of the same nature and wavelength as the primary X-rays. X-rays are merely scattered due to change in direction of incident rays, but without changes in the wavelength (Fig. 6.2).

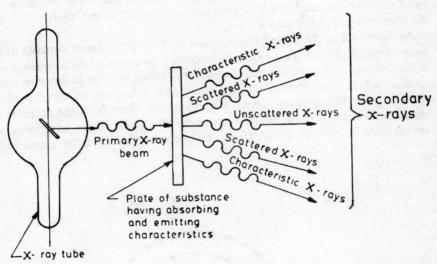

Fig. 6.2 *Secondary x-rays*

6.6 CHARACTERISTIC X-RAYS

These are produced from substances of high atomic weight. The wavelength of such X-rays is equal to or less than that of the primary X-rays and is characteristic of the substance upon which the primary X-rays are incident but do not depend upon the wavelength of the primary X-rays. Characteristic X-rays have less penetrating power than incident X-rays. They are grouped into several series (K L M etc.) from low atomic number to high atomic number elements. The X-rays in the *K*-series of an element have smaller wavelength than in the *L* and *M* series. The higher the atomic weight of the element the more penetrating are the X-rays.

6.7 BETA-RAYS (β-RAYS)

These are fast-moving electrons produced by the photo-electric effect. Beta rays are independent of the nature of the scattering substance but depend upon the quality of the incident or primary X-rays.

6.8 ORIGIN OF X-RAYS

X-rays are produced when high-velocity electrons hit some high melting point and high atomic number material. Many of the electrons that strike matter do nothing spectacular at all. Most of them undergo glancing collisions with particles of the matter and, in the course of these collisions, lose their energy a little at a time and thus merely increase the average kinetic energy of the particles in the material. The result is that the temperature of the target material is increased. It is found that most of the energy of the electron beam (99.8%) goes into heating the target. Only 0.2% of the energy is converted into X-rays.

As discussed in the earlier chapters, an atom of a substance consists of a heavy central positive part called the nucleus surrounded by a number of negatively charged particles called electrons, which revolve in more or less circular orbits. The electrons on the innermost orbits are attracted by the nucleus with the greatest force and, to detach them from the atom, the maximum energy is required. Electrons on the outer orbits experience a comparatively smaller force from the nucleus and a smaller amount of

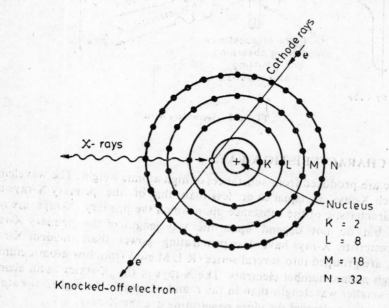

Fig. 6.3 *Origin of X-rays*

energy is required to remove them. The innermost orbit is the K level and the outer orbits are L, M, etc. energy levels.

When an X-ray beam falls on the target, it penetrates deep into the atoms of the target metals and displaces one of the electrons in the innermost orbit, say K orbit, if the X-rays beam possesses a higher energy. Thus a vacancy is created in the electronic structure and this vacancy is filled promptly when an electron from a higher level (L, M, etc.) jumps into the K-level. When an electron jumps or falls into a low-energy level, it releases its energy as radiation as in Bohr's theory (Fig. 6.3).

Although the energy required to ionize an atom by removing an outer electron is much less than 100 eV, the energy required to ionize by removing an innermost electron may be as high as 120,000 eV. When an outer electron falls into such a vacancy, it will radiate a photon of this energy. Such photons are in the X-ray region. This mechanism, which accounts for a significant part of X-ray production, produces X-rays with wavelengths which are characteristic of the target material. Such X-rays are known as characteristic X-rays.

6.9 WAVE NATURE OF X-RAYS OR X-RAY DIFFRACTION

Roentgen failed to determine the nature of X-rays. In 1912, German physicist Max Von Laue conceived of a way of testing the idea that X-rays might be light of very short wavelength and further suggested the use of a natural crystal as a closely spaced three dimensional grating. Laue thought that the atoms of a single crystal might provide the grating needed for the diffraction of X-rays. At Laue's suggestion, Friedrich and knipping directed a narrow beam (pencil) of X-rays at a crystal and set a photographic plate beyond it. The result was a picture like that in Fig. 6.4.

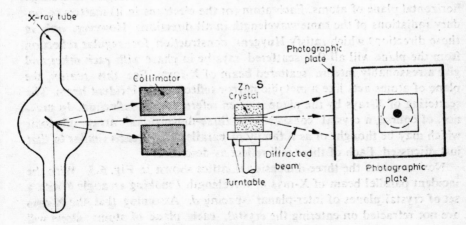

Fig. 6.4 *X-ray diffraction measurement apparatus*

In the experiment, a narrow beam of X-rays from the X-ray tube was allowed to pass through a thin crystal of Zinc Sulphide (Zns) and the beam was received on a photographic plate. Most of the X-rays go directly through the crystal and strike the plate at its centre. To prevent gross over-exposure at thispoint, it is usual to fasten a disc of lead over the centre of the plate. Two lead screens having a pin hole in each were used for collimating a fine pencil of X-rays on the crystal and finally on a photographic plate. Most of the X-rays thus produced a black spot at the centre of the plate. But there were several weak diffracted beams which emerged in different directions and produced a series of dark spots on the same photographic plate. Hence the diffraction pattern consisted of a central dark spot surrounded by a series of smaller spots arranged in a symmetrical order. This diffraction pattern is known as the *Laue pattern*.

The Laue pattern established that X-rays have very short wavelength and confirmed that the crystals have their atoms arranged in a regular structure. Laue patterns or diffraction patterns are widely used in the structural study of a crystal and help in revealing the solid state behaviour during the mechanical working and heat treatment of metals. Later in this chapter we shall discuss the use of X-ray diffraction.

6.10 BRAGG'S LAW

Late in 1912, shortly after the Laue experiment, William L Bragg devised another technique for diffracting X-rays. Instead of observing the effect created by passing the rays through a crystal, Bragg considered how X-rays are scattered by the atoms in the crystal lattice.

Consider an X-ray wave front incident on a surface row of atoms in a crystal plane (Fig. 6.5). Each atom becomes a source of scattered X-radiation. A beam of monochromatic X-rays is incident at an angle θ to a single horizontal plane of atoms. Each atom (or the electrons in it) scatters secondary radiations of the same wavelength in all directions. However, only in those directions which satisfy Huygens construction for regular reflection from the plane will all the scattered rays be in phase with each other and give a reasonably intense scattered beam of X-rays. For this reason, the plane of atoms acts like a metallic mirror reflecting the incident beam. The scattering of X-rays by the plane is often referred to as reflection. In practice, of course, a crystal consists of a three-dimensional array of atoms which may be thought of as a family of parallel planes each similar to that just discussed. Each of these will reflect as described.

Now consider the three-dimensional lattice shown in Fig. 6.5. With the incident parallel beam of X-rays of wavelength λ making an angle θ with a set of crystal planes of inter-planar spacing d. Assuming that the X-rays are not refracted on entering the crystal, each plane of atoms alone will tend to reflect a beam of X-rays at an angle θ. However, these beams may, due to their relative path differences, either interfere destructively or constructively. The criterion for the existence of constructive interference and an intense diffracted beam is that the reflected rays from individual planes should be parallel across the entire wave-front.

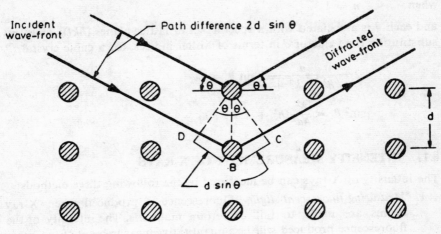

Fig. 6.5 *Bragg's law*

For this to be so, the path difference between two successive rays must be equal to an integral number of wavelengths, i.e.,

$$CB + BD = n\lambda$$
$$d \sin \theta + d \sin \theta = n\lambda$$

where n is an integer

or
$$2d \sin \theta = n\lambda$$

This result is known as Bragg's law and forms the basis of modern X-ray crystallography.

It is very simple to device Bragg's law as stated above but to use Bragg's law in X-ray crystallography wavelength λ can be determined if the interplanar spacing d is known. If diffraction occurs from planes of a lattice at an angle θ_1 then the path difference between the diffracted beams from adjacent plane is 2λ instead of λ; diffraction will occur at an angle θ_2 and the above equation will be satisfied by $n=2$. These reflections are then called first-and second-order reflections respectively

$$\therefore \quad \sin \theta_1 = \frac{\lambda}{2d}$$

$$\sin \theta_2 = 2\frac{\lambda}{2d}$$

and $\quad \sin \theta_3 = \frac{3\lambda}{2d}$

As the order of the spectrum increases the intensities decrease. Thus, in practical crystallography, Bragg's law is of more use in the form

$$2d\,(hkl)\sin\theta = \lambda$$

when $n = 1$

and each θ is associated with a specific set of lattice planes (hkl). Thus, by substituting the value of d in terms of Miller indices, for a cubic crystal,

$$2\,\frac{a}{\sqrt{h^2 + k^2 + l^2}}\sin\theta = n\lambda$$

$$\sin^2\theta = \frac{\lambda^2}{4a^2}(h^2 + k^2 + l^2)$$

6.11 INTENSITY MEASUREMENT OF X-RAYS

The intensity of X-rays can be measured by the following three methods:

1. *Measuring fluorescent light* Fluorescence is produced when X-ray beams are made to fall on certain materials. The intensity of the fluorescence produced will be a quantitative measurement of the intensity of the X-rays directed on such materials.
2. *Using a photographic plate or film* When X-rays are made to fall on a photographic plate or film, it gets blackened. The extent of blackening produced as a result of X-ray exposure can be measured by a densitometer. The intensity of X-rays can thus be measured in terms of the density of X-rays falling on the photographic plate. Modern spectrometers use photographic film instead of a photographic plate.
3. *Measuring ionisation current* A third method is based on the conductivity of air or gas due to X-rays. An ionisation chamber is used to collect and measure these ions quantitatively. The stronger the ionization current the higher the intensity of the X-rays.

We know that air and other gases which are insulators under normal conditions become good conductors under the influence of X-rays due to ionization of their molecules. When X-rays are passed through a gas, the fast-moving electrons of X-ray radiations acting on the molecules and atoms of the gas break them up into positive and negative ions which, by their motion, render the gas conducting and give rise to an ionisation current. The rate of ionisation depends on the intensity of the incident X-ray radiation. Therefore, ionisation current measured by an electrometer connected with the ionisation chamber is very convenient and responds fast to any change in X-ray intensity.

6.12 BRAGG'S CONDITIONS FOR X-RAY DIFFRACTION

The first condition is that the angle which the incident beam makes with the planes must equal that made by the reflected beam.

Let d_1, d_2 and d_3 be the interplanar distances. Then

$$\frac{1}{d_1} : \frac{1}{d_2} : \frac{1}{d_3} = \sin 5.22° : \sin 7.30° : \sin 9.05°$$

$$= 0.091 : 0.13 : 0.157$$

$$= 1 : 1.4 : 1.73$$

which justifies the assumption of the simple cubic lattice.

If the values after reflection from three different reflecting planes are

$$\frac{1}{d_1} : \frac{1}{d_2} : \frac{1}{d_3} = 1 : \frac{1}{\sqrt{2}} : \sqrt{3}$$

it indicates BCC structure of the crystal under study.

Similarly, Bragg's law can be used to determine different types of crystal structures. In the Bragg equation, the interplanar spacing d can also be related to lattice parameters of the unit cell.

In general, the distance between the parallel planes of Miller indices (hkl) in terms of the parameter a for the cubic system is

$$d(hkl) = \left[\frac{a^2}{h^2 + k^2 + l^2}\right]^{\frac{1}{2}}$$

6.17 PRACTICAL X-RAY DIFFRACTION

In order that Bragg's law can be satisfied and X-ray diffraction can occur, the various values of θ and λ must be related. It is necessary, in practice, to vary either the angle of inclination of the specimen (crystal) to the incidence beam or the wavelength (λ) of the radiation. The following three standard methods are used to find the crystal structure experimentally:

(i) In the Laue method a stationary crystal is irradiated by a range of X-ray wavelengths. For a given value of θ the crystal diffracts the beam selectively to satisfy Bragg's law

$$n\lambda = 2d \sin \theta$$

(ii) In the rotating-crystal method, a single crystal is rotated about a fixed axis in a beam of monochromatic X-rays.

(iii) In the powder method, a sample of polycrystalline powder is kept stationary in a beam of monochromatic X-rays. In this method, the angle θ satisfies Bragg's law only for certain orientations.

The nature of X-rays emitted, in each of these three methods, from a standard high-voltage X-ray tube will be as follows. For the Laue method the whole range of wavelengths is used but for the other two methods a single wavelength is required.

6.18 LAUE METHOD

The Laue method is one of the principle methods used for the structural studies of a crystal. As shown in Fig. 6.8 (a) a single crystal is mounted on a rotating table which enables the crystal to be rotated through known angles and maintained stationary in a beam of X-rays of various/different wavelengths. The crystal selects out and diffracts those discrete value of λ, for which crystal planes exist, of spacing d and glancing angle θ satisfying Bragg's equation.

A narrow, parallel beam is collimated on the crystal. Photographic film is placed to receive either the transmitted diffracted beams or the reflected

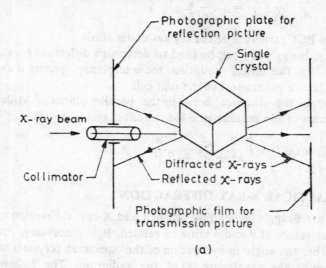

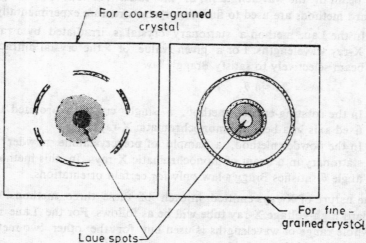

Fig. 6.8 *Laue method for crystal structure examination (a) Experimental arrangement (b) Crystal pattern on photographic film or plate*

beams. The resulting pattern consists of a series of spots. Sharp well-defined spots on the film are indicative of a perfect crystal structure whereas diffuse, broken or extended spots indicate lattice distortion or other imperfections in the crystal (Fig. 6.8 (b)).

This is a quick method of detecting crystal orientation and symmetry but has the following disadvantages:

(i) Due to the wide range of wavelengths used there is overlapping of diffraction images from different crystal planes.
(ii) There is variation in the intensity of the incident X-ray beam due to the large range of wavelengths used in the method.
(iii) Various reflections may make the appearance of the photograph complicated and its measurement impossible.

6.19 ROTATING-CRYSTAL METHOD

In this method, as shown in Fig 6.9 a small (1 mm) single crystal is rotated about a fixed axis in a beam of monochromatic radiation. The resulting variation in the angle θ brings different lattice planes into position for reflection and the diffracted images are recorded on a photographic film placed cylindrically, coaxial with the rotating spindle. The film, when developed, shows a number of spots due to the various planes which have passed through the requisite positions.

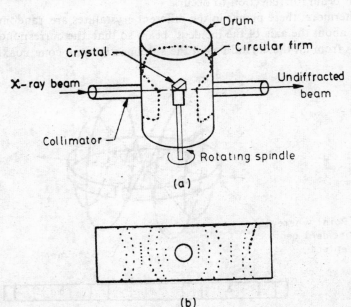

Fig. 6.9 *The rotating crystal technique (a) Rotating crystal, camera arrangement (b) Crystal pattern on cylindrical film*

The explanation for the observed diffracted pattern is somewhat complicated due to changing values of θ between the incident beam and the lattice planes. Secondly, because of the large number of reflections, there is also the problem of overlapping of images. Therefore, to overcome these problems, structural determination, in practice, is done in one of the two ways:

(i) The crystal is rotated through a small angle of about 10° in order to have a limited number of spots on the film.
(ii) By the Weissenberg technique the cylindrical film is moved backwards and forwards parallel to the axis of rotation in exact synchronisation with the rotation of the crystal. The images are then spread out over the whole film.

6.20 POWDER CRYSTAL METHOD

This method is used for crystalline substances available in the form of small crystals or powder. This method is also known as the Debye-Sherrer method after the inventor.

In this method, a monochromatic X-ray beam is allowed to fall on a small specimen of the substance ground to a fine powder and contained in a thin-walled glass capillary tube. Since the orientation of the minute crystal grains is completely random, a certain number of them will line with any given set of lattice planes making exactly the correct angle with the incident beam for reflection to occur.

Furthermore, these planes in the different crystallites are randomly distributed about the axis of the incident beam so that the corresponding reflections from all the crystallites in the specimen lie on a cone coaxial with

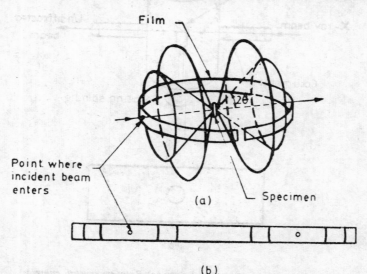

Fig. 6.10 *The powder method*

the axis, with a semi-apex angle twice the glancing angle (Bragg's angle i.e. 2θ). For each set of planes and for each order there will be cone of diffracted rays, as shown in Fig. 6.10. Therefore, a series of concentric circular rings will be formed by the interaction of these cones with a photographic film when it is set with its planes normal to the incident beam, from the radii of which the glancing angle and hence the spacing of the plane can be determined.

This method is used to determine the structure of simple cubic crystals and particularly for the investigation of metal and alloys structures where single crystals are not easily obtainable. Since the diffraction pattern is different for different elements and compounds, this technique provides a fairly rapid method of analysis. In addition, it is a useful metallurgical technique as phase changes can be studied by varying the specimen temperature and recording the associated changes in the diffraction pattern.

6.21 APPLICATION OF X-RAYS

Various applications of X-rays are described under the following main groups:

(i) Industrial applications.
(ii) Medical applications.
(iii) Application of X-rays for scientific research.

6.21.1 Industrial Applications

In industry, X-rays are used for the following purposes:

1. Through X-ray radiography of materials, intrenal defects like imperfection in forgings, weldings and castings can be located.
2. X-rays are used to study the structure of polymers (rubber and plastics) to give valuable information about their atomic arrangement and molecular grouping.
3. Through X-ray crystallography the structure of various alloys is determined by analysing the ingot structure with the help of X-ray diffraction.
4. X-ray fluoroscopy is widely used in inspection work to screen industrial products and locate any hidden defects or impurities on the fluorescent screen. This is a rapid method and is mainly used at international airports to check/detect unwanted goods in air passenger's baggage.
5. X-rays are commonly used to control hot strip steel rolling machines by indicating any variation in the thickness of the hot strip moving out of the rollers. As soon as the hot strip is thicker or too thin than that set by the machine, these will be indicated on the counter, and the distance between rollers is re-adjusted so that the correct thickness of hot strip is produced.

94 Materials Science

6. With X-ray topography, direct observation of dislocations and grain boundaries is possible.
7. X-ray photographs help locate defects like cracks, blow holes and other flaws in the heavy steel plates used in bridges or in defence work. The same technique is used to detect flaws of components used in aeroplanes and rockets.

6.21.2 Applications of X-Rays for Medical Purposes

In medicine X-rays are widely used in radiography and X-ray therapy.

In radiography X-rays are used for the detection of fractures, diseased organs, and foreign matter like bullets in the human body. X-rays are also used for diagnosis of diaseases like ulcers, tuberculosis and for brain scanning, etc. Since the bones, tissues, blood, and flesh have different densities, different X-rays absorption will be registered on the photographic plate which has been exposed to X-rays in testing the human body.

For visual observation, closed circuit TV screens are used to replace photographic plate to give rapid diagnosis informations.

X-ray therapy is used to cure different diseases like cancer, tumours skin diseases, etc., by controlled exposures to X-rays of a suitable quality. Diseased tissues are destroyed by X-rays by exposing diseased portions of the body for a suitable time. Long or over exposure of the human body to X-rays may cause injurious effects like loss of white cells in the blood of the body, sterility and genetic effects.

6.21.3 Applications of X-Rays for Scientific Research

1. To investigate the structure of crystalline solids.
2. To study atomic structure in atomic physics.
3. To study complex organic compounds by analysing their structures.
4. To find the atomic numbers and energy levels and identify elements.

6.22 X-RAY ABSORPTION

The absorption of X-rays by some standard materials such as lead and aluminium has been used for many years as the only quantitative measure of the quality of X-rays. The use of X-rays in radiography is based on their absorption. Lead glass, which is colourless and transparent to visible light, absorbs X-rays almost completely and is, therefore, used to protect the operators using X-ray apparatus. On the other hand, thin foils of aluminium opaque to visible light, absorb very little X-rays.

Among the most remarkable properties of X-rays are their own characteristic absorption lines. A beam of X-rays is weakened when passed through a material, a fraction being scattered away from the beam and a fraction absorbed. The decrease in intensity due to scattering occurs when some rays are deviated from the main direction and therefore leave the parallel beam. The intensity loss due to absorption occurs because some of the X-ray energy is in fact truly absorbed in the substance and transformed into heat.

The experimentally observed primary quantity is the linear absorption coefficient μ defined by the equation, $I = I_0 \exp(-\mu x)$, where x is the thickness of material irradiated, I_0 is the incident intensity and I is the transmitted intensity. The greater the value of μ the greater the diminition of intensity for a given thickness.

X-ray absorption spectra typically display a large decrease in the absorption coefficient with increasing quantum energy and absorption 'edges' which are quantum energies at which the absorption coefficient jumps to higher values. These 'edges' correspond to the series limits for the K, L, M ... series and are correspondingly labelled. The position of the K edge means the work of removing or ejecting an electron from the K-shell where it experiences a field of nearly the unshielded nuclear charge of the nucleus. For an atom to absorb X-ray radiation, an electron must be excited from an inner shell into a less strongly bound state. Since the neighbouring shells are already occupied, discrete, absorption lines due to transitions from one shell to another are scarcely observed. There is, however, a continuum of free states on the otherside of the series limit into which the absorbing electrons can be lifted.

The difference in the values of coefficients of absorption of X-rays of bones and tissues in the human body may be used to locate the defects in them.

REVIEW QUESTIONS

1. Explain Bragg's X-ray spectrometer for the study of diffraction of X-rays in crystals. How is it used?
   (a) to determine the wavelength of X-rays?
   (b) to study the structure of crystals? (AMIE Dec. 74, May 75, Summer 80)
2. Explain how X-rays are produced. What are their properties?
3. Give an account of the production, properties and industrial application of X-rays.
4. What is X-ray diffraction?. Deduce Bragg's law for the diffraction of X-rays by a crystal. What are Bragg's conditions for X-ray diffraction?
   (AMIE May 1975)
5. What are the various experimental methods used in X-ray diffraction to study a single crystal structure?
6. (a) State and derive Bragg Law.
   (b) X-rays with a wavelength of 0.56 Å are used to calculate d_{200} in nickel. The reflecting angle is 9.5°. What is the size of the unit cell?
   (AMIE Summer 1982)
7. Write short notes on:
   (a) X-rays by Coolidge tube (AMIE May 74)

(b) Soft and hard X-rays
(c) Mosley's law   (AMIE May 76)
(d) Industrial applications of X-rays   (AMIE May 74)
(e) Bragg's law
(f) Radiography.

8. Describe the X-ray powder diffraction camera and explain how it is used to determine crystal structure.   (AMIE Winter 79)

9. (a) Explain Bragg's law
   (b) Describe briefly the radiographical method for non-destructive examination of engineering components. How will you decide the exact location of the flaw?   (AMIE Summer 1984)

7

DIFFUSION IN SOLIDS

7.1 INTRODUCTION

It was assumed in the previous chapters that any given atom moves about its mean position in the crystal lattice due to thermal vibration. This vibration caused vacancy motion and contributed to dislocation motion. In this chapter, it will be shown that atomic vibrations and vacancies allow atoms to move through a solid. This motion of matter through other matter is called diffusion. In other words diffusion is the shifting of atoms and molecules to new sites within a material resulting in the uniformity of composition as a result of thermal agitation.

7.2 APPLICATIONS OF DIFFUSION

1. Diffusion is fundamental to phase changes, e.g. γ to α iron
2. Joining of materials by diffusion bonding, e.g. welding, brazing, soldering, galvanizing and metal cladding
3. Important in heat treatment like homogenising treatment of castings, recovery, recrystallization and precipitation of phases
4. Production of strong bodies by powder metallurgy (sintering)
5. Surface treatment of steels, e.g., case hardening
6. Oxidation of metals
7. Doping of semiconductors.

7.3 CLASSIFICATION OF DIFFUSION

1. Self diffusion : Atoms jumping in pure metals,
2. Inter-diffusion : Observed in binary metal alloys such as the Cu-Ni system.
3. Volume diffusion : Atomic movement in bulk in materials.
4. Grain-boundary diffusion : Atomic movement along the grain boundaries alone.
5. Surface diffusion : Atomic movement along the surface of a phase.

7.4 DIFFUSION MECHANISMS

Several atomic mechanisms have been proposed to explain diffusion. All of them are based on the vibrational energy of atoms in a solid. Vacancy mechanism, interstitial mechanism and direct-interchange mechanism are the common diffusion mechanisms.

7.5 VACANCY MECHANISM

This is a very important mechanism for diffusion in (FCC, BCC, HCP) metals. Diffusion can occur by atoms moving into adjacent sites that are vacant. Diffusion by the vacancy mechanism in a pure solid is illustrated in Fig. 7.1 (a). The atoms surrounding the vacant site shift their equilibrium positions to adjust for the change in binding that accompanies the removal of a metal ion and its valency election. Assuming that the vacancies move through the lattice and produce random shifts of atoms from one lattice position to another as a result of atom jumping. Over a period of time such diffusion produces concentration changes. Vacancies are continually being created and destroyed at the surface, grain boundaries and suitable interior positions such as dislocations. The rate of diffusion, therefore, increases rapidly with increasing temperature.

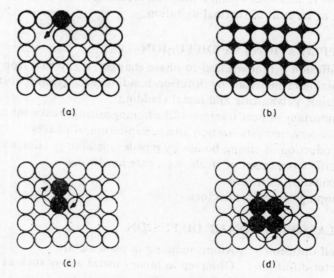

Fig. 7.1 *Diffusion mechanisms (a) Vacancy mechanism (b) Interstitial mechanism (c) Two atoms interchange (d) Four atoms interchange*

If the solid is composed of a single element (pure metal), the movement of atoms is called self diffusion because the moving atom and the solid are the same chemical-element. Copper and nickel are mutually soluble in all proportions in the solid state and form substitutional solid solutions e.g., plating of nickel on copper. Figure 7.2 also shows the vacancy mechanism for atomic diffusion.

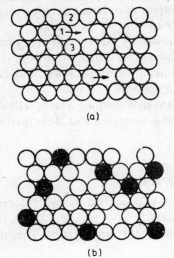

Fig. 7.2 *The vacancy mechanism for atomic diffusion* (a) *Pure solid solution* (b) *substitutional solid solution*

○ A Atoms
● B Atoms

7.6 THE INTERSTITIAL MECHANISM

Interstitial solutions may occur when a solid is composed of two or more elements whose atomic radii differ significantly. The large atoms occupy lattice sites while the smaller ones fit into the voids created by the large atoms. These voids are called interstices. The diffusion mechanism in this case is similar to vacancy diffusion except that the interstitial atoms stay on interstitial sites. This is shown in Fig. 7.1 (b). An activation energy is associated with interstitial diffusion because, to arrive at the vacant site, it must squeeze past neighbouring atoms with energy supplied by the vibrational energy of the moving atoms. Consequently, interstitial diffusion is a thermally activated process. This process is simpler since the presence of vacancies is not required for the solute atom to move.

This mechanism is important in two cases:
(a) The presence of very small atoms in the interstices of the lattice greatly affect the mechanical properties of metals.
(b) Oxygen, nitrogen and hydrogen can be diffused in metals easily at low temperatures.

7.7 DIRECT INTERCHANGE MECHANISM

As illustrated in Fig 7.1 (c) and (d), two or more adjacent atoms jump past each other and exchange positions, but the number of sites remains constant. This may be two-atom or four-atom (Zenner ring) interchange (for BCC). This mechanism results is severe local distortion due to the displacement of the atoms surrounding the jumping pairs. Much more energy is required in this case for atom jumping. The other objection to this mechanism is that a number of diffusion couples of different compositons are produced. This is also called *Kirkendall's effect*.

The inequality of diffusion was first shown by Kirkendall. He used an α-brass/copper couple and showed that zinc atoms diffused out of brass into copper more rapidly than copper atoms diffused into brass. Voids can be observed in brass due to a net loss of Zn atoms.

Theoretically, this effect is very important in diffusion. The practical importance of this effect is in metal cladding, sintering and deformation of metals (creep).

7.8 DIFFUSION COEFFICIENT: FICK'S LAW

Solid state diffusion can be mathematically described by two differential equations called Fick's first and second laws.

Ficks first law describes the rate at which diffusion occurs. This states that

$$dn = -D \frac{dc}{dx} a \, dt$$

where dn = Amount of metal in kg that crosses a plane normal to the direction of diffusion.

dc/dx = Slope of concentration gradient.

D = Diffusion co-efficient

a = Area of plane across which diffusion takes place.

dt = Duration of diffusion

J = The flux or the number of atoms moving from unit area of one plane to unit area of another per unit time. It is proportional to the concentration gradient.

The equation becomes:

$$\frac{dn}{dt} = -D \frac{dc}{dx} a$$

$$\therefore \quad J = -\frac{1}{a} \frac{dn}{dt} = -D \frac{dc}{dx}$$

The negative sign indicates that flow occurs down the concentration gradient. Figure 7.3 shows that the concentration gradient varies with x. A large negative slope corresponds to a high diffusion rate. The B atoms will diffuse from the left side in accordance with Ficks first law. The net migration of B atoms to the right side means that the concentration will

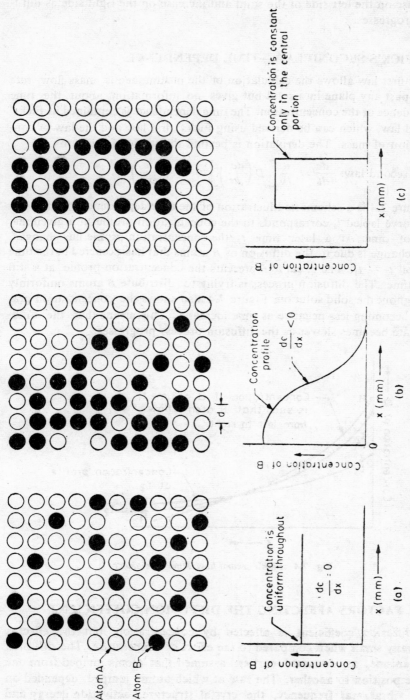

Fig. 7.3 Fick's first law: Model for illustrating diffusion. The concentration of B atoms in the direction indicates the concentration profile

decrease on the left side of the solid and increase on the right side as diffusion progresses.

7.9 FICK'S SECOND LAW—TIME DEPENDENCE

Fick's first law allows the calculation of the instantaneous mass flow rate (flux) past any plane in a solid but gives no information about the time dependence of the concentration. The time dependence is contained in Ficks' second law, which can be derived using Ficks' first law and the law of conservation of mass. The derivation is beyond the scope of this book.

Ficks second law: $\dfrac{dc}{dt} = \dfrac{d}{dx}\left[D\left(\dfrac{dc}{dx}\right)\right] = D\left(\dfrac{d^2c}{dx^2}\right)$

Figure 7.4 is a schematic illustration of the time dependence of diffusion. The curve labled t_1 corresponds to the concentration profile at a given instant of time. At a later time t_2, the concentration profile has changed. This change is due to the diffusion of B atoms that has occurred in the time interval $t_2 - t_1$. The t_3 curve represents the concentration profile at a still later time. The diffusion process is trying to distribute B atoms uniformly throughout the solid solution. Figure 7.4 also shows the concentration gradient becoming less negative as time increases. This means that the diffusion rate becomes slower as the diffusion process progresses.

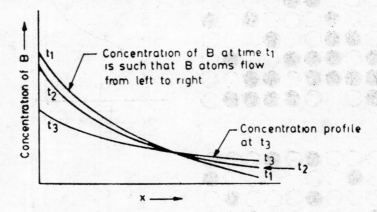

Fig. 7.4 *Fick's second law: Time dependence*

7.10 FACTORS AFFECTING THE DIFFUSION COEFFICIENT

The diffusion coefficient is affected by concentration, but this effect is generally small when compared to the effect of temperature. The diffusion mechanisms, as discussed earlier, assumed that atoms jumped from one lattice position to another. The rate at which atoms jumped depended on their vibrational frequency, the crystal structure, activation energy and temperature. The activation energy depends on the energies of the bonds which are formed between the jumping atoms and their neighbours. The bond energies depend on the concentration.

7.11 SELF-DIFFUSION

We know that diffusion can occur by atoms moving into adjacent sites that are vacant. For an atom to move past other atoms, energy is required. This is the activation energy for vacancy motion, as discussed earlier.

If the solid is composed of a single element (pure copper for example), the movement of the atom is called self diffusion because the moving atom and the solid are the same chemical element.

Self diffusion is very important for annealing and creep. Radioactive tracers have proved to be very useful in determining self diffusion coefficients.

7.12 INTER-DIFFUSION

This takes place in binary metallic alloys, e.g. the copper-nickel system. It is contrary to self diffusion. If nickel had been plated onto the surface of copper, atomic diffusion would bring about nickel homogenization within the copper, after a sufficient time, at elevated temperatures.

7.13 DIFFUSION COUPLE

A diffusion couple consists of two solids of different compositions in contact with one another, so that each tries to diffuse into the other. This is called inter-diffusion and is responsible for diffusion welding.

A diffusion couple (Fig. 7.5) consists of two solids (metal A and metal B) in contact with one another. The rate of diffusion of A into B and B into A for any given value of x can be determined by Ficks' first law.

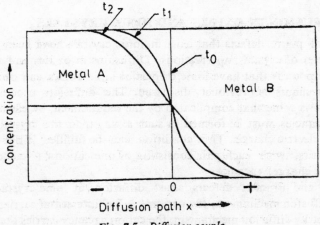

Fig. 7.5 *Diffusion couple*

The concentration-distance profiles at different lengths of diffusion time are shown in Fig. 7.5. Time t_0 is the instant at which diffusion begins and concentration profiles show a step change at the contact surface. A and B diffuse into each other and the concentration profiles change with time as shown by curves t_1 and t_2. The concentration of either A or B, as a function of x and t, can be found from the solution of Fick's second law as applied to the diffusion couple.

The diffusion couple provides one method of experimentally determining the coefficient of diffusion D.

7.14 DIFFUSION WITH CONSTANT CONCENTRATION (CASE-HARDENING)

Case-hardening is a process in which one element (usually in gaseous form) is diffused into another (a solid), the diffusion being limited to a small region near the surface. The properties of this region are changed as a result. Generally, surface regions become harder and brittle whereas the core remains ductile.

This process is governed by Fick's second law. Diffusion of gas atoms into the solid takes place by one of the diffusion mechanisms (usually the interstitial mechanism). The depth to which the atoms of gas penetrate increases with time. Thus, the depth of interstitial alloy increases as diffusion progresses.

For example, nitrogen can be dissolved in the interstitial sites of an iron crystal and the resulting Fe-N alloy is stronger, harder and more brittle than the original iron. The interstitial atoms inhibit dislocation motion and the nitrogen concentration would be highest near the surface in accordance with concentration profiles. This diffusion of nitrogen into iron is called nitriding. Iron and steel can also be case-hardened by carbon, in which case the process is called carburizing.

7.15 DIFFUSION IN OXIDES AND IONIC CRYSTALS

The types of point defects that exist in ionic crystals have been discussed in the chapter of crystal imperfections. Diffusions in oxides and other crystalline compounds that have ionic or partial ionic bonds can occur by the diffusion mechanisms previously discussed. The diffusion process in the present case is somewhat complicated by the requirement of electrical neutrality. Vacancies must be formed in such a way that the crystal does not acquire an electric charge. This condition can be fulfilled if the vacancies form in pairs, with each pair consisting of one cation (+ve charge) and anion (−ve charge) vacancy.

Schottky and Frankel defects assist diffusion in ionic crystals. In the Frankel diffusion mechanism, the cation (+ Ve) interstitial carries the flux. In the Schottky diffusion mechanism, the cation vacancy carries the diffusion flux. The cations diffuse through cation vacancies and anions through anion vacancies. Activation energies are not the same for both processes.

Diffusion in ionic crystals and oxides is very sensitive to the concentration of impurities as they affect the number of vacancies. Consider the addition of a small quantity of cadmium as an impurity to a NaCl crystal. The valance of Cd is $+2$ and that of Na as $+1$. Therefore, a single cadmium ion would have to replace two sodium ions in the crystal in order to maintain electrical neutrality, thus generating a cation vacancy. Therefore, the addition of cadmium cations will increase the diffusion rate.

Impurities also occur in metallic oxides and the rate of oxygen diffusion through an oxide is often increased by the presence of impurities which occupy anion sites.

This diffusion process is quite useful in fabricating parts from high temperature ceramics by the powder metallurgy technique. In the sintering process the powdered particles would bond together by diffusion.

7.16 GRAIN-BOUNDARY AND SURFACE DIFFUSION

Diffusion occurs along the surfaces of a solid and in the grain boundaries of a polycrystal as well as through the volume of a material. Surface atoms form fewer bonds than atoms at the interior of a solid; therefore we would expect surface diffusion to have lower activation energy than volume diffusion. During the solidification process, there is migration of atoms. This is called surface diffusion (Fig. 7.6).

Atoms in the regions of grain-boundaries are not bonded as tightly as interior atoms. Consequently, they diffuse more readily. Surface and boun-

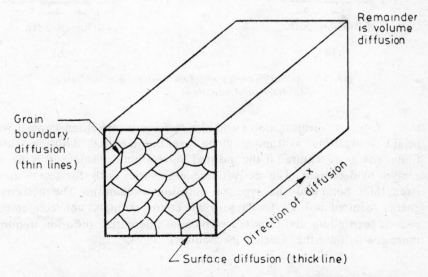

Fig. 7.6 *Showing the areas of volume, grain boundary and surface diffusion*

dary diffusions are produced along paths other than those of crystal lattice diffusion.

7.17 ACTIVATION ENERGY FOR DIFFUSION

We have noted that atomic movement is responsible for the diffusion process. Some energy is required by the atoms for each movement or jump from one lattice position to another. The energy required by the atom to overcome this energy barrier is called the activation energy of diffusion (Fig. 7.7). In the vacancy mechanism, energy is required to pull the atom

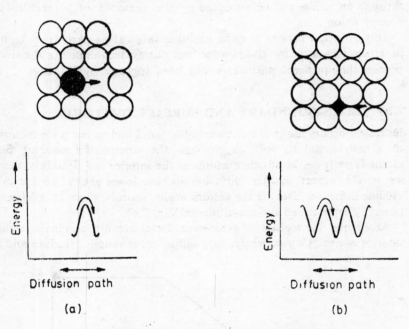

Fig. 7.7 *Activation energy of diffusion* (a) *Vacancy mechanism* (b) *Interstitial mechanism*

away from its nearest atoms and also to force it into closer contact with neighbouring atoms as it moves along them in interstitial diffusion. Additional energy is required if the normal inter-atomic distance is either increased or decreased. The activation energy varies with the size of atom, strength of bond and the type of diffusion mechanism. The activation energy required is high for large-sized atoms, strongly bonded materials such as corrundum and tungsten carbide (as interstitial diffusion requires more energy than the vacancy mechanism).

REVIEW QUESTIONS

1. Explain the engineering importance of diffusion.
2. State and explain Fick's first and second law.
3. Explain briefly the following:
   (a) Diffusion coefficients
   (b) Diffusion couple
   (c) Diffusion with constant concentration.
4. Explain the process of diffusion in the following:
   (a) Grain boundaries and surface diffusion
   (b) Ionic crystals and oxides.

8
DEFORMATION OF MATERIALS

8.1 INTRODUCTION

The behaviour of a solid, when it is subjected to forces, is of fundamental interest to engineers, since most of the things that we design must withstand applied forces. The change in form or dimension under the action

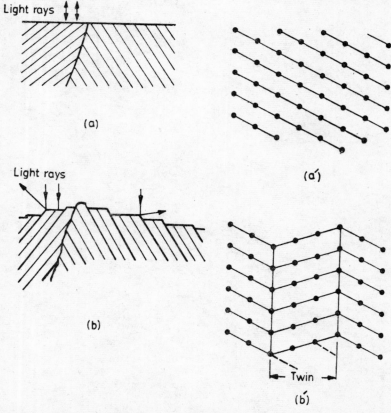

Fig. 8.1 *Effects of deformation* (a), (a') *before straining* (b), (b') *after straining*

of applied forces is called *deformation*. It is also produced by various physical and chemical processes. A deformation can be classified as either recoverable (elastic) or non-recoverable (plastic and visco-elastic). A *recoverable* deformation disappears on removal of the load while a *non-recoverable* deformation remains after removal of the applied load. Both these deformations may be independent or dependent on time. Metals show less elastic but more plastic deformation at even room temperature while steady flow is predominant at higher temperature ranges. Refer Fig. 8.1 for effects of deformation.

8.2 ELASTIC DEFORMATION

Many solid materials behave elastically when subjected to a force or load. The solid deforms when it is loaded but returns to its original position when unloaded. Elastic behaviour is common to metals, rubbers, ceramics and polymers. A change in pressure, or an application of load, results in elastic deformation. Elasticity has its origin in the forces between atoms of the solid, and, therefore, depends on both the chemical bonding and the structure of solid. The elastic properties are independent of crystalline imperfections such as vacancies and dislocations. See Fig. 8.2 for elastic deformation.

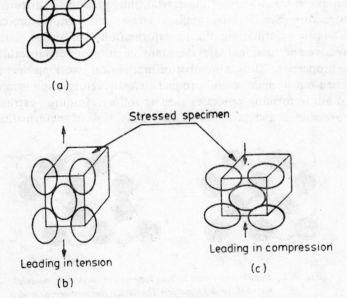

Fig. 8.2 *Elastic deformation*

8.2.1 Ideal Elastic Deformation

Ideal deformation takes places instantaneously upon application of force and disappears completely on removal of the force. Such deformations obey Hooke's law and the elastic strain of the metal is directly proportional to the applied force. Such strain should not be more than one percent of the original length of the material. In other non-crystalline materials such as rubber, elastic deformation is a few hundred percent as compared to about one percent in metals. Ideal deformation occurs with comparatively smaller deformation forces, which can keep working stresses within the elastic limit.

8.2.2 Deviation from Perfect Elastic Behaviour

Hooke's law does not hold good for all engineering applications where deformation is large. Thus, the relationship between stress and strain is no longer applicable. Non-ferrous metals do not exhibit the linear law from the start of the stress-strain diagram. For ductile materials (such as mild steel), at yield point, the rate at which strain changes is not proportional to the stress, due to internal flow or yielding of the metal, and the deformation is non-recoverable.

8.3 PLASTIC DEFORMATION

At stress exceeding the elastic limit, plastic deformation is observed, which always follows elastic deformation. This is accompanied by changes in the internal and external state of the crystal, thus producing distortion of the microstructure. Strain rate, applied stress and temperature control the rate of plastic deformation. Plastic deformation can occur under tensile, compressive and torsional stresses. Many engineering metals exhibit elasto-plastic properties. Thus, a number of mechanical working operations can be carried out to make useful products. Plastic deformation is intentionally carried out in forming processes such as rolling, forging, extrusion, spining, pressing, drawing, stamping, etc. This type of deformation has one

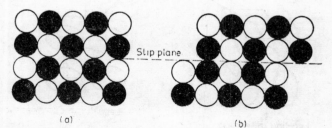

Fig. 8.3 *Plastic deformation of a long-range ordered material (a) Before deformation (b) After the passage of a unit dislocation*

important aspect, the strengthening of metals by cold working. Machinability and wear resistance are entirely dependent on the plasticity of the metals. Refer Fig. 8.3 for plastic deformation.

8.3.1 Mechanisms of Plastic Deformation

As discussed earlier, plastic deformation is associated with dislocation of the atoms within the grains, producing permanent changes in the shape of the material. Plastic deformation of a single crystal occurs by two mechanisms:
  (i) Deformation by slip
  (ii) Twinning

DEFORMATION BY SLIP Slip is a far more common deformation mechanism, where plastic shear moves many interatomic distances with respect to their initial positions. It is also called *shear deformation*. It represents a large displacement of one part of the crystal with relation to another, which may glide more readily along others. It ocrurs on certain planes in the crystal, the slip planes, and in certain directions, the slip directions. Generally the slip plane is the most dense atomic plane and the close-packed direction within the slip plane is the slip direction. In order to produce slip, the criticalre solving shear stress must occur. Figure 8.4 schematically illustrates the slip process in a single crystal under a shear load.

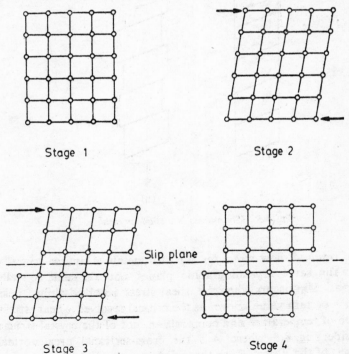

Fig. 8.4 *Slip process in a single crystal*

Slip process in a single crystal As a result of deformation, the thin layers of the crystal are displaced in such a manner that the whole process is non-reversible and there is no return to the original shape of atomic blocks after withdrawal of the external force. Slip is not restricted to the sides of the crystal. Slip causes permanent displacement and removal of the stress will not return the crystal planes to their original positions It occurs readily along certain directions and planes, as illustrated in the Fig. 8.5. The shear stress which produces slip on a crystal plane is called the critical shear stress.

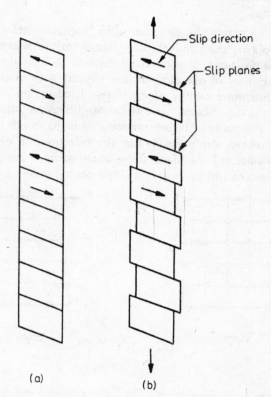

Fig. 8.5 *Slip process in a single crystal*

Critical resolved shear stress for slip All metals of similar crystal structure slip on the same crystallographic planes and the same crystallographic directions. Slip occurs when the shear stress resolved along these planes reaches a certain value known as the critical resolved shear stress. It is a function of temperature and composition, not of the crystal structure.

Consider Fig. 8.6 where A is the cross-sectional area normal to the direction of the force F, so that axial stress is F/A. As the critical shear stress depends upon the direction of force, the slip direction and the angle (λ) between the direction of the force and the slip plane, axial load in the direction of slip is $F \cos \lambda$. Accordingly, the resolved shear stress

$$\tau = \text{Force/Area}$$
$$r = \frac{F \cos \lambda}{A \cos \phi} = \frac{F}{A} \cos \lambda \cos \phi$$

where ϕ = Angle between the direction of the force and the normal to the slip plane

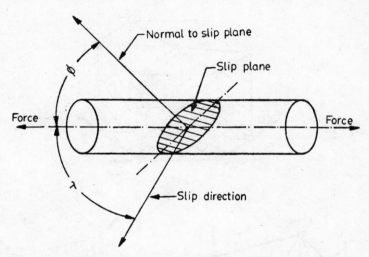

Fig. 8.6 *Relation between tensile stress and critical resolved shear stress*

TABLE 8.1

Metal	Slip Systems		Critical resolved shear stress kg/cm²
	Slip planes	Slip directions	
Copper	{111}	< 110 >	5
Aluminium	{111}	< 110 >	8
Nickel	{111}	< 110 >	
Silver	{111}	< 110 >	
Magnesium	{0001}	< 1120 >	4.5
Cobalt	{0001}	< 1120 >	67.5
Zinc	{0001}	< 1120 >	
β-Tin	{100}	< 0001 >	
	{110}		
Molybdenum	{110}	< 111 >	730
Iron	{110}	< 111 >	280

This equation is known as Schemid's law. Slip occurs with minimum axial force when both λ and ϕ are at 45° and the resolved shear stress is equal to one half the axial shear stress F/A. Slip always begins when the shear stress across the slip planes reaches critical value. The resolved shear stress is less in relation to the axial stress for any other crystal orientation, dropping to zero as either λ or ϕ approaches 90°. Critical shear stress is constant for each material. Table 8.1 gives the critical shear stress for several pure metals. Slip planes and directions for different lattice systems are illustrated in Fig. 8.7.

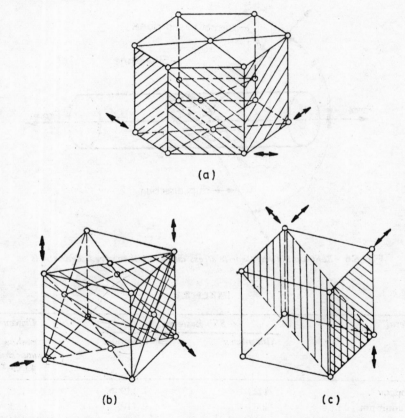

Fig. 8.7 *Slip systems in the common metal lattices (a) CPH lattice (b) FCC lattice (c) BCC lattice*

Factors influencing critical shear stress
1. Purity in metals reduces the critical stress.
2. Surface effects like surface films greatly increase the critical shear stress.
3. Temperature increases the thermal mobility and hence decreases the critical shear stress.

4. Rate of deformation and the extent of initial deformation will also raise critical shear stress.
5. In polycrystalline metals, slip takes place in the preferred directions. The direction of the planes is indicated in such a piece of metal, after deformation, by the presence of slip bands which form on the surface of the metal. Slip bands are made up of several slip planes that are parallel and do not extend through the specimen but stop at the grain boundaries. Slip bands are parallel in a single crystal but differ in orientation from one crystal to another. Figure 8.8 shows slip bands before and after deformation.

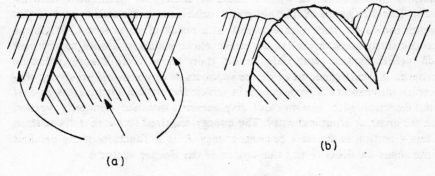

Fig. 8.8 *Slip bands*

Slip systems in common metal lattices A slip plane and a slip direction together constitute a slip system.

(i) The closed-packed hexagonal lattice cell has possible alternative slip-plane systems and directions in addition to the basal plane and its three direction axes (Fig. 8.7 (a)). This is a 3-slip system.

Table 8.2 *Slip Systems Observed in Crystal Structures*

Structure	Slip planes	Slip directions	No. of slip systems
FCC	{111}	< 110 >	4 × 3 = 12
BCC	More common {110}	< 111 >	6 × 2 = 12
	Less common {211}		12 × 1 = 12
	{321}		24 × 1 = 24
HCP	{0001} Basal plane	< 1120 > Close-packed directions	1 × 3 = 3 or 3 × 1 = 3
NaCl	{110}	< 110 >	1 × 6 = 6 or 6 × 1 = 6
Al_2O_3	{0001}	< 1120 >	1 × 3 = 3

(ii) The face centred cubic type of lattice has four sets of easy-slip close-packed planes, the octahedral planes, each with three possible axes of slip. In addition there are nine other possible systems of slightly less easy slip (Fig. 8.7 (b)), each with two possible axes of slip. This makes a total of thirty possible axes of slip, i.e., a 12-slip system.

(iii) The body-centred cubic type of lattice has no close-packed planes, but has six systems of easy slip planes (Fig. 8.7 (c)), each with two possible axis of slip, and three systems of slightly less easy slip with two slip axes, giving a total of eighteen possible slip axes.

Mechanism of slip (dislocation theory) Lattice imperfection is called dislocation. Slip is observed when a material undergoes permanent deformation. The discussion of the observed behaviour in an earlier part of this chapter indicated that slip occurred as a result of shear stress. We know that the shear mechanism cannot cause whole planes of atoms to slip past each other because materials would then show their theoretical shear strength. All experimental evidence supports a mechanism involving dislocation movements. The direction in which the critical shear stress is least is the direction with the shortest displacement distance (Burger's vector) and the greatest atomic density. The energy required to move a dislocation in this direction is the least because energy E is a function of the product of the shear modulus G and the square of the Burger vector b

$$E = f(Gb^2)$$

It has been experimentally proved that, in single crystals of metal, slip starts under a shear stress at least two orders of magnitude lower than the values of 1.5% of E (modulus of elasticity) worked out for a perfect crystal.

Dislocation is one of the important types of crystal imperfections which accounts for differences in the actual and theoretical strength of crystals. The theory of dislocation is helpful in illustrating the stress, strain and other properties of materials. Dislocation in metallic crystals can be observed directly with high-resolution electron microscopy and X-ray diffraction. Slip caused by dislocation is illustrated in Fig. 8.9. The dislocations are edge dislocation, screw dislocation and a mixed one. These have been thoroughly discussed in Ch. 5 or crystal imperfections

Effects of dislocation on cold working If various types of dislocations interact, it can immobilize a dislocation and just reduce its mobility. It is apparent that dislocation mobility decreases as the density of dislocation increases because a density increase causes more interactions. This would imply that the yield stress of a material depends on its dislocation density. The dislocation density can be increased by plastically deforming a crystalline material.

Deformation of Materials 117

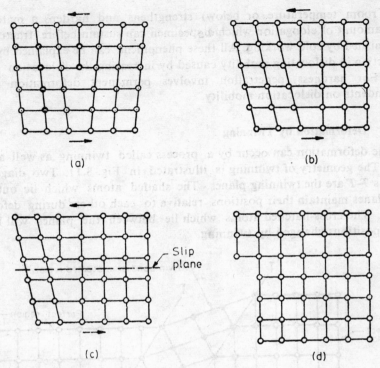

Fig. 8.9 *Progress of dislocation across a crystal*

The effect of deformation on tensile strength, yield point, hardness and elongation is shown in Fig. 8.10. It is seen that cold working (deforming

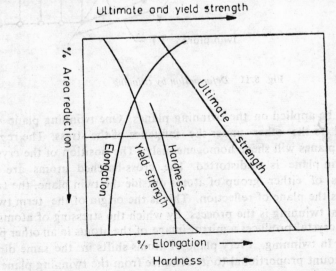

Fig. 8.10 *The effect of cold working on the tensile properties of steel*

it at room temperature or below) strengthens and hardens a material. The amount of elongation which a specimen can sustain before fracture is diminished by cold working. All these phenomena can be explained by the reduction in dislocating mobility caused by increasing the dislocation density. For hardness, penetration involves permanent deformation and dependents on dislocation mobility.

8.3.2 Deformation by Twinning

Plastic deformation can occur by a process called twinning as well as by slip. The geometry of twinning is illustrated in Fig. 8.11. Two diagonal planes T-T are the twinning planes. The shaded atoms which lie outside the planes maintain their positions relative to each other during deformation. The cross-hatched atoms which lie between the planes will have their positions changed by twinning.

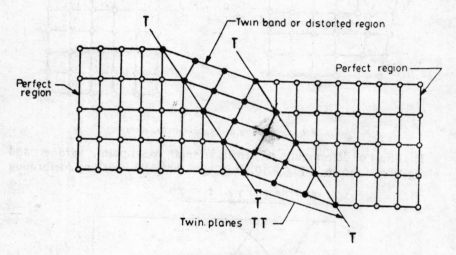

Fig. 8.11 *Deformation by twinning*

Let a stress be applied on the twinning planes. One twinning plane will move parallel to the other under the influence of the stress. The region between twin planes will shear homogeneously. The position of the crystal outside the twin plane is undistorted. The cross-hatched atoms are the mirror images of either group of atoms outside the twin plane, the twin plane acting as the plane of reflection. This is the origin of the term twin. In other words, twinning is the process by which the stressing of atoms in one part of a crystal produces a mirror image of the atoms in an other part of the crystal. In twinning, every plane of atoms shifts in the same direction by an amount proportional to its distance from the twinning plane.

Generally it is not a significant deformation mechanism for cubic metals unless one is concerned with low temperature regions or fast forming processes. Twinning becomes more significant in the case of HCP metals and is observed at room temperature. It occurs on the $(10\bar{1}2)$ plane in the $(10\bar{1}1)$ direction in HCP crystals. Twinning re-orients the crystal region between the twin planes and thus facilitates further slip by placing the planes or potential slip in an orientation that is more favourable for dislocation slip. Twinning also helps in changing the shape of the surface. As twinning is produced suddenly it is accompanied by sound. Thermal treatment, impact and plastic deformation are the causes of twinning. Movements of twin boundaries in heat treatment, as in annealing, is significant. The actual deformation produced by twinning is small, having a maximum of perhaps several percent for ductile HCP crystals. Twins are of two types.

(i) Mechanical twins produced by mechanical deformation.
(ii) Annealing twins produced as a result of annealing following plastic deformation.

NEUMANN BANDS The stress to twin a crystal tends to be higher than that required for slip. Hence, except under certain conditions, slip is the normal deformation mechanism. For example, twinning stress is less sensitive to temperature than the stress for slip; consequently twinning becomes more favourable as the deformation temperature becomes lower. This is the case when BCC iron and its alloys are rapidly loaded at lower temperatures, when thin lamellar twins appear, called Neumann bands (Fig. 8.12).

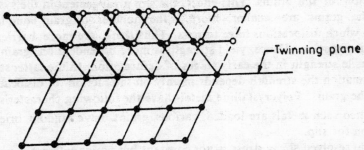

Fig. 8.12 *Neumann bands—The arrangement of atoms on either side of a twin interface*

8.4 DEFORMATION IN POLYCRYSTALLINE METALS

In the previous sections of this chapter, the deformation of single crystals has been considered. Dislocations and the deformation mechanisms of slip and twinning were considered. Deformation in polycrystalline metals like brass are complex, as compared to single crystals, when the effects of grain boundaries and of neighbouring crystals are taken into account. Secondly, common metals are polycrystalline aggregates because they are made up of a number of small crystals or grains.

The slip mechanism operates because of the geometric regularity of a crystal. Deviations from perfect crystal structure tend to reduce the mobility of dislocations. Because a grain boundary is a surface at which two crystals of different orientations meet, the arrangement of atoms in the region of a grain boundary deviates considerably from that found in a single perfect crystal. The effect of a grain boundary on the dislocations is to immobilize them by the repulsive force exerted on the successive dislocation coming down the slip plane. The grain boundary has caused a dislocation pile up. Such pile ups are often observed by electron microscope. Because grain boundaries diminish dislocation mobility, polycrystals are generally stronger and less ductile than single crystals. Static fracture, fatigue and creep strength are increased by changes in grain boundaries.

Structural changes in the deformation of polycrystalline metals are the appearance of slip bands or slip lines on even the slightest deformation. Heavily deformed structures are known as banded or fibrous structures.

As some of the crystals in polycrystalline materials are always oriented with their slip directions in the plane of maximum shear, they tend to behave as single crystals. This results in minute yielding in these crystals when neighbouring crystals are more elastic. Hence yielding is very gradual in the beginning.

Another factor influencing mechanical properties in the early stages of deformation in polycrystalline materials is the grain size. Plastic deformation in the polycrystalline material is never uniform, due to the random orientation of the grains. This effects the slip arrangement in the system. When the grains are smaller, more is the number of grain boundaries through which dislocations have to pass. Thus there are more barriers to dislocation motion. Hence, yield strength is more dependent on grain size than tensile strength in the early stages of deformation. In the latter stages of deformation the strength depends mainly on interaction of dislocations within the grains. Polycrystalline metals have the following characteristics:

(i) When such metals are loaded, various grains have random orientation for slip.
(ii) The resolved shear stress is not constant but varies with grain orientation; because of random orientations greater stress is required.
(iii) There are a number of slip planes within a crystal.
(iv) Considerable structural changes result in deformation.
(v) Under load, all crystals are perfectly elastic.
(vi) They show elastic after effects which can not be obtained in single crystals. Application of this effect is in work hardening.
(vii) After elastic deformation, a significant amount of permanent deformation and an appreciable amount of yielding can be obtained in the crystals.

8.5 WORK HARDENING

We have discussed the mechanism of deformation in a single crystal and polycrystalline materials. Now the effects of deformation in structural changes and in properties of metals is outlined in this section. Since the change in hardness is of prime importance, it is considered first.

Ductile materials show increase in strength and hardness when plastically deformed at temperatures lower than the crystallization temperature. This is called work hardening or strain hardening. A little above the elastic limit, a small plastic deformation results in a relatively large increase in strain hardening. The rate of strain hardening decreases rapidly beyond the elastic limit and becomes a constant value until fracture occurs. The strain hardening process is of great significance in many metal-forming and fabrication operations in industries. As a result of work hardening, electrical conductivity can be decreased and the rate of chemical action increased. Work hardening might cause development of internal stresses, increase in corrosion and crack formation, elastic distortion and fragment of crystals.

The mechanism of work hardening is based on the interaction of dislocations with each other and with barriers which immobilize them when crossing through them. The simultaneous action of deformations in the various slip planes is continuous, causing deformation of metals. These dislocations pile up at the boundaries in the crystals causing density dislocations. Applied stress on the slip planes is opposed by the newly developed back stress produced by pile-ups, the greater the amount of dislocation density the more being the effect of work hardening or straining. Another mechanism of work hardening is based on the intersection of various dislocations moving through the active slip planes. In both cases the resistance

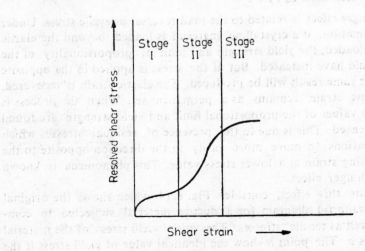

Fig. 8.13 *Three stages of work hardening*

of the metal to deformation is increased to work or strain the metal. This implies an increase in strength as well as hardness. Work hardening reduces ductility.

Three stages of work hardening for a work hardenable metal are shown in the stress-strain curve in Fig. 8.13.

(i) Corresponding to stage I, the work hardening rate is practically constant.
(ii) At stage II, the work hardening rate is higher than that of stage I due to the predominant action of dislocation.
(iii) The material continues to work harden in stage III but at a decreasing rate due to the lower mobility of dislocation.

8.6 SEASON CRACKING

Season cracking causes disintegration of the metal. Season cracking results from the combined effects of internal stress and intercrystalline corrosion. This term is commonly applied to metals such as brass. Internal stresses of very high intensity may be left in an object after cold working. These stresses are susceptible to corrosion in parts of metals if these are stored for a long period. Internal stresses make alpha brass more susceptible to inter-granular cracking in atmospheres containing ammonia. Development of cracks is accelerated by corrosive agents like sea water, industrial atmosphere, etc. This defect is overcome by annealing the brass at 200-300 °C.

The tendency of brass cracking also increases with increased zinc or alloying content which results in active response to inter-granular attack along the grain boundaries.

8.7 BAUSCHINGER EFFECT

The Bauschinger effect is related to the load reversal or cyclic stress. Under plastic deformation, if a crystalline material is loaded beyond the elastic limit and unloaded, the yield strength and limit of proportionality of the material would have increased. But if the stress is applied in the opposite direction, the same result will be produced. The elastic strain is recovered, but the plastic strain remains as a permanent set. When the process is repeated, the values of the proportional limit and yield strength are found to have decreased. This is due to the presence of residual stresses which cause dislocations to move more easily in the direction opposite to the original, causing strain at a lower stress value. This phenomena is known as the Bauschinger effect.

To illustrate this effect, consider Fig. 8.14 which shows the original stress (σ), strain (ϵ) diagram for a ductile meterial subjected to compression as well as tensile stresses. The initial yield stress of the material in tension is a. The point b show the identical value of yield stress if the material is tested in compression. This strength will indicate point c on the path *OaC* if it further stressed in this order. On unloading, the path of the tensile stress will be along *cd*. If at this stage compressive stress is applied, it will be observed that yield strength (compressive) has been decreased. It is now *e* instead of *a*. While the yield stress in tension was increased by strain hardening or work hardening from *a* to *c*, the yield strength in compression was decreased. This phenomenon occurs because the subsidual stress caused by the previous (tensile) strain increased the stress under the reversed load as when the stress is reversed. This is Bauschingers effect. It is given by the strain *B* called Bauschinger's strain.

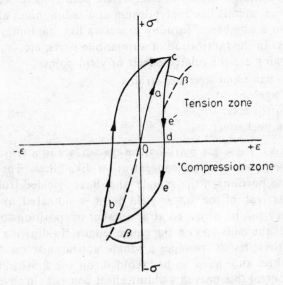

Fig. 8.14 *Bauschinger's effect and hysteresis loop*

This phenomenon is observed in the cold working of metals. Cold rolled steel for instance, is strengthened by work hardening. Rollers compress the mild steel stock and cause elongation, thus increasing the tensile yield stress. However, lateral compression causes a decrease in tensile yield stress across the bar.

8.8 YIELD POINT PHENOMENON AND RELATED EFFECTS

Figure 8.15 shows the stress-strain diagram obtained during tensile deformation when a sample of mild steel was loaded. It behaves elastically upto a certain load and then suddenly yields plastically. The first higher point at which yielding starts is called the upper yield point and the other lower value having more strain is known as the lower yield point. The stress required to start plastic flow is higher than to maintain plastic flow, as is clear in the diagram, after the yield point is obtained.

Due to dislocations there is work hardening in the sample. This causes the stress-strain curve to acquire a smooth and gradual rise. This is called the *yield point phenomenon* which states that the elongation that occurs at constant stress has an important bearing in the yielding strength value. There is a large strain energy associated with the dislocation of a crystal lattice which will try to reduce it during rearrangement of the crystal structure. Impurities also reduce dislocation movements. Thus, a higher stress is needed.

In this way, yield point has significance in the practical use of various metals. In addition to iron and steel, yield points have been observed in polycrystalline metals like molybdenum and alluminium alloys. It is also important for a number of forming processes like stampings and drawing of thin sheets in the fabrication of automobile parts, etc.

The following are the related effects of yield point:
(i) Luders band and stretcher strains
(ii) Strain ageing
(iii) Blue brittleness
(iv) Orange peel effect

LUDERS BANDS   These are markings on the surface of a tensile test sample, formed at the points of stress concentration like fillets. These bands distinguish those portions of the sample that have yielded from those which have not. Arrival of the upper yield point is indicated by the formation of these. They can be observed at a number of positions on the sample, e.g., at both the ends and on the gauge length. In drawing and stamping operations, these bands produce a rough appearance on the surface of the metal sheet and have to be avoided on the finished products. The remedy to control this uneven yielding effect consists in overstraining the sheets prior to pressing operations, by means of temper roll pass, so that the yield point phenomenon is eliminated.

STRAIN AGEING   This refers to changes in the properties of an over-strained alloy with time. If a test sample which has been overstrained to remove the yield point is allowed to rest or age after plastic deformation, the yield point returns with a higher stress (Fig. 8.15 (c)). If the specimen is unloaded and retested without ageing then the yield point will occur (Fig. 8.15 (b)) due to weak dislocation effects from the atmosphere of carbon and nitrogen atoms. Strain ageing or strain-age-hardening is accompanied by hardening due to increase in stress value. This is observed when the sample, after the test of Fig. 8.15 (a), is unloaded and permitted to rest or age for some time and then retested, when the yield point as in Fig. 8.15 (c) reappears. It is evident in commercial iron and steel. It has an important influence on the study of fatigue and creep behaviour of some metals.

BLUE BRITTLENESS   This is the effect produced when the temperature range of the test is raised; the yield point becomes less pronounced. Failure of the sample occurs at low strain with high stress. The section of the stress-strain diagram shows a steeper curve with rapid formation of successive yield points due to quick diffusion and dislocation effects. This effect appears at about 160 to 300 °C (Fig. 8.15(d)).

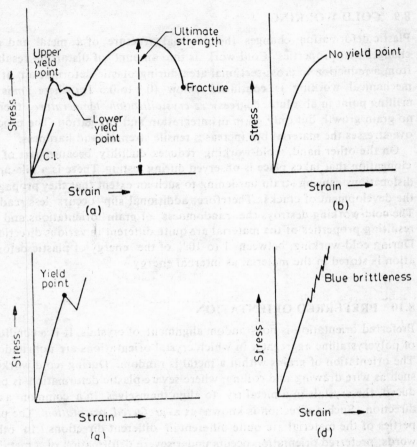

Fig. 8.15 *Yield point phenomenon and related effects (stress-strain diagram) (a) Anchored atmosphere (first test) (b) Free atmosphere (unloaded retest) (c) Strain aging (anchored atmosphere)*

ORANGE PEEL EFFECTS These markings appear on metals during stretching. It is also a surface roughening phenomenon commonly seen in coarse-grained metals at low temperature-forming operations. Grain refinement eliminates this effect.

The sharp yield point observed in many metals is a result of the Cotterell theory which states that there are always special atmospheres to reform the interaction between the dislocations and the solute atoms (e.g. carbon, nitrogen, etc.). Such atmospheres anchor the dislocations and higher stresses are required to free a dislocation. The value of this stress should be corresponding to upper yield points. Free dislocations lower the yield point. Stress-strain diagrams shown also indicate that a Cotterell atmosphere gives yield point and blue brittleness.

8.9 COLD-WORKING

Plastic deformation changes the internal structure of a metal and also changes its properties. Cold-work is the amount of distortion resulting from a reduction in cross-sectional area during plastic deformation. If the mechanical working is continued below (0.3 to 0.5 Tm where Tm is the melting point in absolute degrees) *re-crystallisation temperature*, there is no grain growth but only grain disintegration and elongation. The process overstresses the material but increases tensile strength and hardness.

On the other hand, cold-working reduces ductility because part of the elongation that takes place is observed during testing. There is a pile-up of dislocations causing strain hardening to such an extent that they propagate the development of cracks. Therefore, additional slip occurs less readily. The cold-working destroys the randomness of grain orientations and the resulting properties of the material are quite different in various directions. During cold-working, between 1 to 10% of the energy of plastic deformation is stored in the material as internal energy.

8.10 PREFERRED ORIENTATION

Preferred orientation is non-random alignment of crystals. It is a condition of polycrystalline aggregates in which crystal orientations are not random. The orientation of grains within a metal is random. During cold working such as wire drawing and rolling, where severe plastic deformation is produced, the crystals in a metal try to align themselves in a common axial direction. Such a direction is known as a *preferred orientation*. The properties of the material are quite different in different directions. In other words, preferred orientation occurs under severe deformation of a metal in which certain crystallographic planes tend to orient themselves in a preferred number with respect to the direction of maximum strain (Fig. 8.16).

Preferred orientation resulting from plastic deformation is strongly dependent on the slip and twinning systems. The slip planes of all crystals are caused to rotate into more favourable directions with respect to the direction of critical shear stress. These are determined by X-ray methods. The X-ray pattern satisfies the condition for Bragg reflections.

Preferred orientation has the following engineering applications:
1. Directional properties such as strength and magnetic permeability are valuable in crystals of some metals (silicon steels), but undesirable in others.

2. The manufacture of transformer cores of iron sheets helps in orienting the core iron in the direction of the magnetic field, thus saving a large amount of electric energy dissipated as heat in the iron core by eddy currents.
3. The formation of a strong preferred orientation will result in an anisotropy in mechanical properties, i.e., during fabrication, different mechanical properties can be obtained in different directions.
4. It results in the loss of ductility in semifinished products like wires and sheets.
5. In cast metals having columnar growth.

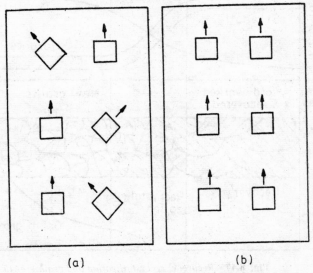

Fig. 8.16 *Grain orientation (a) Random (b) Preferred*

8.11 RECOVERY, RECRYSTALLISATION AND GRAIN GROWTH

Recovery, recrystallisation and grain growth are the phenomena intimately associated with the annealing of a plastically deformed crystalline material. Plastic deformation breaks up the structure of the crystal lattice and causes distortions, which make the cold-worked metal unstable thermodynamically. Heating the metal and releasing the internal stresses will return it to its undeformed state by lowering its energy. If the temperature is raised sufficiently, the metal tries to approach equilibrium through three stages

(i) Recovery
(ii) Recrystallization
(iii) Grain growth.

Figure 8.17 indicates these three stages and their effects on material properties.

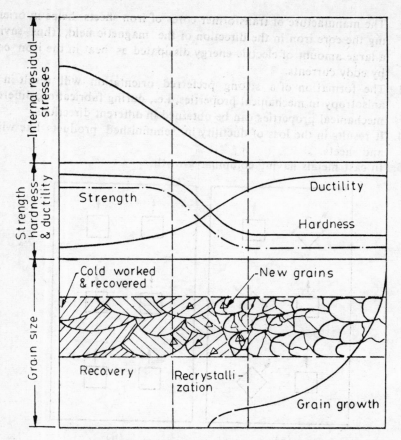

Fig. 8.17 *Recovery, recrystallization and grain growth*

8.11.1 Recovery

This refers to the improvement of properties of a material which occurs before recrystallisation and therefore does not affect grain growth. The extent to which a property, such as hardness or strength, is modified by recovery is usually small compared to the change caused by recrystallisation. Recovery occurs at moderate temperature; some metals exhibit recovery below room temperature. The process of recovery is thought to be the result of dislocations interacting with each other at high dislocation densities, by virtue of attractive and repulsive forces. The interaction is enhanced by the unpinning of dislocations; climb appears to be an important unpinning mechanism in the process.

Recovery softens a material by reducing the dislocation density. Low angle grain boundaries are formed by the climb mechanism which allows the movement of dislocations. The formation of low-angle grain boundaries during recovery, called polygonization, helps in the softening of metals or ceramics. Low-angle grain boundaries often form polygons within the grain; hence the name polygonization (Fig. 8.18).

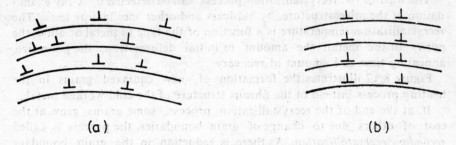

Fig. 8.18 *Recovery movement of dislocation to produce polygonization (a) Uncovered (b) Recovered*

There is improvement in the elastic limit and ductility during recovery but there is no change in microstructure at this stage. The rate of recovery is fastest initially, but drops off with time. It is also a function of temperature and is the first stage of annealing. Recovery eliminates voids or vacancies which were created by the interaction of dislocations. The recovery process is used to relieve stresses trapped during welding, casting, forging and extrusion operations without affecting the strength obtained during work-hardening.

The major property changes are—a noticeable increase in electrical conductivity and the reduced distortions produced by residual stresses, as in casting and weldings.

8.11.2 Recrystallization

Recrystallization is the process of nucleation and growth of new crystals which replace all the deformed crystals of the cold-worked material. It occurs on heating to a temperature above the recovery range. Recrystallization follows recovery. It is a process in which new, strain-free grains are nucleated and grow until they have consumed all the work-hardened material. New grains having a low dislocation density appear as a result of recrystallization, and the hardness and strength of the material are significantly reduced while ductility increases. The largest changes in physical properties occur during the recrystallization phase of annealing. The fraction of a material that recrystallizes during annealing depends on both time and temperature.

Recrystallization lowers the energy of a solid by eliminating dislocations; therefore, it is the strain energy of the dislocations that causes recrystallization to occur. A crystal which has been severely work hardened has a high dislocation density and therefore, releases more energy during recrystallization than one that is less work hardened. As a result, heavily cold-worked specimens recrystallize at higher rates. The recrystallization rate is also affected by all the parameters which influence dislocation mobility, such as impurities and grain size.

The start of the recrystallization process can be detected by X-ray examination of the microstructure, by hardness and other mechanical tests. The recrystallization temperature is a function of the type of metal or alloy, the purity of the metal, the amount of initial deformation, the grain size, annealing time and amount of recovery.

Figure 8.17 illustrates the formation of new equiaxed grains in the heating process instead of the fibrous structure of the cold-worked metal.

If, at the end of the recrystallization process, some grains grow at the cost of others due to change of grain boundaries, the process is called *secondary recrystallization*. As there is reduction in the grain boundary energy which causes reduction in grain boundary area, this phenomenon is also known as *abnormal grain growth*. After completion of this stage (recrystallization), the metal reprocesses its original structure and physical properties.

8.11.3 Grain Growth

This is the third stage of annealing that follows recrystallization. Heating beyond recrystallization temperature range causes the size of the recrystallized grains to increase. Some of the grains grow by consuming others. Grain growth lowers the energy of a solid because surface energy is associated with grain boundaries. As in the other annealing processes, high temperature gives atoms of the crystals sufficient mobility to arrange themselves in low-energy configurations. Some softening is associated with grain growth because grain boundaries impede dislocation slip. Grain size affects the surface appearance of the metal.

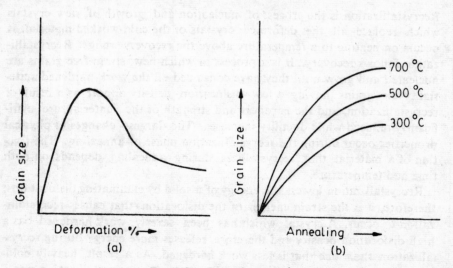

Fig. 8.19 *Grain size control*

All crystalline materials, metals and non-metals, exhibit these characteristics of grain growth. An increase in temperature enhances the thermal vibration of the atoms, which in turn facilitates the transfer of atoms across the interface from small to large grains. A subsequent decrease in temperature slows down this process but does not reverse it (Fig. 8.19 (b)).

The only way to decrease grain size (refine) is to cold-work and plastically deform the existing grains and start new grains (Fig. 8.19 (a)). The process of grain growth (Fig. 8.20) depends on the following:

(i) Annealing temperature and time
(ii) Rate of heating
(iii) Insoluble impurities
(iv) Alloying elements
(v) Rate of cooling
(vi) Extent of initial deformation.

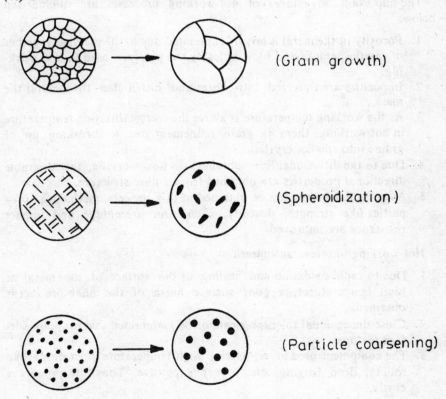

Fig. 8.20 *Process of grain growth*

8.12 HOT-WORKING

The plastic deformation of metals or alloys has large applications in the production and forming of metals to any desired shape. It must be noted that the cold-working process is carried out below the recrystallization temperature whereas hot working processes are carried out at temperatures above the recrystallization range. In hot working, hardening due to deformation and softening due to annealing occur simultaneously. The work-hardening process decreases when the temperature is increased constantly due to the active slip system. If plastic deformation is carried out at a higher rate than softening, the metal will become harder. Thus, at a temperature where hardening and softening processes balance each other, the rate of deformation is termed temperature deformation. Temperatures above this will result in hot working and below in cold working. The rate of deformation is an important factor in hot-working operations. Slow rates allow annealing to occur, thus reducing the power required for hot working. High rates of deformation are undesirable as they inherit the characteristics of the cold working process.

The important advantages of hot-working processes are summarised below:

1. Porosity in the metal is largely eliminated due to the pressing together of cavities, cracks and blow holes by the pressure used in hot working.
2. Impurities are squeezed into fibres and distributed throughout the mass.
3. As the working temperature is above the recrystallization temperature in hot working, there is grain refinement due to breaking up of grains into smaller crystals.
4. Due to the directional flow obtained in hot working, the desirable directional properties are obtained from a fibre structure.
5. As grain refinement can be obtained in this process, mechanical properties like strength, ductility, elongation precentage and impact resistance are improved.

Hot working has several limitations:

1. Due to rapid oxidation and scaling of the surface of the metal at high temperatures, a poor surface finish of the final product is obtained.
2. Close dimensional tolerances cannot be maintained due to poor surface texture.
3. The equipment used to withstand high-temperature operations like rolling, drop forging, etc. is very expensive. Thus the process is costly.

REVIEW QUESTIONS

1. Explain deformation of metals. How does it take place? State its effect.
2. Compare elastic and plastic materials when they are subjected to tensile loads.

3. State the difference between elastic and plastic deformation. Explain each in detail.
4. Explain the terms slip and twin. How does slip occur? Explain slip directions and slip planes with diagrams.   (AMIE May 1978)
5. Explain the difference between slipping and twinning. How does twinning occur in metals? Name and explain two types of twins.   (AMIE Dec 77)
6. Write short notes on:
   (a) Season-cracking of brass   (AMIE Winter 1986, Summer 1987, Winter 1987,
   (b) Recrystallization
   (c) Grain growth
   (d) Preferred orientation
   (e) Dislocation theory
   (f) Elastic and plastic deformation in steel
7. What is slip? On what crystallographic planes and in what directions it is likely to occur in BCC, FCC and HCP metals?
8. Explain the mechanism for dislocation in plastic deformation with neat sketches.   (AMIE 1974)
9. What is critical resolved shear stress? On what factors does it depend?   (AMIE Summer 1987)
10. What is strain hardening? State the effects of strain hardening.
11. Write a short note on the Bauschinger effect.   (AMIE Winter 1982, Winter 1987)
12. (a) What is the effect of cold working on the mechanical properties of metals and alloys?
    (b) Discuss the changes in properties when a severely cold-worked metal is annealed at successively higher temperatures.   (AMIE 1977)
13. Distinguish between the term 'recovery' and the 'recrystallization' involved in the process of heating cold-worked metals.   (AMIE 1974, Winter 1987)
14. (a) Define Burger's vector and illustrate it on the sketch of an edge dislocation.
    (b) What is the significance of dislocations in the plastic deformation of metals?   (AMIE 1977)
15. What is hot and cold working? How do they differ? Explain their effects on the properties of Materials.   (AMIE Summer 1984, 87)
16. Explain the yield point phenomenon in materials in terms of dislocation.
17. Explain Bouschinger effect as applied to the deformation of metals.

9
THEORY OF ALLOYS: CONSTITUTION AND EQUILIBRIUM DIAGRAMS

9.1 INTRODUCTION

Most commercial metals have the property of unlimited mutual solubility in the liquid state. Metals generally form homogeneous liquid solutions in the liquid state. Few metals (lead and iron) are completely insoluble in each other and separate according to their specific gravity in the liquid state. Limited solubility in the liquid state is much more frequent. In such cases a homogeneous liquid solution will be obtained if the amount of second metal added to the first metal does not exceed its maximum solubility in the first metal at the given temperature. If the maximum solubility of any two of the metals is increased, the liquid will separate into two layers.

In the solid state the alloy may be present in one or more of the following forms:

1. As a solid solution
2. As an intermediate phase or intermetallic chemical compound
3. As a finely divided mechanical mixture of the metals
4. As a finely divided mechanical mixture of solid solution.
5. As a finely divided mechanical mixture of chemical compounds of metals, the individual metals and solid solutions

Alloys are more widely, used in industry than pure metals because their physical and chemical properties can be easily varied to suit the exact individual requirement. This can be achieved by making alloys of different metals. The characteristics of any series of alloys can be studied suitably by using alloy diagrams. From such diagrams it is possibe to get an idea of the structure and the physical and chemical properties of any given alloy.

9.2 BASIC TERMS

In the study of solid phases and phase diagrams the following terms are frequently used.

A *system* is explained as the whole complex of phases of one or several components at different pressures and compositions. It may be composed of solids, liquids, gases or their combinations and may have metals and non-metals separately or in any combination. It is so isolated from its surroundings that it is unaffected by these and is subjected to the change in the overall composition, temperature, pressure or total volume, only to the extent allowed by the investigator.

The *components* are substances, either chemical elements or chemical compounds, whose presence is necessary and sufficient to make up a system. A pure metal is a one-component system, an alloy of two metals is a binary or two-component system, etc.

A *phase* is a physically and chemically homogeneous portion of a system, separated from other portions by a surface and an interface, each portion having different composition and properties. In an equilibrium diagram, liquid is one phase and solid solution is another phase. The number of phases in a system is the number of different substances that exist in it in a homogeneous sense.

9.3 SOLID SOLUTIONS

Some metals used commercially for engineering purposes are pure but, in many cases, other elements are added in order to get specific properties. If such additions become an integral part of the solid phase, the phase obtained is called a solid solution, by analogy with familiar aqueous solutions such as sugar in water. Solid solutions form readily when solvent and solute atoms have similar sizes and electron structures, so that it is chemically homogeneous and the component atoms of the elements cannot be distinguished physically or separated mechanically. There is a homogeneous distribution of two or more constituents in the solid state so as to form a single phase or solid solution.

Solid solutions are of two types:
(i) Substitutional
(ii) Interstitial

9.3.1 Substitutional Solid Solution

A solute atom may occupy two alternative positions in the lattice of solvent (matrix) metal. If the two atoms are of comparable size the solute atom will substitute at random for one of the matrix atoms in the crystal lattice. This kind of structure is called a substitutional solid solution. For example, brass is an alloy of copper and zinc which forms solid solutions most readily as the atoms of these elements have similar sizes and electron structure. The solution of copper and nickel to form monel is another example because the nickel atoms substitute for copper atoms in the crystal structure.

Substitutional solid solutions are of two types.
(i) Random substitution solid solutions
(ii) Ordered substitutional solid solutions.

When there is no order in the substitution of the two elements (as shown in Fig. 9.1(a)), the chance of one element occupying any particular atomic site in the crystal is equal to the atomic percent of that element in the alloy. In such a case the concentration of solute atoms can vary considerably throughout the lattice structure. The resulting solid solution is called a random or disordered substitutional solid solution. If the atoms of the solute material occupy similar lattice points within the crystal structure of the solvent material, this is called an *ordered solution* (Fig. 9.1 (b)). Such ordering is common at lower temperatures since greater thermal agitation tends to destroy the orderly arrangement. If the random conditioned alloy is cooled slowly to undergo rearrangement of the atoms, diffusion takes place which results in uniformity and definite ordering of the atoms in the lattice structure.

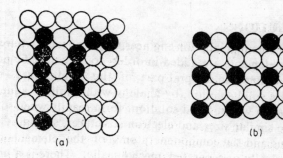

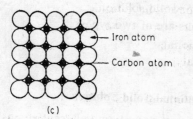

Fig. 9.1 (a) *Solid solutions* (b) *Substitutional solid solutions*
(c) *Interstitial solid solution*

9.3.2 Interstitial Solid Solutions

There are few relatively small atoms that can be accommodated in the interstices between solvent atoms to form an interstitial solid solution. Carbon in iron is the example of such a type of solution which is the basis of steel hardening (Fig. 9.1 (c)). Other elements like:

(i) Nitrogen for maintaining stainless steel in the austenitic condition for nitriding process.
(ii) Hydrogen when introduced into steels during welding operations, acid-cleaning and plating results in hydrogen embrittlement and causes a sharp decrease in ductility.

Interstitial solid solutions normally have very limited solubility and are generally considered of secondary importance. In some alloys, both interstitial and substitutional solid solutions are formed to an appreciable extent.

9.4 HUME-ROTHERY'S RULES

In the course of an alloy development, it is frequently desirable to increase the strength of the alloy by adding a metal that will form a solid solution. In the choice of such alloying elements, a number of rules govern the formation of substitutional work of Hume-Rothery. Unfortunately, if an alloying element is chosen at random, it is likely to form an objectionable intermediate phase instead of a solid solution.

Hume-Rothery's Rules are described below.

CHEMICAL AFFINITY FACTOR The greater the chemical affinity of two metals the more restricted is their solid solubility. When their chemical affinity is great, two metals tend to form an intermediate phase rather than a solid solution.

RELATIVE VALENCY FACTOR If the alloying element has a different valence from that of the base metal, the number of valence electrons per atom, called the electron ratio, will be changed by alloying. Crystal structures are more sensitive to a decrease in the electron ratio than to an increase. Therefore, a metal of high valence can dissolve only a small amount of lower valance metal, while the lower valence metal may have good solubility for a higher valence metal.

RELATIVE SIZE FACTOR If the size of two metallic atoms (given approximately by their constants) differs by less than 15 percent, the metals are said to have a favourable size factor for solid solution formation. So far as this factor is concerned, each of the metals will be able to dissolve appreciably (to the order of 10%) in the other metal. If the size factor is greater than 15%, solid solution formation tends to be severely limited and is usually only a fraction of one percent.

LATTICE-TYPE FACTOR Only metals that have the same type of lattice (FCC for example) can form a complete series of solid solutions. Also, for complete solid solubility, the size factor must usually be less than 8 percent. Copper-nickel and silver-gold-platinum are examples of binary and ternary systems exhibiting complete solid solubility.

A qualitative estimate of the solid solubility of one metal in another can be obtained by considering these four factors. It should be noted that an unfavourable relative size factor alone is sufficient to limit solid solubility to a low value. If the relative size factor is favourable, then the other three factors should be considered in deciding the probable degree of solid solubility. It must be emphasized that numerous exceptions to these Hume-Rothery rules are known.

9.5 INTERMEDIATE COMPOUNDS OR INTERMEDIATE PHASES

In many binary alloy systems (Fig. 9.2), when the chemical affinity of metals is great, their mutual solubility becomes limited and intermediate phases are formed rather than solid solutions. The copper–zinc system is an example of this behaviour. When the solubility of copper in zinc is exceeded, a Zn-rich beta phase appears with the Cu-rich alpha phase. Intermediate phases may range between the ideal solid solution and the ideal chemical compound. The crystal structure and properties of such phases are different from that of the solid solution. The intermediate phases may have either narrow or wide ranges of homogeneity, and may or may not include a composition having a simple chemical formula. Properties of the intermediate phases that obey the valency law like chemical compounds (Al-Sb), as would be expected from their ionic or covalent bonding characteristics, are essentially non-metallic. Moreover, they are brittle and have poor electric conductivity. Such phases are known as intermediate compounds of fixed composition. The other types of intermediate phases of variable

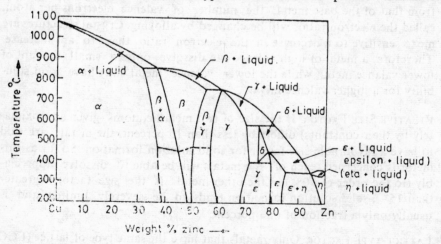

Fig. 9.2 *Copper-zinc phase diagram showing intermediate phases. This diagram represents a series of alloys known as brasses (1) Upto 38% wt zinc, copper forms a substitutional solid solution and α-brasses. (2) Zinc in excess of 38% wt leads to the formation of β-brasses which are harder than α-brasses.*

composition which do not obey the valency law are called electron phases or compounds. Definite compositions of such phases appear in certain binary equilibrium diagrams, which depend on the ratio of electrons to atoms at those compositions. Many of these compounds are grouped in three classes according to their electron ratio:

(i) 3 : 2 ratio produces BCC beta phase in copper and zinc (CuZn)
(ii) 21 : 13 ratio produces complex cubic gamma phase in copper and zinc (Cu_5Zn_8)
(iii) 7 : 4 ratio produces a close-packed hexagonal epsilon phase ($CuZn_3$)

9.6 PHASE DIAGRAMS

The solidification of metal alloys is clearly indicated by means of phase diagrams, also known as equilibrium or constitution diagrams. In phase diagrams, graphic representations of changes in state due to variations in concentration and temperature enables the phase content of the alloy to be determined at any temperature and composition. They enable the phase transformations while heating or cooling the alloy under equilibrium conditions to be followed, i.e., when all processes in the given system are reversible. This means that changes taking place in a system as the result of a process proceeding in one direction are fully compensated by changes due to the reversal of the process in the system. All phase diagrams have temperature as the vertical scale (ordinate) and percentage composition by weight as the horizontal scale (abscissa).

However, although many useful engineering materials consist predominantly of one phase, a greater number are mixtures of phases. The mixture of two or more phases in one material permits interaction among the phases and the resulting properties are usually different from those of individual phases. It is possible to rectify these properties by altering the distribution of the phases or the shape.

A phase diagram permits the control and study of metallurgical processes like solidification of alloys and metals, crystal growth, phase separation, purification of materials, structural changes produced in welding, forging, casting and heat treatment, etc. In principal, equilibrium diagrams represent quite definite types of equilibrium—complete or partial solubility, presence or absence of chemical compounds, etc., and may be drawn on the basis of theoretical considerations. Equilibrium diagrams are valid only under the condition that all processes which may occur in the given system are of the equilibrium type.

Phase diagrams are of three types:
(i) Single-component or unary phase diagrams
(ii) Two-component or binary phase diagrams
(iii) Three-component or ternary phase diagrams.

Binary diagrams are extensively used and are considered in the succeeding sections.

9.7 THE PHASE RULE/GIBB'S PHASE RULE

All changes which take place in a system consisting of several phases in accordance with external conditions (like temperature and pressure) conform to the so called phase rule, also called Gibb's phase rule. This states that the maximum number of phases P which may co-exist under equilibrium conditions is equal to the sum of the number of components C and between the number of degrees of freedom in the system (the number of variable factors).

It is expressed mathematically as follows:

$$P + F = C + n$$
$$P + F = C + 2$$
∴ n = no of external factors
= 2 (temperature and pressure)

The number of degrees of freedom is the quantity of independent external or internal variables, like temperature, pressure and concentration, which may be changed without causing the disappearance of a phase or the formation of a new phase in the system. In studying chemical equilibrium, temperature and pressure are considered as external factors determining the state of the system. In applying the phase rule of to metal systems the effect of pressure is neglected, leaving only one variable factor—temperature. The equation will then be:

$$F = C + 1 - P$$

In equilibrium all factors have definite values, hence the degree of freedom cannot be less than zero

$$C - P + 1 \geqslant 0$$
then $\quad P \leqslant C + 1$

which shows that the number of phases in a system cannot exceed the number of components plus one. Therefore, no more than three phases may be in equilibrium in a binary system. In a ternary system no more than four phases may be in equilibrium.

A pure metal at solidification temperature, for example, is a one-component system consisting of two phases of identical composition

$$F = 1 + 2 - 2 = 0$$

Thus, the number of phases and the number of degrees of freedom equals zero ($F = 0$). This is called non-variant equilibrium. If the number of phases is less than the maximum possible number by one, the number of degrees of freedom will also increase by one ($F = 1$). Such a system is said to be *monovariant*.

In an alloy of two metals which is a two-phase and two-component system at solidification, $F = 1$. When $F = 2$ the system is known as *divariant*. Therefore the system may be in equilibrium at different temperatures and concentrations

All transformations occurring in alloys and depending on temperature and concentration ($C = 2$) can be clearly represented by equilibrium diagrams. These diagrams are plotted with concentration as the abscissa and temperature as ordinate. A number of examples of practical applications of the phase rule, for studying phase diagrams of two-component alloys ($C = 2$), are given in this chapter.

9.8 TIME-TEMPERATURE COOLING CURVES

The alloy diagrams are constructed from a series of time-temperature cooling curves. When a metal or alloy solidifies from the state of fusion, it will be observed that latent heat is given out due to changes taking place in the structure of the metal or alloy at the time of solidification. Due to this, at certain stages the fall in temperature of the metal or alloy is totally arrested for a specified time. Such points are called *critical points*. These critical points are determined by the abrupt inflexion of the curves due to thermal effects in transformations. Three types of cooling curves are commonly used in plotting phase diagrams.

COOLING CURVE FOR PURE METAL Liquid metal cools from A to B. From B to C heat is given out and the temperature remains constant. This is also called the critical point. Between B and C the mass is marshy (partly liquid and partly solid). On further cooling from C to D the metal reaches room temperature. The slips of AB and CD lines indicate specific heats of liquid and solid metals, respectively (Fig. 9.3 (a)).

COOLING CURVE OF A BINARY SOLUTION Curve AB is the same as for pure metals. The freezing line BC drops until the whole mass is solid at point C. This is due to the fall in temperature recorded during freezing in a

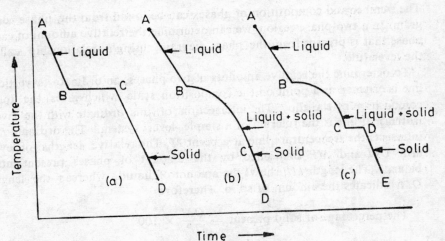

Fig. 9.3 *Cooling curves for (a) Pure metal (b) binary alloys or solid solution (c) binary eutectic system*

binary alloy. The solid further cools along line CD to reach room temperature (Fig. 9.3 (b)).

COOLING CURVE OF A MULTIPHASE ALLOY The cooling curve of a multiphase alloy or a binary eutectic system is shown in Fig. 9.3 (c). In this system the two components are completely soluble in the liquid state but entirely insoluble in the solid state. They are liquid along AB upto point B. At B the temperature drops along BC and crystallization of one component starts. At point C the components solidify simultaneously at constant temperature, the lowest for a given system, and are called eutectic alloys. A mechanical mixture of two (or more) phases which solidify simultaneously from the liquid alloy is called an eutectic (From the Greek Word *eutektos* which means easily melted). Cooling from D to E is as usual.

9.9 CONSTRUCTION OF A PHASE DIAGRAM

Figure 9.4 shows how to construct a phase diagram for alloys of two metals A and B in their various compositions, using the data from the cooling curves. The arrestment points have been joined by dashes to get the actual phase diagram. In this diagram there is complete intersolubility in the liquid and solid phases. The upper line corresponds to the temperature at which the alloys begin to solidify the liquidus. The lower line shows the complete solidification and is called the solidus. The area between the liquidus and solidus represents the alloys in a semisolid state. Greek letters α, β, γ, etc. are commonly used in phase diagrams to designate solid solutions.

9.10 THE LEVER RULE

The number and composition of phases can be found from the phase diagram. In a two-phase region, we can determine the relative amount of each phase that is present from the phase diagram, using a relationship called the lever rule.

To determine the relative amounts of two phases, an ordinate or vertical line is errected at a point on the composition scale which gives the total composition of the alloy. The intersection of this ordinate with the given isothermal line is the fulcrum of a simple lever system. The ordinate KL intersects the temperature line at a point M. The relative lengths of lever arms OM and MP multiplied by the amount of phases present must balance. The length MP shows the amount of liquid, whereas the length OM indicates the amount of solid. Therefore,

$$\text{The percentage of solid present} = \frac{OM}{OP} \times 100$$

$$\text{The percentage of liquid present} = \frac{MP}{OP} \times 100$$

Theory of Alloys: Constitution and Equilibrium Diagrams 143

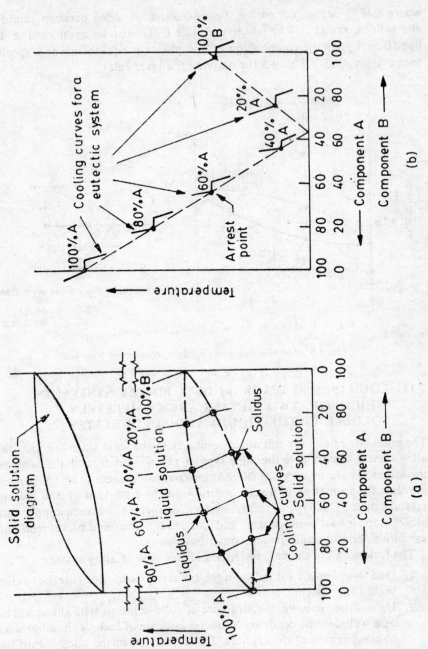

Fig. 9.4 Series of cooling curves giving rise to a phase diagram

where $OM + MP = OP =$ Total composition of alloy between liquidus and solidus, say at t_p. The isothermal (line OMP) can be considered a tie line since it joins the composition of two phases in equilibrium at a specific temperature t_p (See Fig. 9.5 for deriving the lever rule).

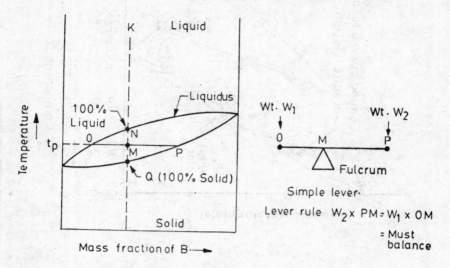

Fig. 9.5 *The phase diagram used for deriving the lever rule*

9.11 EQUILIBRIUM DIAGRAM OF A BINARY SYSTEM IN WHICH THE TWO METALS ARE COMPLETELY SOLUBLE IN THE LIQUID AND SOLID STATES

The main conditions for unlimited solubility in the solid state are that the alloy system should have the same type of crystal lattice and the sizes of the atoms should be very similar. Atom size difference over 15 percent prevents the formation of a solid solution due to distortion of the solvent crystal lattice. This type of alloy system is formed between copper and nickel, copper and gold, copper and platinum, iron and nickel iron and vanadium, bismuth and tin and cobalt base alloys.

The following are the characteristics of this type of alloy system:

1. Due to unlimited solubility in the binary system, no eutectic exists within the system.
2. Upon faster cooling, the structure of solid-solution-type alloys will be of a well-defined dendritic character because of lack of homogeneity of composition of the crystals. This formation of the alloy within the crystals is called coring or dendritic segregation. This can be removed by annealing.
3. The variation in mechanical properties in this system is small with respect to variation in alloy composition. The variation in physical properties (electrical and corrosion resistance) and fabrication pro-

perties (rolling, forging and stamping) may change greatly with only slight variations in composition.
4. Alloys which form homogeneous, solid solutions are easily worked and are widely used as engineering materials.

Referring to Fig. 9.6 the cooling curve for pure copper (100% copper) contains a flat portion during which complete solidification takes place at constant temperature. When 20 wt% nickel is added the flat portion is no longer present but the solid solution solidifies over a range of temperature. With nickel wt% increasing similar behaviour occurs, but when 100% nickel is reached the liquid transforms into solid and again contains a flat portion indicating the solidification of pure metal. From these cooling curves the phase diagram is constructed as shown in the same figure. The liquidus and solidus lines, liquid phase, solid solution fields, etc., are also marked in the diagram.

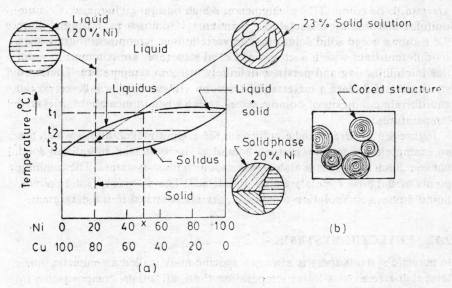

Fig. 9.6 (a) *Equilibrium diagram for a solid solution (copper-nickel system)*
(b) *Cored structure showing non-uniform distribution of metallic constituents*

Now let us consider the solidification of the alloy containing any composition, say x, 50% copper and 50% nickel which will be typical of the solidification of any alloy in the system. Solidification now starts at temperature t_1 as it reaches the liquidus, and the tie line shows that the first solid is of composition x_1% nickel. If a series of tie-lines is drawn at lower temperatures in the two-phase regions, it can be noted that the proportion of solid increases with decreasing temperature and the composition of the solid forming at any given temperature changes with temperature along the solidus. Similarly, the composition of the remaining liquid follows the liquidus. When the cooling rate is slow enough to maintain equilibrium, the solid format changes composition by diffusion so that, at any temperature, all the solid is of the same composition. The first dendrites deposited are richer in nickel, the remaining melt will have a composition which is richer in copper. Thus, when the temperature has fallen to t_2, the first dendrites should absorb copper from the liquid to change composition. Solidification continues in this manner, the liquid vanishes and the solid has a uniform composition. The microstructure of the alloy so formed will consist of grains of homogeneous solid solution of composition $x\%$ nickel.

Commercially, the rate of cooling is very rapid and the composition of the solid solution as discussed above rarely occurs due to the rapid cooling. This results in a non-equilibrium structure which prevents the first dendrites from absorbing the proper amount of copper as the temperature falls to t_2. The grains obtained are heterogeneous due to variation in composition and are said to be cored. This phenomenon which occurs giving rise to non-uniform distribution of metallic constituents is known as coring. Figure 9.6 shows a cored solid solution. The variations in composition are shown by different lines within each grain. Cored structures are common in practice in chill casting and persist indefinitely at room temperature. The cored structure is therefore a metastable structure. However, it will revert to the equilibrium condition of homogeneous grains when annealed at elevated temperatures.

Figure 9.6 illustrates the equilibrium for copper-nickel alloys which are an example of complete mutual solubility in both the liquid and solid phases. Such systems are also called isomorphous systems. The melting points of the pure metals are also indicated. The liquidus, solidus lines, liquid phase, solid solution fields, etc., are also marked in the diagram.

9.12 EUTECTIC SYSTEMS

In eutectic systems there is always a specific alloy, called an eutectic mixture, that freezes at a lower temperature than all other compositions. In such systems, two distinct solid phases or a mixture of more phases is formed when a liquid solution of fixed composition solidifies at a constant temperature without an intermediate pasty stage. The process is reversible on heating.

The alloys in which the components solidify simultaneously at a constant temperature, the lowest for a given system, are called eutectic alloys. The eutectic temperature and composition determine a point on the phase diagram called the eutectic point. The eutectic reaction can be expressed in the following form:

$$\text{Liquid} \underset{\text{Heating}}{\overset{\text{Cooling}}{\rightleftharpoons}} \text{Solid}_A + \text{Solid}_B$$

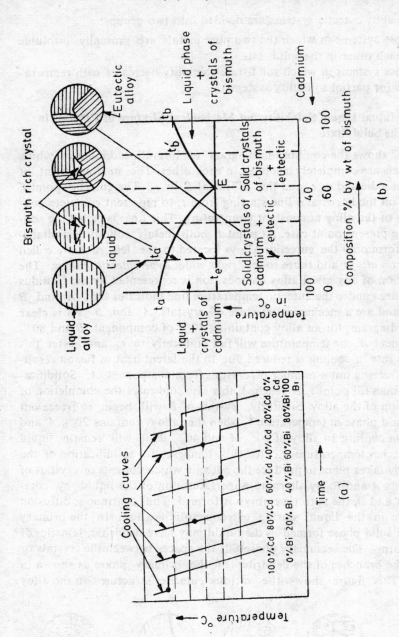

Fig. 9.7 *Equilibrium diagram for eutectic systems (a) Time-temperature cooling curves for a series of alloys between cadmium and bismuth. (b) Equilibrium diagram of the cadmium-bismuth system, showing point E as eutectic point*

Binary alloy eutectic systems are divided into two groups:
(i) Those systems in which the two pure metals are mutually insoluble in each other in the solid state.
(ii) Those systems in which solid-state solubility decreases with temperature (or partial solubility systems).

9.12.1 Mutual Liquid Solubility but Mechanical Mixture of Crystals in the Solid State

Figure 9.7 shows the equilibrium diagram of bismuth-cadmium in which both metals are completely insoluble in each other. The melting point of metal A is indicated as t_a and that of metal B by t_b. The eutectic point E is shown on the horizontal line starting from t_e to represent complete solidification of the alloy at constant temperature. This line is called the solidus in the present point case. At point E both metals solidify simultaneously to form only the eutectic. Alloys located to the left of E are called *hypoeutectic alloys* and those to the right side *hypereutectic alloys*. The solidification of any given alloy composition is represented by the liquidus line. The area under the eutectic temperature line indicates that A and B are solid and are a mechanical mixture of crystals A and B. As is clear from the diagram, for an alloy containing 20% of component B and 80% of component A, the temperature will fall uniformly to t_a and, after this point, the rate of cooling is reduced due to the latent heat of fusion resulting from freezing out a number of crystals from component A. Solidification continues till point t_e is reached; this also indicates the completion of solidification of the alloy. Similarly, crystals of B will begin to freeze out of the liquid phase at temperature t_b when the alloy contains 20% A and 80% B. On cooling an alloy of 60% A and 40% B, it will remain liquid until it reaches temperature t_e. At this temperature solidification of the entire alloy takes place to produce the eutectic which consists of crystals of components A and B. By alternate supersaturation of the liquid by components A and B, the eutectic embryo is formed. Thus continuous diffusion takes place in the liquid during eutectic solidification. In the primary phase, the solid phase formed in the liquid may have dendritic, lameller or nodular forms. The secondary phase fills the spaces between the crystals or between the branches of the dendrites of the primary phase as shown in Fig. 9.8. This figure shows the various eutectic structures in the alloy system.

Nodular   Chinese script   Lamellar   Divorced   Accicular

Fig. 9.8 *Common eutectic structures in alloy systems*

9.12.2 Mutual Liquid Solubility but Partial Solid Solubility

As described in the bismuth-cadmium system, very few alloys which are soluble in all proportions in the liquid state crystallise into two pure metals, since most metals show some solubility for each other in the solid state. This group of alloys is the most common and contains many important industrial alloy systems. Metals lead-tin, iron-carbon and lead-silver, form the binary alloys that can accommodate only a certain percentage of solute metals, resulting in limited solubility in the solid state. The solid solubility decreases with falling temperature. In some such alloy systems eutectic reaction takes place. Figure 9.9 shows the solidification characteristics of alloys of the lead-tin system.

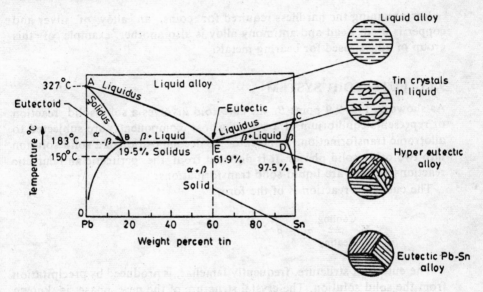

Fig. 9.9 *Lead-tin equilibrium diagram showing point E as eutectic formation and point B as eutectoid point*

1. Lines AE and EC are the liquidus. Crystals of a solid solution of tin in lead (α) begin to precipitate from the liquid along line AE; the solid solution of lead in tin (β) precipitates along line CE.
2. Points B and D indicate the points of maximum solubility at the eutectic temperature. Lines AB, BE and CD are the solidus.
3. At the eutectic point E, the solid solution of α and β are separated simultaneously from the liquid phase to form a mechanical mixture $\alpha + \beta$.
4. The melting point of lead and tin are also indicated.
5. No pure metals exist in a solid alloy of any composition.
6. At 183°C an eutectic alloy is formed, containing 61.9% tin and 38.1% lead.

7. The maximum solid solubility of lead in tin is 2.5% and that of tin in lead is 19.5%.
8. On fast cooling to room temperature, the saturated solid solution is converted into a supersaturated solid solution. There is separation of surplus phase due to decomposition of the supercooled solid solution, thereby, affecting considerable changes in the properties of alloys. An increase in hardness is the characteristic for this process and if the alloy is heated, the precipitation rate is increased and it will take less time to harden and strengthen the alloy. This process is called *age-hardening* or *precipitation hardening* and is made use of in non-ferrous alloys.

For obtaining the hardness required for coins, an alloy of silver and copper is used. Lead and antimony alloy is also another example of this group of alloys, used for bearing metals.

9.13 EUTECTOID SYSTEM

As shown in Fig. 9.9 point B, the eutectoid involves a solid-solid reaction or represents equilibrium of a system whose components are subjected to allotropic transformation. It indicates the decomposition of a solid solution into two other solid phases. It is different from the peritectic and eutectic reactions as both are liquid-solid transformations.

The eutectoid reaction is of the form:

$$Y \underset{\text{Heating}}{\overset{\text{Cooling}}{\rightleftharpoons}} \alpha + \beta$$

The eutectoid structure, frequently lamellar, is produced by precipitation from the solid solution. The crystal structure of the new phase is known as the *Widmanstatten structure*. The eutectoid reaction is observed in many alloy systems like Al-Cu, Cu-Zn, etc.

This reaction occurs in the iron-carbon diagram in which austenite (a solid solution of carbon in γ-iron) is decomposed into pearlite. Heat treatment of steel is the most important example of the use of this reaction.

9.14 PERITECTIC AND PERITECTOID SYSTEM

Equilibrium diagrams of alloys whose components have complete mutual solubility in the liquid state and limited solubility in the solid state are also termed representations of alloys with peritectic transformation.

In the eutectic system, the crystals of beta solid solution precipitated at the beginning of solidification react with the liquid alloy of a definite composition to form new crystals of alpha solid solution. In the peritectic reaction, two phases are used to produce one different phase with reaction just the opposite of the eutectic reaction (Figs 9.10 and 9.11).

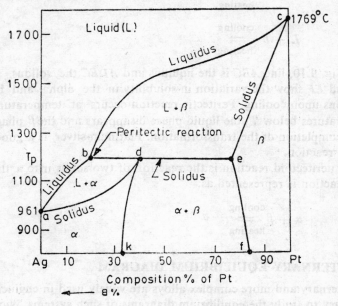

Fig. 9.10 *The silver-platinum peritectic system*

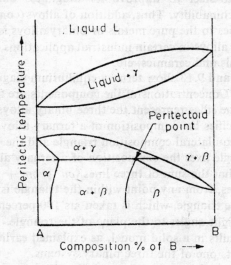

Fig. 9.11 *Peritectoid diagram*

The peritectic reaction is also like other systems of solidification of different metals, but is comparatively less common. This reaction also occurs at constant temperature.

The peritectic reaction is written as:

$$\text{Liquid} + \text{Solid 1} \underset{\text{heating}}{\overset{\text{cooling}}{\rightleftharpoons}} \text{New Solid 2}$$

i.e.
$$L + \beta \underset{\text{heating}}{\overset{\text{cooling}}{\rightleftharpoons}} \alpha$$

In Fig. 9.10, line ABC is the liquidus and $ADEC$ the solidus. The lines DK and EF show the variation in solubility in the alpha and beta solid solutions upon cooling. Peritectic reaction occurs at temperature T_d. At temperatures below T_d the liquid phase disappears and the β phase appears upon completion of the transformation. Platinum-silver is a good example of this reaction.

The peritectoid reaction is the reaction of two solids into a third solid. The reaction is represented as:

$$\alpha + \beta \underset{\text{heating}}{\overset{\text{cooling}}{\rightleftharpoons}} \gamma$$

9.15 TERNARY EQUILIBRIUM DIAGRAM

Since ternary and more complex alloys are widely used in engineering, it is necessary to study the equilibrium diagrams of such systems. Nickel is frequently added to steel to improve its toughness and lead to brass to improve its machinability. Thus, addition of alloys (components) possessing certain properties to the pure metals or binary alloys is very significant for development of alloys in certain industrial applications, e.g., stainless steels, high speed steels and ceramics, etc.

Figures 9.12 and 9.13 show ternary equilibrium diagrams of a three-component system. Concentrations of the components are shown in the triangle corners while the sides represent the three binary alloys. Any point within the triangle specifies the composition of a ternary alloy. The base of the diagram is an equilateral composition triangle and the location of a point within the triangle fixes the composition of a ternary alloy. This geometrical rule states that the sum of three lines ($on + om + op$) drawn parallel to the three sides, from any point within the triangle is equal in length to one side l of the triangle, which is taken as 100 percent. The temperatures are plotted at right angles to the plane of the triangle. The development of the diagram results in a solid model, as explained earlier; each side of the model represents one of the three binary systems.

This system may form a ternary eutectic with lower melting point than that of the binary eutectic. Sometimes the third element may combine with impurities in the alloy to produce desirable effects such as the combining of manganese added to steel with sulphur to make manganese sulphide which is ejected from the steel during forging. Addition of manganese to steel results in better forging properties.

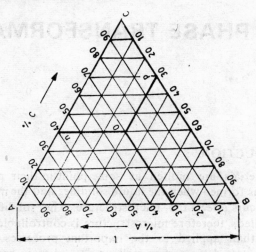

Fig. 9.12 *Ternary equilibrium diagram*

REVIEW QUESTIONS

1. What do you understand by a solid solution? Explain with neat sketches the substitutional solid solution and interstitial solid solution. Give two examples of each in the alloy system.
2. Write short notes on the following:
    (i) Hume-Rothery's rules
    (ii) Interstitial compounds
    (iii) Gibb's phase rule
    (iv) Cooling curves
3. Draw the phase diagram for a binary system showing complete solubility in liquid and solid state.
4. In the copper-nickel system, determine the fraction of phase in an alloy of NiCu at 1300 °C.
5. A lead-tin alloy containing 10% tin is cooled from 200 to 100 °C. How much β (wt %) precipitates from the alloy as a result of the cooling?

10

PHASE TRANSFORMATIONS

10.1 INTRODUCTION

Most solid materials are molten at some point in their processing. The grain structure is formed when the liquid solidifies and the manner in which solidification occurs affects the microstructure. The solidification process can be controlled, therefore microstructure is controllable. Solidification principles are than applied to the important processes of casting, heat treatment, solid state devices, recrystallization and grain growth.

Metallurgical processes often involve the transformation of a metal from one phase to another. Every transformation can be considered to be dependent on two factors:

(i) A thermodynamic factor that determines whether the transformation rate is possible.
(ii) A kinetic factor that determines whether transformation is possible at a practical rate.

10.2 RATE OF TRANSFORMATION

The control of structure is achieved by producing or suppressing deliberately a structural transformation. Before discussing the kinetics and mechanism of the transformations, it is necessary to know how the above-mentioned factors affect the transformation. The time taken to complete transformation and the available driving force equal to the magnitude of the free energy change during transformation are also important. As the transformation occurs on cooling a material from an elevated temperature, so the cooling rates may be exteremely fast, very fast, normal, slow or extremely slow.

An extremely fast rate of transformation may take micro-seconds, as in the manufacture of solid state devices from amorphous materials.

Very fast transformations may take a fraction of a second, by drastic quenching.

Normal transformations take place in the range of a few seconds or minutes. Slow transformations take a few hours to days, whereas extremely slow transformations may take a few days to years.

10.3 MECHANISM OF PHASE TRANSFORMATION OR NUCLEATION

Consider a simple transformation where a crystal β that is in contact with the liquid of the same will grow as the liquid freezes. As a starting point, consider a container holding both the liquid and solid phases of a pure substance at its melting point. As heat is removed from the container, the liquid solidifies. The change from the liquid to the solid state reflects the tendency of the substance to maintain itself in its most stable (or equilibrium) state. The liquid and the solid can co-exist in equilibrium at the meeting point, therefore both states are equally stable at this temperature. At lower temperatures the solid state is the stable one, while equilibrium at higher temperatures requires the liquid state.

The Gibbs free energy ΔF is a useful measure of the tendency for a transformation to occur. At the melting point (T_m), the liquid and solid phases are in equilibrium and there is no tendency for the transformation to occur (enthalpy change) i.e.,

$$\Delta F_v = \Delta H_v - T_m \Delta S_v = 0$$

or $\quad\quad \Delta S_v = \dfrac{\Delta H_v}{T_m} =$ change in entropy $\quad\quad\quad\quad (1)$

If it is assumed that ΔF_v and ΔS_v do not vary with temperature, then the free energy change accompanying the solidification at any temperature T can be determined as

$$\Delta F_v = \Delta H_v - T \Delta S_v$$
$$= \Delta H_v \left(\dfrac{T_m - T}{T_m}\right) = \dfrac{\Delta H - \Delta T}{T_m}$$

where ΔT is the degree of super cooling

Since heat is given off during solidification i.e., ΔH_v is negative and ΔF_v is negative, the solidification reaction is possible provided the temperature T is less than the melting point (T_m)

Figure 10.1 shows the variation of free energy during nucleation as a function of the particle radius r at different temperatures. Now if a drop of liquid of radius r is to form in the spherical new phase particle, ΔF that accompanies this formation is

$$\Delta F = 4\pi r^2 \gamma + 4/3 \, \pi r^3 \, \Delta F_v \quad\quad\quad\quad (2)$$

= Energy required to create interface + energy released by volume of condensing phase

since γ is the surface energy per unit area of the interface separating the initial and product phases. The surface energy (the first term) is always positive but the second term will be negative when ΔF_v is negative for the phase transformation being considered. The sum ΔF goes through a maximum at some critical radius denoted by the superscript* whose value is

$$r^* = 2\gamma/\Delta F_v \qquad (3)$$

$$\therefore \quad F^* = \frac{16\pi\gamma^3}{3(\Delta F_v)^2} \qquad (4)$$

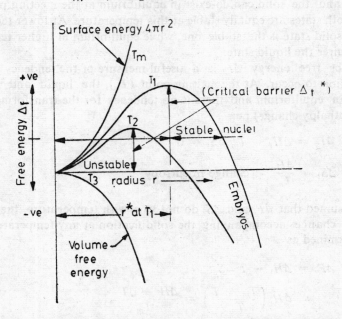

Fig. 10.1 *Variation of the free energy during solidification as a function of particle radius r at different temperatures*

The drops of liquid described by the curve ΔF versus r arise because of successive random collision of atoms in pairs. Thus there will be drops of the whole range of sizes per unit volume of the parent phase n. Then, from Boltzmann's equation, the number of critical sized particles per unit volume is given by

$$n^* = n e^{-\Delta F^*/KT} \qquad (5)$$

where K is the Boltzmann's constant.

If the radius of the drop or sphere is less than the critical radius (r^*), the sphere will revert to the liquid phase. This sphere is called an embryo. If the sphere or drop has a radius larger than the critical radius, it is called a nucleus and serves as a nucleation site. The size of the critical radius depends on the degree of supercooling of the liquid.

The rate of nucleation (I_v) or the rate of formation of stable nuclei can be determined from the rate at which these nuclei are struck by additional atoms (z). This is given by the equation

$$I_v = z \times n^* \, 4\pi \, (r^*)^2 \tag{6}$$

The rate of nucleation increases with decreasing temperature as ΔF^* is decreasing.

10.4 HOMOGENEOUS NUCLEATION

Nucleation can be homogeneous or heterogeneous. In a homogeneous process, the probability of nucleation occurring at any given site is identical to that at any other site within the assembly. Homogeneous nucleation was explained in the above discussion in terms of free energy distribution among the liquid drops. It was observed that, for this, extremely large amounts of supercooling was required to obtain a practical rate of nucleation of drops in the volume of liquid.

10.5 HETEROGENEOUS NUCLEATION

It is often observed that when a substance changes its phase, the phase change begins at a particular or preferred location. In the heterogeneous process, the probability of nucleation occurring at certain preferred sites in the assembly is much greater than at other sites. A common example of this is water boiling in a pot. The steam bubbles usually originate at a particular location on the bottom of the pot. These are called heterogeneous nucleation sites.

In practice, before homogeneous nucleation has a chance to occur, drops of liquid form on any foreign material that serves as a nucleation catalyst. Suitable materials for this purpose are the walls of the container, dust particles, grain boundaries, stacking faults and dislocations.

The mechanism of heterogeneous nucleation is similar to that described for homogeneous. In additon to the difference in the ΔF^* term discussed above, the number of sites will be preferred sites only.

10.6 NUCLEATION AND GROWTH

The effect of a liquid solidifying on several nuclei is illustrated in Fig. 10.2. The dark spots in Fig. 10.2 (a) represent nuclei or very tiny particles identical in structure to the β crystal. These form in the liquid phase and grow to larger sizes till the transformation is complete. As the temperature of the melt diminishes, the liquid solidifies on the nuclei and crystals grow. One crystal grows from each nucleus. The parallel lines in Fig. 10.2 (b) represent the same direction in each crystal. It is observed that the crystals are not in alignment with each other. Their orientation is random. As solidification progresses, the crystals join each other and form **boundaries**, as shown in Fig. 10.2 (c). Each boundary marks a discontinuity in the crystal structure. The individual crystals are called grains and the boundaries are called grain boundaries. If solidification begins at a large number of nuclei, the resulting solid will have many grains and is said to be polycrystalline. Most engineering materials are polycrystalline.

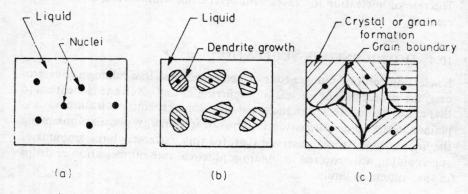

Fig. 10.2 *Solidification process of a liquid containing nuclei*

10.7 APPLICATIONS OF PHASE TRANSFORMATIONS

Phase transformations are of great technological importance. These are usually observed in microstructural changes in cooling or freezing (dendrite formation), in castings, in amorphous structures (solidification phenomenon in glassy structures), in heat treatment or in the binary phase diagram of $Fe-Fe_3C$ systems and TTT diagrams for eutectoid steels and in recrystallization and grain growth during mechanical working.

10.7.1 Dendrite Formation

Freezing is accompanied by the release of latent heat of fusion. Therefore, temperature gradients must exist within the solid or liquid or both. This influences the solid structure which is formed. As freezing progresses, the solid-liquid interface moves away from the nucleus. Solidification of metals often results in the appearance of branched, trace-like particles called dendrites. An example of dendritic structure in solid lead is shown in Fig. 10.3. The lead begins to solidify in a particular direction shown by the straight lines in the figure. Other dendrites then branch out from the main one. Crystal growth is preferentially along the three mutually perpendicular directions. The liquid between the dendritic arms solidifies last and this region can be distinguished from the dendritic arms themselves by an etching process which brings out the compositional differences caused by coring.

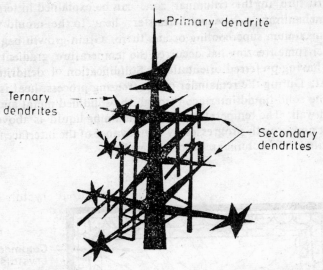

Fig. 10.3 *Dendritic growth showing the first stage of solidification*

10.7.2 Casting

Casting is a process of great practical importance. Many items are fabricated by pouring a molten material into a mould and allowing it to solidify. Thermal gradients and temperature changes can become complex during the process. These affect the microstructure of the cast metal or casting. Let us consider the cast metal or ingots formed by pouring the melt into a cylindrical mould. Heat is taken out through the walls of the mould during freezing.

Figure 10.4 shows an equal-sized grain structure in the region next to the mould wall. This region is followed by the columnar zone in which the grains are long, thin and oriented in the radial direction. How much of each zone appears in a particular ingot depends upon the temperature of the liquid and the mould at the time of pouring.

The freezing mechanism of this structure, which is unique in alloys, can be explained as follows:

In most casting processes, the mould is considerably cooler than the melt before pouring (the mould is often at room temperature). When the molten metal is poured, the portion of the liquid near the mould wall chills quickly and supercools. Nucleation (usually heterogenous) occurs at many points in the chill zone. The nuclei grow until the grains join each other, resulting in a large number of randomly oriented grains. This is the structure of the chill zone. If the liquid is highly supercooled, many nuclei will grow and the structure will be fine grained. Conversely, a small amount of supercooling results in a coarse grained structure. As melt temperature increases, so does grain size.

The structure in the columnar zone can be explained in terms of two freezing mehanisms. As the melt transfers heat to the mould wall after pouring, maximum supercooling occurs there. Grain growth begins at the wall. After some freezing has occurred the temperature gradient will form dendrites having preferred orientations. Solidification of dendrites releases latent heat. During the remainder of the freezing process, heat is conducted from the solid-liquid interface through the solidified metal and through the mould wall. The temperature of the remaining liquid is above the freezing point and growth progresses by the advance of the interface formed by the dendrites. The columnar structure results.

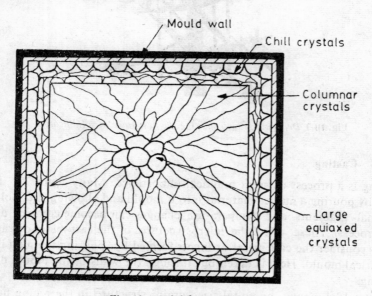

Fig. 10.4 *Solidification of a casting*

10.7.3 Amorphous Structures—Glass Transition Temperature

In all the processes discussed so far, the liquid atoms have been fairly free to move about; they have been mobile. Mobility was required so that liquid atoms could move to a solid surface or could aggregate together to form nuclei. The basic building unit was the atom. If we consider polymers or silicates, the situation changes somewhat. The building block of many polymers is the polymer chain, while silicate structures involve chains or sheets. These molecules are larger than the atoms we have been considering and their mobility is lower. When a melt containing large chain molecules is cooled, solidification can occur by either (or both) of two mechanisms. These are crystallization and vitrification.

A substance which vitrifies forms an amorphous glass when it cools. Let us start with a substance which is capable of either crystallizing or vitrifying. Figure 10.5 shows the effect of temperature on the specific volume of the material. The specific volume decreases slowly with temperature in the liquid region. At the melting point (T_m), the substance can follow either of two routes. If nuclei are available for solidification and the liquid molecules are mobile, crystallization will occur. The freezing of crystalline solids occurs at well-defined temperatures and is accompanied by the release of latent heat and, usually, a reduction in specific volume. An alternate cooling path is also available and leads to a vitreous or glassy solid. Suppose that the mobility of the molecules in the melt was very slow when the melt reached T_m, so that crystallization could not occur readily. Decreasing the temperature below T_m will result in a super cooled liquid. The molecular mobility can be related to the viscosity of the liquid. Viscous liquids have low mobility and liquids become viscous as the temperature is lowered. Further cooling of the supercooled liquid results in the formation of a glassy material. This material has an amorphous structure like the liquid but its high viscosity causes it to behave very much like a solid. The upper curve in Fig. 10.5 changes its slope in the neighbourhood of T_g. T_g is called the glass transition temperature. Significant changes occur in physical properties in this temperature region.

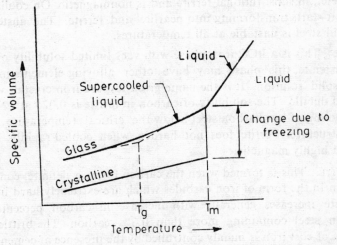

Fig. 10.5 *Solidification of crystalline and amorphous (glassy) materials showing changes in specific volume. T_g is the glass transition temperature*

10.7.4 Phase Transformation In Iron-Carbon Systems

The iron-carbon system is the most important subject in the study of steels and cast irons. This is also very important to modern industry due to the extensive applications of ferrous metallurgy in fabrication, casting and production processes. Theoretically, it is based on phase transformations in metals. For example, in iron manufacture there is a transformation on cooling liquid iron to the solid state, as in casting of metals. Another example of phase transformation is in heat treatment processes where different mechanical properties can be attained by changing the crystal structure.

The iron-carbon system provides the most important knowledge on heat treatment, based on polymorphic transformation and eutectoid decomposition. The primary constituent of the iron-carbon system is metal iron. Common alloys of the iron-carbon system include steels and cast irons. Allows with a carbon content upto 2% are called steels and those with carbon contents exceeding 2% are called cast irons.

10.8 MICRO-CONSTITUENTS OF IRON-CARBON SYSTEM

Decomposition of austenite produces the various constituents of iron and steel which can be observed under a microscope. The important micro-constituents are classified below.

AUSTENITE This is a solid solution of ferrite and iron carbide in gamma iron which is formed when steel contains carbon upto 1.8% at 1130 °C. This FCC structure contains carbon in interstitial manganese, nickel chromium, etc., in substitutional ferrite and is nonmagnetic. On cooling below 723 °C it starts transforming into pearlite and ferrite. The austenite in a eutectoid steel is unstable at all temperatures.

FERRITE This is a BCC iron phase with very limited solubility of carbon. Like austenite, this phase may have other alloying elements in substitutional solid solution. It is the name given to pure iron crystals which are soft and ductile. The solubility of carbon in ferrite is 0.025 at 723 °C. The slow cooling of low carbon steel below the critical temperature produces ferrite structure. Ferrite does not harden when cooled rapidly. It is very soft and highly magnetic.

CEMENTITE This is formed when the carbon forms a definite combination with iron in the form of iron carbides which are extremely hard in nature. Cementite increases generally with increase in carbon percentage. It is found in steel containing more than 0.8% carbon. The brittleness and hardness of cost iron is mainly controlled by the presence of cementite in it. It is magnetic below 200 °C.

PEARLITE This is a mechanical mixture of about 87% ferrite and 13 percent cementite. This mixture derives its name from the fact that it shows in oblique lighting under the microscope, the rainbow colours of mother-of-pearl when properly polished and etched. A steel with 0.8 percent carbon is wholly pearlite. Less than 0.8% carbon is hypoeutectoid and more than 0.8% carbon is hypereutectoid steel. The former contains ferrite and pearlite and is soft while the latter contains pearlite and cementite which are hard and brittle. Pearlite is composed of alternate plates of cementite and ferrite.

BAINITE This is a ferrite cementite aggregate that is formed by the growth of a ferrite nucleus. It is the product of isothermal decomposition of austenite. It is present in two forms, the feathery bainite obtained in the upper part of the temperature range and needle-like or accicular bainite produced by lower reaction temperatures.

MARTENSITE This is a body-centered, tetragonal iron phase produced by entrapping carbon on decomposition of austenite when cooled rapidly. It is the main constituent of hardened steel. It is magnetic and is made up of a needle-like fibrous mass. It has a carbon content upto 2%. It is extremely hard and brittle. The decomposition of austenite below 320 °C starts the formation of martensite.

TROOSTITE This is another constituent of steel obtained by quenching tempering martensite. It is composed of the cementite phase in a ferrite matrix that can not be resolved by a light microscope. It is less hard and brittle than martensite. It is also produced by cooling the metal slowly until transformation begins and then rapidly to prevent its completion. It has a dark appearance on etching. It is weaker than martensite.

SORBITE This is also produced by the transformation of tempered martensite. It is produced when steel is heated at a fairly rapid rate from the temperature of the solid solution to normal room temperature. It has good strength and is practically pearlitie. Its properties are intermediate between those of pearlite and troostite.

10.9 THE ALLOTROPY OF IRON

The melting point of iron is 1539 °C. Iron is relatively soft and ductile. The important allotropic forms of iron are alpha, beta and gamma iron. The existence of phases depends upon the temperature to which the iron is heated. Figure 10.6 shows the cooling curve for pure molten iron plotted in time v. temperature coordinates. When the molten iron solidifies from the liquid state at 1539 °C, this transformation is accompanied by an evolution of heat. This is represented by the horizontal line on the temperature curve at 1539 °C. For a considerable amount of time, the temperature remains steady at this point. When the transformation of liquid into solid is complete, the iron is in β (beta) form (BCC structure). After some time the temperature falls to 1404 °C and there is evolution of heat represented by the horizontal line at 1404 °C. This heat allows the temperature to remain steady. Here an allotropic modification of iron takes place, from beta to gamma (FCC structure). The third temperature effect on further cooling, occurs at 910 °C. Here gamma iron changes into alpha iron (α-iron BCC structure). This transformation is non-magnetic whereas the earlier two transformations were paramagnetic in nature. The fourth and last allotropic change occurs on further cooling till a temperature of 768°C is attained and corresponds to alpha iron. Alpha iron here is highly magnetic and exists upto room temperature. At this temperature the maximum solubility of carbon is about 0.025%. These transformations are reversible, i.e. the same changes are observed while heating from room temperature to the molten state of iron (about 1500 °C).

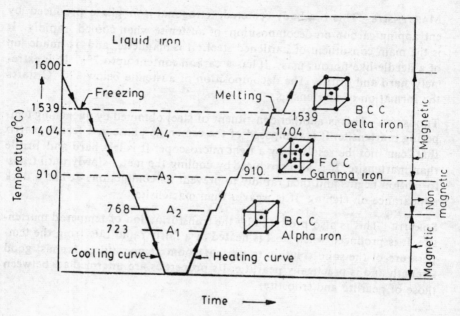

Fig. 10.6 *Allotropic transformations in iron*

10.10 IRON-CARBON EQUILIBRIUM DIAGRAM

This diagram represents the entire range of iron-carbon alloys, i.e., plain carbon steels and cast irons. The iron-carbon equilibrium diagram is shown in Fig. 10.7. It indicates transformations that take place in an alloy of iron-carbon, from pure iron to cementite (6.67 % carbon) The temperature is represented on the vertical axis and carbon percentages on the horizontal axis. Several areas of the diagram are indicated by the names of the phases existing at temperatures and concentrations determined by the lines (boundaries) of these areas.

All alloys represented by the compositions and temperatures in the region above *ABCD* are completely liquid. The curve *ABCD* is called the liquidus. The melting point of pure iron (1539 °C) is marked as point A in the diagram. Similarly, point *D* is the melting point (1550 °C) of iron carbide or cementite. As the temperature of the liquid falls along the line *ABC*, crystals of austenite separate from the liquid. In the same way, crystals of iron carbide separate from the liquid along the line *CD*. High-temperature transformations (gamma iron ⇌ delta iron) take place at the

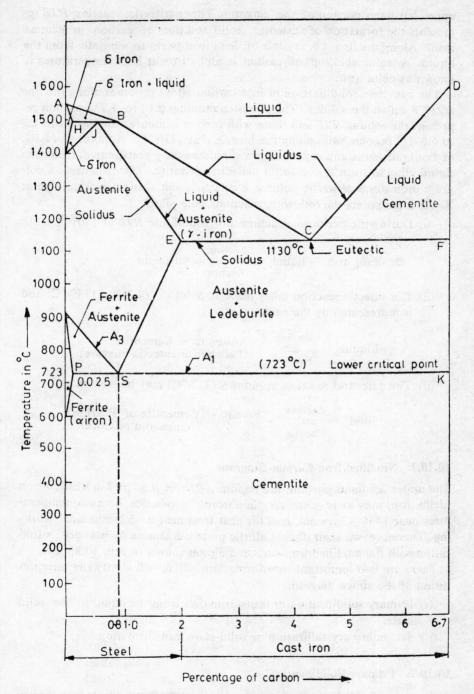

Fig. 10.7 *Iron-carbon equilibrium diagram*

upper left-hand portion of the diagram. The peritectic reaction HJB represents the formation of austenite (solid solution of carbon in gamma iron). Along the line AB, crystals of delta iron begin to separate from the liquid. A solid solution of carbon in alpha iron at high temperatures is known as delta iron.

The complete solidification of iron-carbon alloys proceeds along the line HJECF called the solidus. The alloys containing 0.18 to 1.7% carbon represent the solidus HJE and those with carbon contents ranging from 1.7 to 6.67% become solid along the line ECF at 1130 °C. At point C (4.3% carbon), austenite and cementite are simultaneously precipitated from the liquid alloy to form the eutectic called Ledeburite. The iron-carbon diagram indicates a peritectic point J, a eutectic point C and a eutectoid point S. At these points the following reactions take place:

(i) During the peritectic reaction (horizontal line HJB at 1401 °C)

$$\text{Delta } (\delta) \text{ iron} + \text{liquid} \underset{\text{Heating}}{\overset{\text{Cooling}}{\rightleftharpoons}} \text{Austenite}$$

(ii) The eutectic reaction takes place at point C, (4.3%, 1130 °C) and is represented by the equation

$$\text{Liquid} \underset{\text{Heating}}{\overset{\text{Cooling}}{\rightleftharpoons}} \text{Austenite} + \text{Cementite} \\ \text{(Ledeburite eutectic mixture)}$$

(iii) The eutectoid reaction at point S (723 °C) may be written as

$$\text{Solid} \underset{\text{Heating}}{\overset{\text{Cooling}}{\rightleftharpoons}} \text{Ferrite} + \text{Cementite or Pearlite} \\ \text{(Eutectoid mixture)}$$

10.10.1 Modified Iron-Carbon Diagram

The upper left hand portion, the region, ABJN of Fig. 10.7 in which δ-iron (delta iron) may exist is not very important in practice because temperatures near 1401 °C are not used for heat treatment or in mechanical working. Therefore, we shall discuss all the practical aspects of primary solidification with the modified iron-carbon diagram shown in Fig. 10.8.

There are two important transformations which will clarify the interpretation of the above diagram.

(i) Primary solidification or transformation from the liquid to the solid state.

(ii) Secondary crystallization or solid-state transformation.

10.10.2 Primary Solidification

Let us consider the course of events when liquid alloys of various carbon contents are cooled to a temperature just below the eutectic temperature

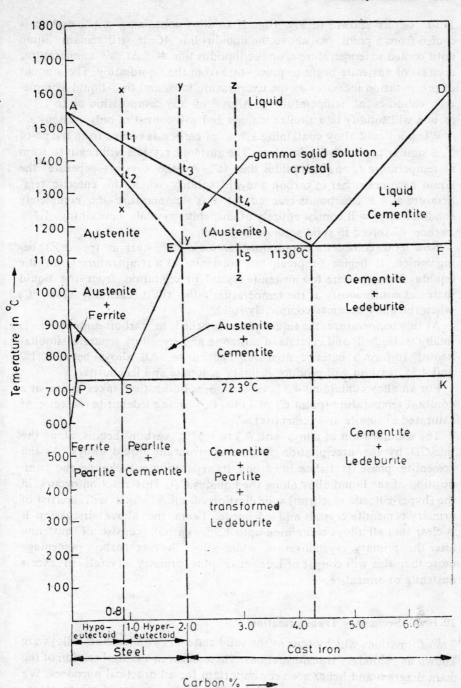

Fig. 10.8 *Modified iron-carbon equilibrium diagram*

1130 °C, as shown in Fig. 10.8. If an alloy containing 0.8% carbon is cooled from a point (x) above the liquidus line AC, it will remain liquid until cooled to temperature t_1 on the liquidus line AC. At this temperature, crystals of austenite begin to precipitate from the liquid alloy. The amount of precipitation increases as the temperature falls, and the liquid completely solidifies at temperature t_2. Alloys of any composition up to 1.7% carbon will solidify in a similar manner and will consist of only austenite.

When a liquid alloy containing 1.7% of carbon is cooled from the point Y, a similar process will take place. The austenite crystals will begin to form at temperature t_3 on the liquidus line AC. As the crystals seperate, the liquid becomes richer in carbon and will solidify when the eutectic temperature at 4.3% carbon is reached. At this temperature the completely solidified alloy will consist entirely of austenite crystals containing 1.7% carbon dissolved in solid solution in γ-iron.

Now let us consider a liquid alloy containing 3% carbon (point Z) being cooled. It begins to precipitate austenite at a temperature t_4 on the liquidus line AC. Here too austenite crystal precipitation from the liquid increases continuously as the temperature falls, till it reaches t_5 (1130 °C) when the alloy becomes completely solid.

At this temperature, the liquid phase enriched in carbon upto 4.3% is finally solidified and crystals of austenite and cementite separate simultaneously to form a eutectic mixture, Ledeburite. All alloys between 1.7 and 4.3% carbon will produce primary austenite and ledeburite.

For an alloy containing 4.3% of carbon, solidification takes place at a constant temperature (point C) of 1130 °C, forming ledeburite (eutectic at saturated austenite and cementite).

The solidification of alloys with 4.3 to 6.67% carbon begins along the line CD by the precipitation of cementite from the liquid alloys. As the cementite phase is richer in carbon, its separation will change the composition of the liquid alloy along the liquidus CD. This reaction occurs in the (hyper-eutectic cast iron) solidification of alloy which will consist of primary cementite crystals and ledeburite. From the above discussion it is clear that all alloys containing upto 1.7% carbon consist of austenite after the primary crystallisation while alloys having carbon percentage more than this will consist of ledeburite plus primary crystals of excess austenite or cementite.

10.10.3 Secondary Transformation

Transformations which occur in the solid state in iron-carbon alloys are known as secondary transformations. These occur in the steel section of the main diagram and hence are very important for all practical purposes. We have already discussed in Sec. 9.8, that the temperatures at which structural changes in steel on heating and cooling occur are called arrest points or critical points. These are designated by the symbols A_c (for heating)

A_r (for cooling where A stands for arrest, C stands for *chauffage* (a French word used for heating) and r stands for *refroidissment* (cooling).

In iron-carbon alloys, secondary transformations are accompanied by (1) the decomposition of austenite and (2) the transformation of gamma iron to alpha iron. The point G at 910 °C corresponds to the transformation of alpha iron into gamma iron and vice-versa. Decomposition of austenite takes place along the line GS and the crystals of ferrite separate from the austenite during this process. The excess carbon from the austenite decomposition also forms the cementite. At point S (0.8% carbon) at 723 °C a solid phase reaction takes place in which simultaneous separation of ferrite and cementite forms a new structure called the pearlite. Point S is known as the euctectoid point.

The critical points along the line GS are designated as A_{c3} in heating and A_{r3} in cooling. Along the line SE, cementite is separated during austenite decomposition and all temperatures along this line are marked as A_{cm} points. The temperature at which pearlite is formed in cooling is represented as A_{r1} and A_{c1} corresponds to the transformation of pearlite into austenite in heating. The pearlite structure consists of alternate thin plates of ferrite and cementite. For all iron-carbon alloys, the formation of pearlite takes place along the line PSK.

The region GPQ indicates the area of ferrite formation and the point P represents the solubility of carbon in alpha iron at 723 °C, which is 0.025 % carbon. On cooling, this solubility is reduced, as shown by the line PQ. The composition of ferrite for a given carbon content between the points P and Q can be determined by applying Lever's rule. In this region, the excess cementite which is separated is always present along the boundaries of the ferrite grain. Similarly, in gamma iron, the reduction of carbon solubility during cooling can be observed along the line SE.

Let us consider, the changes that occur in steels containing 0.025 to 0.8 % carbon when they are cooled from the solid solution or austenite region. Changes occur only when the temperature drops to the line GS. In this case, the alloys have a single-phase austenite structure. Below this line, Ferrite precipitates from the austenite; thus a two-phase state exists at temperatures below line GS, i.e. gamma-iron austenite and alpha-iron ferrite. The carbon concentration of the austenite will increase continuously along the line GS on cooling and at a temperature of 723 °C it will reach 0.8 % as indicated by point AC_1, the eutectoid point. Austenite of the eutectoid composition will decompose at the constant temperature of 723 °C and pearlite will be formed (ferrite + cementite mixture).

Thus, steels containing less than 0.8 % carbon are known as hypoeutectoid steels and those containing more than 0.8 % hypereutectoid steels. Hypoeutectoid steels consist of ferrite and pearlite when they are completely cooled. The higher the carbon content, the more pearlite and less ferrite they will contain. For hypereutectoid steels, when temperature falls below line ES, the austenite is saturated with carbon and precipitates as cementite upon further cooling. Thus, below line *ES*, a two-phase state exists consisting of austenite and secondary cementite for all completely cooled hypereutectoid steels (0.8-1.7 % carbon).

Now let us consider the secondary solidification characteristics of cast irons. Completely cooled hypoeutectic cast irons will have a structure consisting of pearlite, bedeburite (pearlite + cementite) and secondary cementite. In hypereutectic cast irons, the structure consists of ledeburite (pearlite + cementite) and primary cementite. Eutectic (4.3 % C) consists of lebeburite alone.

10.11 FORMATION OF AUSTENITE

Formation of austenite from pearlite is the transformation that occurs on heating steel slightly above the equilibrium temperature (A_{c1}), the free energy of pearlite is more than that of austenite. It is based on the nucleation and subsequent austenite crystal growth process. Therefore, this transformation is of the diffusion type and is dependent upon the movement of carbon atoms over considerable distances. On heating above the equilibrium temperature A_{c1}, regions of austenite are formed due to the dissolution of cementite and disappearance of ferrite. The higher temperature, greater is the rate of carbon diffusion and thus the higher the pearlite to austenite transformation. Therefore, this process consists of three stages, formation of the austenite nuclei and their growth by taking up cementite and ferrite, dissolution of the cementite and homogenisation of austenite or equalising the austenite composition throughout the crystals.

10.11.1 Austenite Grain Size and Grain Size Control

Grain size is a very important factor in relation to the various physical properties of steel and is of prime importance in the behaviour of metals under different loads. Grain size generally refers to austenite grain size. In selecting the rate at which steel is to be heated, the effect of austenite grain growth on increase in temperature must be taken into consideration. Increase in size of the austenite grain has the following effects:

(i) Improves strength of steel after heat treatment, in fine grains.
(ii) Improves machining finishes and machinability.
(iii) Mechanical properties like tensile strength, creep strength and hardenability are increased.
(iv) Quenching cracks and distortion are reduced in fine grains.
(v) Improves the fabrication properties of steel, i.e. a fine-grained steel, may be heated to higher temperature without the fear of over heating, i.e. appreciable coarsening of grains.

The following factors govern the grain size:
1. Nature and amount of deoxidizers—The tendency towards grain growth is determined by the method of deoxidation of the steel and the deoxidisers used. A deoxidiser is added to molten steel to eliminate trapped gases and to reduce the iron oxide. This process produces a fine-grained steel. Steel is de-oxidised by ferro-manganese and ferro-silicon (for the coarse-grained group) steels containing vanadium and other alloying elements (of fine-grained type) and also by aluminium.
2. Composition of steel.
3. Metallic and non-metallic inclusions.
4. Mechanical working processes like rolling, forging, etc.
5. Heat treatment processes, working temperatures and environment.
6. Grain size, time of heating and cooling and tendency to grain growth.

10.11.2 Grain Size Measurement

The American society for testing and materials (ASTM) has standardised a grain size index which is widely used for determining the austenite grain size in steels. For this the following relationship may be used:

$$n = 2^{N-1}$$

where n = number of grains per square mm as seen in a specimen viewed at a magnification of 100

N = ASTM grain size number

ASTM has recommended the following methods for measuring grain size:

(i) Comparison method
(ii) Intercept method
(iii) Planimetric method

In the comparison method, the grain size is measured by comparison, under a microscope with a magnification of × 100 (after etching), with standard grain size ASTM charts. Grain sizes of steel are usually graded into eight classifications. Steels with grain size numbers from 1 to 5 inclusive are coarse grained while numbers 6, 7 and 8 are fine-grained. By trial and error absest match is determined and the grain size of the steel is noted from the index number of the machining chart.

In the intercept method a photo-micrograph is used. On any straight line, the length of the line drawn in mm divided by the average number of grains intercepted by it gives the grain size. This method is suitable for non-equiaxed grains.

The plantimetric method, when used on a photomicrograph, indicates the number of grains per square mm in a drawn area of a circle or triangle.

For most steels, the specimen is first carburised at 930 °C for 8 hours. On slow cooling or normalisation, the austenite grain size will reveal a cementite network due to the transformation of austenite into pearlite. As the heating temperature of the standard test does not exceed 930 °C, the grain size before and after the test remains unchanged.

10.12 ISOTHERMAL TRANSFORMATIONS—TTT DIAGRAM

A TTT (Time-Temperature-Transformation) diagram is also called an S-curve, C-curve isothermal (decomposition of austenite) diagram and Bain's curve. TTT diagrams are extensively used in the assessment of the decomposition of austenite in heat-treatable steels. As the iron-carbon phase diagram does not show time as a variable, the effects of different cooling rates on the structures of steels are not revealed. Secondly, equilibrium conditions are not maintained in heat treatment. The iron-carbon equilibrium diagram reveals on the phases and corresponding microstructures under equilibrium conditions but many useful properties of the steels are obtained under non-equilibrium conditions such as variable rates of cooling as produced during quenching and better transformation of austenite into pearlite and martensite.

To construct a TTT diagram, a number of small specimens of steel are heated to a temperature at which austenite is stable and then rapidly cooled to a number of temperatures like 650 °C, 600 °C, 500 °C, 250 °C, etc. The specimens are held at these temperatures for different periods of time (isothermally) until the austenite is completely decomposed. It will be observed experimentlly that, at the start of the cooling as marked by points B_1, B_2, B_3 and B_4, there is no decomposition of austenite. This time period is called the incubation period. After this, austenite begins to decompose into the ferrite-cementite mixtures. After a certain period of time, the process of decomposition of austenite is stopped, as designated by points E_1, E_2, E_3 and E_4. It will also be observed that the rate of decomposition of austenite is not constant but is rapid initially and gradually slows down.

Figure 10.9 shows the isothermal transformation of the austenite in steel containing 0.8 % carbon. In such diagrams, for the sake of convenience, the time scale is logarithmic as the decomposition of austenite takes from a fraction of a second to hours. By plotting the starting and end points of the decomposition of austenite, two different curves are obtained. Austenite transformation takes place in the area between these two curves. The important products of austenite decomposition are labelled at the respective positions on the TTT-diagram.

The austenite is stable above 723 °C and its stability first rapidly decreases with increase in cooling rate. It is least stable at about 500-550 °C and then starts to increase below this temperature range. At temperatures between 250 and 50°C, we observe a diffusionless transformation of austenite into a hardened steel structure called Martensite. This is a supersaturated solid solution of carbon in α-iron.

At temperatures near A_{r1} (700 °C) a ferrite-cementite mixture with coarse pearlite is obtained. From 700 °C to 550 °C, the product of auste-

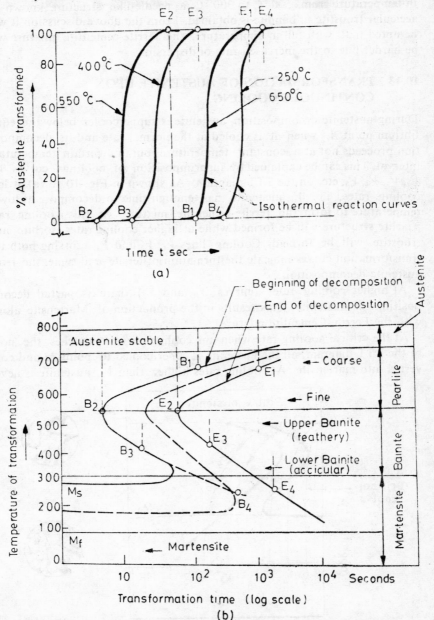

Fig. 10.9 *The TTT diagram* (a) *Isothermal reaction curves* (b) *Construction of TTT diagram*

nite decomposition (a mixture of ferrite-cementite), lamellar in structure, and finer pearlite is obtained. This structure is known as sorbite. When the temperature is lowered to between 550 °C to 500 °C an evenly dispersed mixture of ferrite-cementite called troostite is obtained. With further fall in temperature from 500 °C to 300 °C, a needle-like structure known as accicular troostite or bannite is obtained. From the above discussion it will be noted that, with fall in temperature, the ferrite-cementite mixture will be harder due to the increased rate of dispersion.

10.13 TRANSFORMATION OF AUSTENITE UPON CONTINUOUS COOLING

During austenite decomposition, austenite is suppercooled below the equilibrium point A_{r1} when it is cooled at the normal rate and its decomposition proceeds not at a constant temperature but in a certain temperature interval. This can be explained by superimposition of coolings curves V_1, V_2, V_3, V_4, V_5, etc. on the TTT diagram. As shown in Fig. 10.10 (a) at low cooling rates, i.e. V_1, austenite has enough time to decompose at lower temperature to form the ferrite-cementite mixture. At this cooling rate pearlite structure will be formed while at higher cooling rates sorbite and troostite will be formed. Cooling lines V_1, V_2 and V_3 crossing both the transformation curves indicate the formation of ferrite and cementite from austenite decomposition.

At higher cooling rates (curves V_4 and V_5) there is partial decomposition of Austenite. This results in the production of Martensite along with the Ferrite-Cementite mixture.

At the critical cooling rate when the cooling curve V_c touches the nose of the TTT diagram, all the austenite is super cooled to point M and converted into martensite. At cooling rates higher than V_c, austenite is never

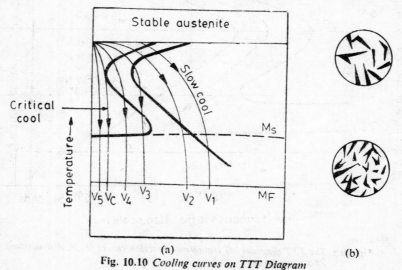

Fig. 10.10 *Cooling curves on TTT Diagram*

completely decomposed into martensite but a certain amount is untransformed. This is called retained austenite.

10.14 MARTENSITIC TRANSFORMATION

Martensite is a metastable phase of steel formed by transformation of austenite below MS temperature. Martensite is the supersaturated solution of carbon in α-iron and, in a sense, may be regarded as a product after pearlite and bainite. Its structure is broken down by tempering (Fig. 10.10 (a) and (b)).

Martensite is an interstitial supersaturated solid solution of carbon in iron with a body-centred tetragonal lattice and accicular of needle-like structure. The principal difference between martensitic and pearlitic transformations is that the former is diffusionless.

Martensitic transformation may be considered as similar to allotropic transformation, i.e. a phase transformation in a single-component system. Therefore the general laws of phase transformation are valid in this case. Martensitic transformation develops by nucleation and subsequent crystal growth like any other transformation.

When austenite is cooled at extremely high rate from above the upper critical point, as shown by the cooling line V_5 in Fig. 10.11, there is no ferrite-cementite transformation of austenite and all the austenite is transformed into a new phase, martensite. In other words, martensitic transformation may occur only at cooling rates not less than the critical value, which provides for supercooling austenite to martensite point M where no processes occur.

REVIEW QUESTIONS

1. Describe the allotropic transformations in iron and discuss their importance in practical applications.
2. What do you understand by critical points and critical range and how are they related to heating and cooling of steel? Discuss their importance in phase transformations.
3. Explain the terms: eutectoid, hypereutectoid and hypoeutectoid.
4. Describe the process of decomposition of austenite to the following:

    (a) Pearlite
    (b) Ferrite
    (c) Cementite
    (d) Bainite
    (e) Martensite
    (f) Retained Austenite.

    Compare their properties and structures.
5. Describe the working of iron-carbon diagrams and list the advantages and limitations of these diagrams when applied to heat treatment.
6. Explain the working of TTT diagrams and what information is supplied by them?
7. Describe the process of austenite decomposition of alloyed steels with TTT diagrams.

11
HEAT TREATMENT

11.1 INTRODUCTION

Heat treatment may be defined as an operation involving the heating of solid metals to definite temperatures, followed by cooling at suitable rates in order to obtain certain physical properties which are associated with changes in the nature, form, size and distribution of the micro-constituents. Heat treatment is a very important process in the various fabrication and manufacturing operations. The purpose of heat treatment is to achieve one or more of the following objectives:

(i) To relieve internal stresses set up during cold-working, casting, welding and hot-working operations.
(ii) To improve machinability
(iii) To change grain size
(iv) To soften metals for further treatment as wire drawing and cold rolling
(v) To improve mechanical properties
(vi) To modify the structure to increase wear, heat and corrosion resistance
(vii) To modify magnetic and electrical properties
(viii) To remove trapped gases
(ix) To remove coring and segregation.

The theory of heat treatment is based on the principle that when an alloy has been heated above a certain temperature, it undergoes a structural adjustment or stabilization when cooled to room temperature. In this operation, the cooling rate plays an important role on which the structural modification is mainly based. For steel the eutectoid reaction in the iron-carbon diagram involves the transformation and decomposition of austenite into pearlite, cementite or martensite. Temperature ranges and heat treatment processes are indicated in Fig. 11.1. Fig. 11.2 Shows common microstructure of steel obtained during heat-treatment.

Heat Treatment 177

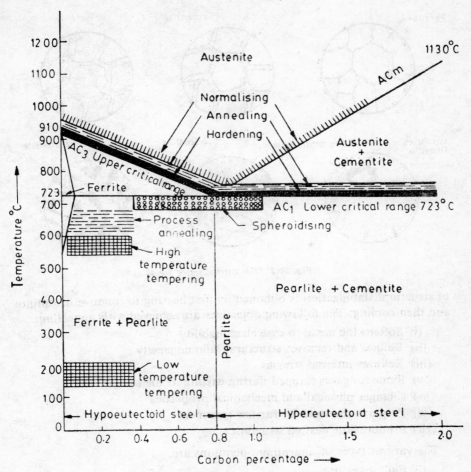

Fig. 11.1 *Heat treatment ranges and processes*

11.2 HEAT-TREATMENT PROCESSES

In order to obtain the desired objectives, one or more of the following heat-treatment processes may be used:

- (i) Annealing
- (ii) Normalising
- (iii) Hardening
- (iv) Tempering
- (v) Case hardening
- (vi) Surface hardening.

11.3 ANNEALING

This is one of the most important heat-treatment operations. The condition

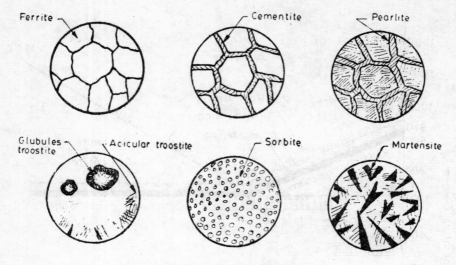

Fig. 11.2 *Microstructures of steel*

of structural stabilization is obtained by first heating to remove instability and then cooling. The following objectives are achieved with annealing:

(i) Softens the metal to ease machinability
(ii) Refines and removes structural inhomogeneity
(iii) Relieves internal stresses
(iv) Removes gases trapped during casting of metals
(v) Changes physical and mechanical properties
(vi) Prepares the steel for further treatment
(vii) Produces the desired structure.

The various types of annealing operations are:

(i) Full annealing
(ii) Process annealing
(iii) Spheroidise annealing
(iv) Diffusion annealing.

11.3.1 Full Annealing

This removes all structural imperfections by complete recrystallisation. The purpose of annealing is to soften the metal, relieve stresses and refine the grain structure. Full annealing consists of

(i) Heating the steel to about 50°-75°C above the upper critical temperature for hypo-eutectoid steel and by the same temperature above the lower critical temperature for hyper-eutectoid steels
(ii) Holding it at this temperature for a sufficient time depending upon the thickness of work and
(iii) Slowly cooling it in the furnace.

After the annealing temperature is obtained, the work is held at that temperature for a certain period of time to allow structural changes to take place. The coarse structure produced during solidification breaks down to a finer size and the structure as a whole becomes more homogeneous in chemical composition. The hardness of the steel is greatly reduced whereas there is increase in ductility. Measures should be taken to avoid over-heating, decarburisation and heavy scaling. The approximate annealing temperatures for steel are given in Table 11.1. Figure 11.3 shows the effect of annealing temperatures on mechanical properties.

Table 11.1 Annealing Temperatures

% Carbon content	Annealing temperature in °C
Less than 0.12 (Dead MS)	875-924
0.12 to 0.45 (MS)	840-970
0.45 to 0.80 (Medium carbon steel)	815-840
0.5 to 0.80 (Medium carbon steel)	780-810
0.8 to 1.50 (High carbon steel)	760-780

Holding time at the annealing temperature is approximately 3 to 4 minutes for each millimetre thickness of the largest section.

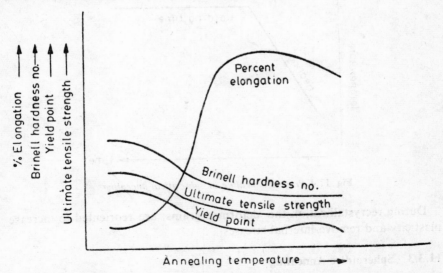

Fig. 11.3 *The effect of annealing temperature on the properties of a cold-rolled high-purity aluminium sheet*

The metal is either slowly cooled in the furnace or removed at high temperature and packed in sand, lime, mica, or charcoal to avoid decarburisation. The rate of cooling varies with the composition. The cooling time can be prolonged when packed in ashes. The rate of cooling varies from 30 to 200 °C per hour depending upon the stability of austenite. When slow cooling is done, austenite is decomposed into pearlite and ferrite structures in hypo-eutectoid steels. In eutectoid steel austenite is converted into pearlite. In a properly annealed steel the pearlite consists of coarse plates of cementite and ferrite. Hyper-eutectoid steels may undergo full annealing after hot working like rolling. Austenite decomposes into pearlite and cementite in case of hyper-eutectoid steels.

11.3.2 Process Annealing

Process annealing is usually carried out to remove the effects of cold working and to soften the steel to make it suitable for further plastic deformation, as in the sheet and wire industries. It is the recrystallization of cold-worked steel by heating below the lower critical temperature. The exact temperature depends upon the extent of cold working, grain size, composition and time held at heat. This process is very useful in mild steels and low carbon steels (Fig. 11.4).

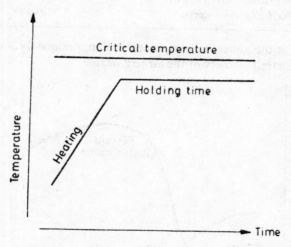

Fig. 11.4 *Process annealing or sub-critical annealing*

During recrystallization the deformed grains are reoriented to increase plasticity and remove internal stresses.

11.3.3 Spheroidise Annealing

This is a form of annealing in which cementite in the granular (globular) form is produced in the structure of steel. This process causes the agglomeration of all carbides in the steel in the form of small globules or spheroids. This process is usually applied to high-carbon steels which are difficult to machine.

The process consists of heating the steel slightly above the lower critical point (730-770 °C) holding at this temperature, and then cooling slowly to a temperature of 600 °C. The rate of cooling in the furnace is from 25 to 30 °C per hour (Fig. 11.5). Varieties of heat treatment can be used to produce a spheroidized structure, but all of them are relatively long and costly. Another method is to use, a high temperature isothermal transformation of the austenite.

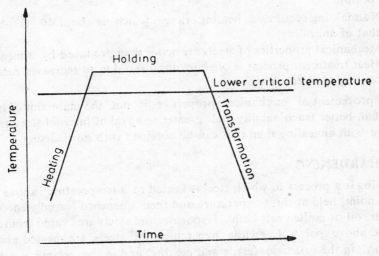

Fig. 11.5 *Spheroidize annealing*

11.3.4 Diffusion Annealing

In order to remove the heterogeneity in the composition of heavy castings, diffusion annealing is used. This process homogenises the austinite grain when heated to above the upper critical point. This process is always followed by full annealing for fine-grained structure in the castings.

11.4 NORMALISING

Normalising is frequently applied as a final heat treatment process to products which are subjected to relatively high stresses. The process consists of heating steel to a point 40 to 50 °C above its upper critical temperature, holding at that temperature for a short duration and subsequently cooling in still air at room temperature. This is also known as air quenching. This process is suggested for manufacturing operations like hot rolling and forging which are carried out on steels in the austenite range. It is also useful for eliminating coarse-grained structure in castings, removing internal stresses that may have been caused by hot or cold working and improving the mechanical properties of the steel by eliminating the carbide network at the grain boundaries of the steels.

Normalising produces microstructures consisting of ferrite and pearlite for hypoeutectoid steels and pearlite and cementite for hypereutectoid steels. The alloy steel structure consists of ferrite and sorbite. Normalising raises the yield point, ultimate tensile strength and impact strength values of steels. Normalized steels are harder and stronger but less ductile than annealed steels with the same composition.

If normalising is compared with annealing, the following points are worth noting:

(i) Normalising requires a heating range which is about 40 °C above that of annealing
(ii) Mechanical properties of steels are better than produced by annealing
(iii) Heat treatment process is of short duration due to increased rate of cooling of metal in air.

If improvement of mechanical properties is not the main aim of heat treatment, better machinability and greater removal of internal stresses is possible with annealing than that can be obtained with normalising.

11.5 HARDENING

Hardening is a process in which steel is heated to a temperature above the critical point, held at this temperature and then quenched (rapidly cooled) in water, oil or molten salt baths. Hypoeutectoid steels are heated from 30° to 50 °C above point A_{c1} while hypereutectoid steels are heated above point A_{c1}. In the first case ferrite and pearlite, and in the second pearlite and cementite are transformed into austenite upon heating. A considerable part of the cementite is retained.

Cooling at a rate higher than the critical value enables the austenite to supercool to the martensite point. Hardened steel is in a stressed condition and very brittle and cannot be used for practical purposes. After hardening, steels must be tempered to

(i) reduce the brittleness
(ii) relieve the internal stresses caused by hardening
(iii) obtain predetermined mechanical properties.

The purposes of hardening followed by tempering are:
(i) To increase the hardness and wear resistance.
(ii) To improve ductility, strength and toughness.

Hardening is applied to all tools and machine parts made from carbon steels and alloy steels. The process is carried out in three stages

(i) Heating the work to a temperature above the critical point
(ii) Holding the work at that temperature for a definite period
(iii) Quenching in a suitable medium.

The hardening process is based on a very important metallurgical reaction of decomposition of eutectoid. This reaction is dependent upon

(i) Adequate carbon content to produce hardening
(ii) Austenite decomposition to produce pearlite, bainite and martensite structures
(iii) Heating rate and time
(iv) Quenching medium
(v) Quenching rate
(vi) Size of the part
(vii) Surface conditions

ADEQUATE CARBON AND ALLOY CONTENT Steels with low carbon contents will not respond to hardening treatment satisfactorily. In order to produce hard structures like martensite, some carbon must be present. The carbon content is held near 0.5% in many engineering steels; as the carbon increases the possible hardness also increases but ductility decreases rapidly. For wear resistance the carbon content may be increased to over 1.0%, for example in tool and die steels. There are some alloys which remain dissolved in the austenite and become very hard during the decomposition of austenite into martensite.

AUSTENITE DECOMPOSITION The use of equilibrium diagrams in heat treatment is restricted to fixing the austenitizing temperatures predicting the phases that are eventually obtained at a given composition and temperature. The principal source of information of the actual process of austenite decomposition under non-equilibrium conditions is the TTT diagram which relates the transformation of the austenite to the time and temperature conditions to which it is subjected. This diagram explains in detail austenite decomposition to various structural components like pearlite, bainite and martensite. It is also known as the S-curve or C-curve.

HEATING RATE AND HEATING TIME If the heating rate is too great there will be a temperature gradient set up between the outer surface and the inner core. This will set up internal stresses. The heating rate and time depend on the composition of the steel, its structure, form and the size of the part to be heated. The exact heating rate depends upon the heat-producing capacity of the furnace, capacity of the work to be heated and range of heating. By preheating the work, heating rate is generally reduced.

The best way of heating for hardening is to first heat the articles to the required temperature, then hold at that temperature until the entire volume attains the same temperature. The lower the heating temperature, the longer the holding time. The heating time for carbon tool steels and medium alloy structure steels should be 25-40% more than that for carbon structure steels. The heating time for high alloy structures and tool steels should be 50-100% more than that for carbon structure steels. The more uniform the heating is, the higher the permissible rate may be. For this reason, heating in salt baths may be conducted more rapidly than heating in box furnaces.

QUENCHING MEDIA   The quenching media must provide for a cooling rate above the critical value to prevent austenite decomposition in the pearlite and intermediate regions. In martensite transformation temperature range, cooling should be slower to avoid high internal stresses, distortion and cracking. The most widely used quenching media are water, oil, air and molten salts.

Water and aqueous solutions are most frequently used as quenching media in hardening carbon and low carbon steels with high critical rates of cooling. It tends to form a pronounced vapour blanket or bubble which causes high structural stresses and even quenching cracks. Aqueous solutions like brine are less sensitive to heating; they lower the cooling rate but also cause distortions, as water. Mineral oils are more suitable for quenching alloy steels than plain carbon steels. This prevents quenching defects due to uniform cooling. The quenching capacity of oil is limited to a small temperature range from 20° to 150 °C. The disadvantages of oil quenching are

(i) the comparatively low cooling rate
(ii) the high inflammability of the oil and
(iii) its tendency to thicken (become gummy) in the course of time. Air blasts may also be used to cool alloy steels. High speed steels are often quenched in molten salts to harden them.

QUENCHING RATE   The rapidity with which heat is absorbed by the quenching bath has a considerable effect on the hardness of the metal. Different quenching media give the steel different quenching rates. In order to increase the quenching rate, the parts may be moved around the quenching medium either by hand or by a mechanical conveyor. The whole operation of quenching can be done these days with automatic quenching equipment.

THE SIZE OF THE PART   Small pieces can be more easily hardened than larger ones because less time is required for the quenching. In order to have a good hardening effect in the mass of the piece the following rules should be observed:

(i) Long articles (both cylinders and of other cross sections) should be immersed with their main axis normal to bath surface.
(ii) Thin and flat articles should be immersed on edge.
(iii) Recessed article should be immersed with recess upward.
(iv) Heavy articles should be held stationary in the bath and the liquid should be agitated.

SURFACE CONDITION   As impurities decrease the hardening effect, the surface of the article should be cleaned before heat treatment. Hot water, wire brushes and sand blasting may be employed to remove surface impurities.

11.5.1 Hardenability

Hardenability is a measure of the ease with which steel can be hardened by heat treatment and is strongly affected by alloying elements. Hardness is a measure of resistance to plastic deformation by indentation. The hardenability of steel determines the depth and distribution of hardness induced by quenching. It is the ability of the material to become uniform or to harden in the depth direction. Fully hardened articles will have the same properties throughout their cross section. Variation in structure in incomplete hardenability will lead to a corresponding variation in properties.

The depth of hardening is usually taken as the distance from the surface to the semi-martensitic zone, i.e. 50% martensite plus 50% pearlite. The hardness of the semi-martensitic zone, also called the 50% martensite zone, depends upon the composition of the steel. Full hardening of carbon steels is observed in articles of diameter or thickness upto 20 mm. Alloy steels harden to a considerably larger depth due to the high stability of the supercooled austenite and the correspondingly lower critical cooling rate.

11.5.2 Factors Affecting Hardenability

1. The composition of steel and the process of manufacture
2. The quenching media and the method of quenching
3. The size of the austenite grains
4. The size and shape of the piece
5. The presence of non-metallic inclusions and undissolved carbides.

The hardenability in steel increases as the percentage of carbon increases and it is further improved by the addition of such alloying elements as manganese, nickel, chromium and vanadium. Cobalt is the only one known to decrease hardenability. As earlier discussed, the rate of cooling is controlled by the quenching media and the method of quenching. Fine-grained steels have lower hardenability. The larger the austenite grain size prior to quenching, the higher the degree of hardenability. The presence of impurities tends to reduce the hardenability. Figure 11.6 shows the variation or properties due to variation in hardening.

11.5.3 DETERMINATION OF HARDENABILITY

There are a number of methods available to measure the hardenability or hardening response of the steel to heat treatment

(i) By the appearance of the fracture
(ii) By the distribution of hardness along the cross-section
(iii) By an end quench test or Jominy Test.

The hardenability depends upon the depth of hardness obtained or the rate of critical cooling. Figures 11.6 show the hardness distribution curves for hardening steel bars of different diameters with water and oil quenching. This method is not very satisfactory because the time taken and material consumed are pretty large.

The most widely standard hardenability test is the Jominy test. The apparatus used for the test is shown in Fig. 11.7 (a). The specimen to be tested is a cylindrical bar of 25 mm diameter and 100 mm length.

186 *Materials Science*

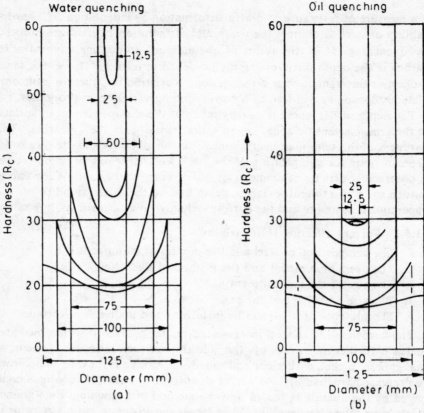

Fig. 11.6 *Hardenability of specimens of steel containing 0.45% C of different diameters.* (a) *Water quenched* (b) *Oil quenched*

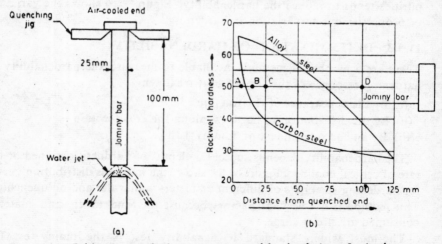

Fig. 11.7 (a) *The standard arrangement used for hardening a Jominy bar* (b) *The hardness of a Jominy bar made of steel after being quenched in parts labelled A, B, C & D*

The specimen is heated to a suitable austenitising temperature and held there long enough to form a uniform austenitic structure. It is then placed in a jig and one end of the specimen is quenched by water. When the entire bar has cooled to room temperature, Rockwell hardness readings are taken every 1.5 mm from the quenched end. Fig. 11.7 (b) shows hardness as a function of distance from the end of the bar for carbon steel and alloy steel. It is seen that the hardness falls rather quickly with distance for this carbon steel. For alloy steels having higher hardenability, the curve will not fall as fast.

The hardenability of steel can be increased by supressing the austenite-pearlite transition. If this was done, martensite could form at lower cooling rates and a piece of steel could be hardened to a greater depth when quenched. The alloying elements used with steel generally increase the hardenability by increasing the time required for the austenite pearlite transformation. Alloying elements also increase the temperature required for tempering martensite. Particular alloying elements are used for enhancing particular properties. Nickel, for example, generally increases toughness, while chromium aids wear properties by forming very hard carbides. It is common to use several alloying elements in a steel in order to get a desired combination of properties.

11.5.4 Hardening Methods

The various hardening methods are:
  (i) Quenching in a single medium
  (ii) Quenching in two media
  (iii) Hardening with self tempering
  (iv) Stepped quenching or martempering
  (v) Isothermal quenching or austempering.

The most extensively used method is conventional hardening by quenching in a single medium. The disadvantage of this method is that the cooling rate in the martensitic transformation range will be very high. Therefore cracks, distortions and other defects may occur in this method.

In order to avoid the above-mentioned defects, the quenching is carried out in two media. First water quenching is done quickly followed by a second quenching in a less intensive quenching medium (oil or air) where the article is held until completely cooled. This method is commonly used in heat treating carbon steel tools.

Hardening with self-tempering can be obtained by not allowing the article to be cooled completely in the quenching medium. Instead it is withdrawn in the course of quenching to retain a certain amount of heat in the core. This heat is used for the self tempering. This method is widely used in heat treatment of hammers, punches and chisels in order to have a high surface hardness, especially for induction hardening **operations**.

STEPPED QUENCHING OR MARTEMPERING This is a hardening method that produces martensite. This method is also known as hardening by interrupted quenching. First the steel is heated to the hardening temperature then quenched in a medium (salt bath) having a temperature slightly above the point where martensite starts to form (usually from 150° to 300 °C). The article is held until it reaches the temperature of the medium and then cooled further to room temperature in air or oil. The holding time in quenching medium or bath should be sufficient to enable a uniform temperature to be reached throughout the cross section but not long enough to cause austenite decomposition. Austenite is transformed into martensite during the subsequent period of cooling to room temperature. This treatment provides a structure of martensite and retained austenite in the hardened steel.

Martempering has the following advantages over conventional quenching

(i) Less volume changes occur due to the presence of large amounts of retained austenite and the possibility of self tempering of the martensite
(ii) Less wraping since the transformation occurs simultaneously in all parts of the article and
(iii) Less danger of quenching cracks appearing in the article.

This process is limited to carbon steel thin articles upto 8 mm in thickness. Alloy steel articles may be considerably thicker. Martempering forms martensite without transforming it into bainite.

AUSTEMPERING In isothermal quenching or austempering the steel is heated to the required hardening temperature in the same manner as in martempering but the quenching time in the salt bath is longer. Another difference in this method is that the salt bath temperature for austempering is above the martensite point to ensure a sufficiently complete austenite decomposition into bainite (acicular troostite). After which it is allowed to cool to room temperature, the rate being immaterial. Bainitic structures produced in this way are free from cracks, softer than martensite and possess good impact resistance. Tempered bainite, however, has mechanical properties inferior to those of tempered martensite.

The austempering process is based on austenitic transformation at constant temperature. The difference between martempering and austempering has been represented diagrammatically by TTT curves or S-curves in Fig. 11.8.

It must be noted, however, that hardening with quenching in a hot medium is not suitable to all grades of steel and for articles of all sizes. Improper procedures may substantially reduce the mechanical properties. The austempering process is shown in Fig. 11.9.

Molten salts are usually used as medium in martempering and austempering. The lower the temperature of the salt bath, the higher the cooling rate it provides. Since cooling in a salt bath is achieved only by conduction

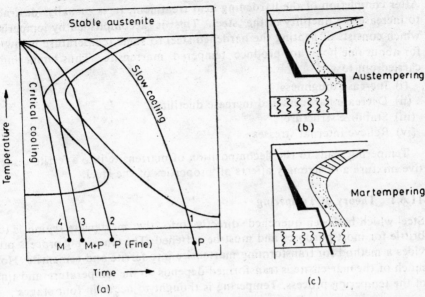

Fig. 11.8 *S-curves or TTT curves (a) for 0.8% carbon steel with continuous cooling curves 1, 2, 3, 4 superimposed M-Martensite; P-Pearlite (b) For austempering (c) for martempering*

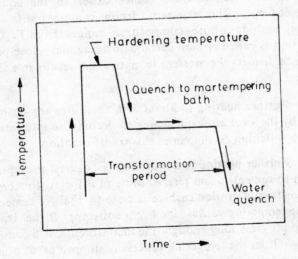

Fig. 11.9 *Austempering process*

cooling capacity is increased to a great extent by agitation. Steel is not oxidised when heated in chlorides. The thin film of chlorides covering the article protects it from oxidation.

11.6 TEMPERING

After completion of the hardening heat treatment, it is usually desirable to increase the ductility of the steel. This is accomplished by tempering, which consists of heating the hardened steel to some temperature below A_1 for about one hour to produce tempered martensite. Thus tempering is carried out to:

(i) Increase toughness
(ii) Decrease hardness and increase ductility
(iii) Stabilize structure
(iv) Relieve internal stresses

Tempering leads to the decomposition of martensite into a ferrite-cementive mixture and strongly affects all properties of the steel.

11.6.1 Theory of Tempering

Steel which has been quenched directly into the martensite region is too brittle for most purposes and must be softened. The tempering process provides a method for transforming martensite into ferrite and cementite. How much of the martensite is transformed depends on the temperature and time of the tempering process. Tempering is thought to occur in four stages.

STAGE 1 Heating a hardened steel upto a temperature of 200 °C causes the martensite to reject some of the interstitial carbon. In doing this, the tetragonal martensite structure comes closer to the equilibrium B.C.C. structure of ferrite. This rejected carbon combines with some martensite to form a carbide whose composition ranges from Fe_2C to Fe_3C. This precipitation is called epsilon carbide or hexagonal close-packed carbide. Its presence distorts the martensite matrix and results in a slight hardening of the steel.

STAGE 2 Further heating to about 300 °C, causes any austenite that was retained by the steel after quenching to decompose into ferrite and cementite. Some softening accompanies this transformation.

STAGE 3 Further heating to about 400 °C causes the epsilon carbide to transform to cementite and ferrite. Most of it forms to cementite because the composition of epsilon carbide is close to that of cementite. This portion of the tempering causes significant softening. If the transformation is allowed to progress long enough, the final structure will consist of cementite and ferrite. Often the tempering process is stopped at a point where steel contains cementite, ferrite and martensite.

STAGE 4 (400 °C—A_1) If tempering is done at a temperature just below the lower critical point or the eutectic point, the cementite forms spheres or spheroidized steel. In steels containing one alloying addition, cementite forms first and the alloy diffuses to it.

A temperature range greater than 300 °C should be avoided as tnere is a possibility of reduction in impact strength or temper brittleness due to the precipitation of carbides from the martensite in tempering.

In assigning a tempering procedure it must be noted that longer tempering times are required for heavy parts than for small ones to obtain the same results. Some tempering operations take several hours to complete the structural changes. Properties of steels after structural improvement, i.e., hardening followed by high tempering, are always higher than those of annealed steels. This is due to the difference in structure of the ferrite-cementite mixture.

Tempering is the principal method for relieving residual streeses in hardened steel. The higher the tempering temperature, the more completely the internal stresses caused by quenching will be relieved. The cooling rate after tempering has a great effect on the residual stresses. Articles of a complex form, of carbon and certain alloy steels, should be cooled slowly, after tempering at a high temperature, to prevent the appearance of new stresses.

11.6.2 Tempering Procedures

Tempering procedures are classified in accordance with the heating conditions.

LOW TEMPERATURE TEMPERING This is done in the range of 150°-250 °C and its purpose is to reduce internal stresses and increase toughness without any appreciable loss in hardness. This process is used in the heat treatment of carbon tool steels and alloy tool steels as well as of parts that have been surface hardened or case carburised.

Temperature control in low tempering is achieved by observing the so-called temper colours which appear on the ground surface of steel at temperatures from 200-300 °C.

TABLE 11.1 *Approximate Tempering Temperatures and Temper Colours*

Temper Colour	Temperature (°C)	Articles
Pale yellow	220	Hacksaw
Straw	230	Planning and slotting tools
Dark straw	240	Milling cutters, drills, shear blades
Light brown	250	Pen-knife blades, taps, metal shears, punches
Brownish purple	260	Stone-cutting tools, twist drills, reamers
Purple	270	Axles, press tools, augers, surgical tools
Dark purple	280	Cold chisels for steel and cast iron
Bright blue	290	Screw drivers, cold chisels for wrought iron
Dark blue	300	Springs, wood saws

The temper colour is the colour of the thin layer of oxides that forms on the surface of the heated steel. The colour depends on the thickness of the oxide layer.

MEDIUM-TEMPERATURE TEMPERING   This method requires heating the work to 350 °C to 400 °C. It is used to toughen the steel at the expense of hardness. Martensite and any austenite decompose into ferrite which is precipitated as extremely fine particles of cementite (secondary troostite). This is used for coil and laminated springs and provides the highest attainable elastic limit with sufficient toughness.

HIGH-TEMPERATURE TEMPERING   This is done in the range of 500°-650 °C. It almost completely eliminates internal stresses and provides the most favourable ratio of strength to toughness for structural steels. The tempered steel has a sorbite structure after treatment.

11.7 SUB-ZERO TREATMENT OF STEEL

A certain amount of retained austenite may always be found in hardened steel. Retained austenite reduces hardness, wear resistance and thermal conductivity and makes the dimensions of the article unsuitable. A subzero treatment has been devised to reduce the retained austenite in hardened steel. It consists of cooling the metal being treated to subzero temperatures. Such treatment is suitable only when the temperature at which the martensitic transformation is completed (M_f) is below zero.

Cooling to M_f transforms the retained austenite into martensite. This increases the hardness of the part and its dimensions become more stable. There is no purpose in cooling below M_f since no additional transformation of retained austenite occurs below this temperature. Subzero treatment is usually carried out in the temperature range $-30°$-120 °C. The holding time at this temperature is 1-1½ hours.

Subzero treatment is most frequently used for high-speed steel tools, measuring tools, carburised gears and other machine parts of alloy steels.

11.8 CASE HARDENING

There are many important components like gears, bearing surfaces, cam shafts and wear resistant faces on members which must also be tough, shock resistant and capable of carrying high stresses. The steel used for this purpose is usually a low-carbon steel and does not respond appreciably to heat treatment. The process consists of increasing the carbon content of the case so that it will respond to hardening and keeping the core soft and ductile. The carbon is introduced by the process of diffusion from carbon monoxide gas which is brought into contact with the surface at an appropriate temperature (870°-950 °C) by some controlled means. The process is called case-carburizing.

Case hardening is also known as a process of chemical heat treatment in which the saturation of the surface of low-carbon steel with a certain element (carbon) by diffusion of this element from the surrounding medium at a high temperature takes place. Any case hardening operation comprises three elementary processes:

(i) Processes that take place in the external medium and result in the liberation of the diffusing element in an *ionic* state, e.g., the decomposition of carbon monoxide to produce ionic carbon.
(ii) Forming of chemical bonds with the diffusing atoms and atoms of the matrix metal.
(iii) Diffusion.

There are three methods of carburizing:
(i) Pack carburizing
(ii) Liquid carburizing
(iii) Gas carburizing.

11.8.1 Pack-Carburizing

Pack carburizing involves the following reactions:

(i) The carburizing material contained in the box releases CO when heated to 900 °C
(ii) Dissociation of CO with the evolution of carbon ions:
$$4 CO \rightarrow 2 CO_2 + 2C$$
(iii) Case enrichment with carbon:
$$2 CO + 3 Fe \rightarrow Fe_3C + CO_2$$

It is one of the oldest processes of manufacturing high-carbon steel. Iron parts which are to be pack-carburized are placed in heat-resistant boxes containing a loose mixture of carbonaceous materials such as charcoal, wood, bone or charred leather combined with barium carbonate and soda ash (Fig. 11.10). Carbonates are added to accelerate the carburizing process. The packed boxes are properly sealed and placed in a furnace to heat. Iron at a temperature (900 °C) near its critical point has an affinity for carbon. Thus carbon enters the metal to form a solid solution with iron and makes the outer surface into a high-carbon steel. The box is then held at that temperature for about 5 hours. Large boxes may require more than 24 hours.

After pack-carburizing treatment, the parts with a high-carbon case (up-to 0.9 % carbon) and a low-carbon core are heated in different ways depending upon the properties required. First grain refining of the core is done by heating up to the critical point and then cooling. The case is then heated and quenched to produce the desired hard structure. Carbon steels are quenched in water and alloy steels are usually quenched in oil to produce hard fine-grained cases. The core stucture remains soft.

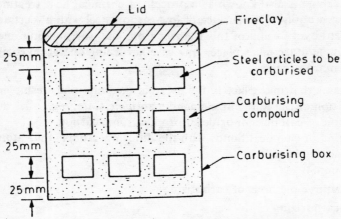

Fig. 11.10 *An arrangement for the solid carburizing process*

11.8.2 Liquid Carburizing

Liquid carburization is a similar process but, as its name suggests, it takes place in a liquid medium. The carburizing process is performed in baths of molten salts containing 75% sodium carbonate, 15% sodium chloride and 10% silicon carbide. The carburizing effect in this bath is due to the CO evolved according to the formula:

$$SiC + 2Na_2CO_3 \rightarrow \underline{Na_2SiO_3 + Na_2O} + 2CO + C$$
$$\text{Slag}$$

The principal advantages of salt-bath carburizing are,
 (i) uniform heating
 (ii) possibility for direct quenching from the bath
 (iii) small deformation of the work
 (iv) better than cyaniding because cyanide case depth is 0.25 mm and liquid carburizing allows as 6.25 mm case depth
 (v) a wide range of parts can be carburized
 (vi) time of heat treatment of steel is reduced
 (vii) very suitable for mass production of medium sized parts.

Its disadvantage is the use of poisonous salts which require careful operation in the presence of fumes and risk of explosions.

11.8.3 Gas Carburizing

Gas carburizing is a process of comparatively recent development in which directly applied carbon monoxide gas reacts with the surface of a steel to give a much more direct and rapid absorption of carbon. This is achieved by holding the component in an atmosphere of a mixture of CO, CO_2, hydrogen and other gases so proportioned that the maximum rate of carbon absorption is attained.

Components of simple shape are suspended from hooks in the atmosphere-controlled gas furnace over a quenching tank. By suitable release or lowering of suspension, the components may be quenched directly from the furnace without exposure to atmospheric oxygen. This permits a quality of surface finish impossible to obtain by other methods.

The gaseous atmosphere can be generated in several ways:

1. One very common method is to use ordinary coal gas modified so that a high proportion of CO is developed before feeding into the treatment furnace.
2. Another method uses a neutral gas which will not affect the metal surface, into which are introduced CO, propane, butane or methane vapours. The latter may be done by allowing a liquid hydrocarbon to drip, at a controlled rate, into the furnace atmosphere so that it vapourizes and releases its constituents which can react with the metal surface. This process is also known as drip carburizing.

The main reaction which supplies active carbon to the steel part is given below:

$$CH_4 \rightarrow C_{at} + 2H_2 \quad \text{(dissociation of methane)}$$

ADVANTAGES OF GAS CARBURIZING

1. Case depth can be obtained accurately
2. Process is rapid as less time is required than in pack carburizing
3. Gas carburizing operation may be feasibly mechanised as carburizing may be combined with subsequent quenching.
4. Many laborious operations are excluded
5. Less floor space is required than pack carburizing
6. Results are reliable as skilled personnel are required to maintain the necessary controls.

11.8.4 Nitriding

It may be desired to finish-machine an intricate part to its final dimensions and then to obtain a hard surface either over the whole part or on some selected portion without causing distortions that might easily arise in the carburizing process, e.g., in the manufacture of dimensional gauges or cylinder liners of IC engines. This is obtained by nitriding.

Nitriding is the process of saturating the surface of steel with nitrogen by holding it for prolonged periods at a temperature between 480 °C and 650 °C in an atmosphere of ammonia. Nitriding increases the hardness of the surface to a very high degree. It also increases the wear resistance, fatigue limit and resistance to corrosion in such media as the atmosphere, water, steam, etc.

Nitriding produces a hard case without quenching or any further heat treatment.

At the nitriding temperature, ammonia gas dissociates as follows:

$$2NH_3 \rightarrow 2N + 3H_2$$

The atomic nitrogen thus formed diffuses into the iron. Nitriding is usually applied to medium carbon alloy steels which acquire high wear resistance as a result.

11.8.5 Advantages and Disadvantages of Nitriding

Advantages
1. Very high surface hardness may be obtained
2. Distortion and cracks are minimum as quenching is eliminated
3. Good wear resistance
4. Nitrided parts retain hardness upto 500 °C
5. Machining and finishing is generally not required
6. Economical for mass production
7. Better process than carburizing.

Disadvantages
1. Long operation times, upto 100 hours for 0.038 mm depth
2. Limited alloy steels containing only aluminium, chromium, vanadium and molybdenum can form good nitrides
3. Not economical unless specialized equipment is available
4. Oxidation due to prolonged heating time.

11.8.6 Cyaniding

In order to produce hard cases on the surfaces of low or medium carbon steels, the cyaniding process may be used. In cyaniding, carbon and nitrogen are added to the surface layer of the steel to increase its hardness, wear resistance and fatigue limit. The steel is heated in a molten cyanide salt bath maintained at 950 °C, followed by water or oil quenching.

Salt bath compositions may very according to the temperature of the salt, thickness of the case to be obtained, type of steel to be heat treated and period of operation. Case thicknesses from 0.075–1.5 mm can be obtained in process.

Molten salts like sodium chloride, sodium carbonate, sodium cyanide, soda ash and barium chloride are commonly used in this process.

Carbon monoxide is formed by the decomposition of sodium cyanate. It evolves atomic carbon which is diffused in the steel. At the same time, nitrogen is also dissociated from sodium cyanate and introduced in the steel.

11.8.7 Advantages and Disadvantages of Cyaniding

Advantages
1. Bright finish of parts can be retained

2. Cracks and distortions can be minimised by uniform heating maintained by the salt
3. Most suitable for parts subjected to high loads.

Disadvantages
1. Risk of spiltering of poisonous salts
2. Unhealthy fumes formed, requiring careful handling operation
3. Liquid carburizing can produce better surface hardnesses.

11.8.8 Carbonitriding

Parts made of various tool and low-carbon steels are given the same treatment as cyaniding, not by liquid salt baths but by gas atmospheres. In carbonitriding, both carbon and nitrogen simultaneously saturate the surface of the steel but the process is slower. At low temperatures of about 550 °C, this process is applied to high-speed steel tools. Complex machine parts are heated at 800 °C, followed by subsequent tempering. This produces distortion-free heat treatment. To save time, temperatures as high as 950 °C can be used, as in gas carburizing. This increases wear resistance of the steel.

Carbonitriding is the process of heating parts in an atmosphere of natural gas and ammonia in a 3 : 1 proportion at the critical temperature, followed by quenching and tempering. It is a highly progressive method of heat treatment of structural steels. The lower temperature of the process increases the service life of the furnace and its accessories and also reduces fuel consumption.

11.9 SURFACE HARDENING

This is also a method of heat treatment in which the surface layers of a metal are hardened to a certain depth while the core is maintained relatively soft. It differs from case hardening in that the chemical composition of the surface is changed. Secondly, heating and quenching of the metal is rapid and thus the core of the metal remains unaffected. There are two methods of heat treatment in this group:

(i) Induction hardening
(ii) Flame hardening.

11.9.1 Induction Hardening

Induction hardening is a process of surface hardening in which the surface to be hardened is surrounded by a copper inductor through which a high-frequency current of about 2000 cycles per second is passed. The inductor acts as a primary coil of a transformer. The work to be hardened is placed in the inductor in such a way that it does not touch the inductor (Fig. 11.11).

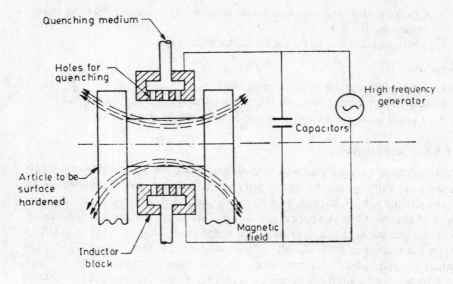

Fig. 11.11 *Arrangement for high-frequency induction heating*

The inductor block has a number of holes to spray water for quenching. The heating effect in the work is produced by the induced eddy currents and hysteresis loss in the surface of the work. Steels containing 0.35-0.55 % carbon are most frequently induction hardened. The hardening temperature is above 768 °C (Curie point) in order to increase the depth of current penetration. The heated areas are quenched immediately by water sprays directed from the inductor.

Induction hardening is, at present, extensively used in many industrial plants e.g., camshafts, crankshafts, gears, axles and many other automobile parts, tractor parts and similar wearing surfaces. This process may be applied by three methods.

SINGLE OPERATION The entire surface of the work is simultaneously heated and quenched. This method is applicable for hardening small surfaces like pins, drills and small shafts.

NTERMITTENT OPERATION Parts like teeth of gears or cam surfaces are consecutively heated and quenched, intermittently, and the process is repeated for all locations one by one.

PROGRESSIVE OPERATION This method of induction hardening is most suitable for long shafts, axles and other similar parts. As the work passes the inductor coil, its top surface is heated red hot. The work keeps moving forward on a conveyor and, as soon as it comes out of the inductor heating range, is quenched by powerful water sprays. Thus, heat treatment is performed directly in the machining production line without interrupting the technological sequence of operations.

11.9.2 Advantages and Disadvantages of Induction Hardening

Advantages
1. Heating times are extremely short (between 1 and 5 seconds)
2. No sealing or surface oxidation
3. Deformation due to heat treatment is considerably reduced
4. Permits automation of heat-treatment processes
5. Close tolerances on dimensions are maintained due to induction heating of the surface and articles which were finished before heat treatment remain finished after treatment
6. Eliminates the use of costly alloy steels for manufacture of the above-mentioned products.
7. Induction hardened steels have high hardness, higher wear resistance, higher impact strength and higher fatigue limit in comparison with ordinary, hardened steels.

Disadvantages
1. Not economical for mass production
2. Cost of equipment is high
3. Restricted to medium-carbon and low-alloy steels only.

11.9.3 Flame Hardening

Like induction hardening, this process is also based on rapid heating and quenching in order to produce a hard surface and soft core in the work. An oxy-acetylene flame is used to heat the work above its critical temperature and quenching is done by means of a spray of water directed on the surface. The torch for heating the work may be stationary or may move progressively over the work which may or may not spin (Fig. 11.12).

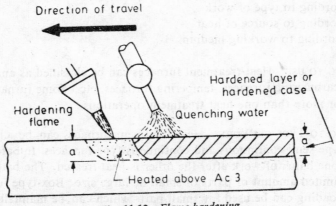

Fig. 11.12 *Flame hardening*

This method is applied for hardening cast gears, mill rolls or worms. The chief advantage is the low cost of equipment and simplicity of operation. A disadvantage is the danger of overheating and uncontrollable hardness produced at two different locations due to uncontrolled temperatures.

11.10 HEAT-TREATMENT FURNACES

Heat-treatment furnaces are special chambers in which the heat treatment processes are carried out. The primary objectives of heat treatment furnaces are:

  (i) Uniform heating and cooling of components
  (ii) Temperature should be readily controllable
  (iii) Loaded furnace should be heated to the required temperature in a reasonably short time.
  (iv) Should occupy the minimum working space
  (v) Should maintain a controlled atmosphere to prevent decarburisation, oxidation, erosion and formation of scale on the surfaces of work components.

The basic requirements in the selection of heat treatment furnaces are:

  (i) The size and shape of work that can be heat treated
  (ii) The volume of production, i.e., batch-type or continuous-type, that can be heat treated
  (iii) The type of heat-treatment process that can be done by the furnace, e.g. annealing, tempering and nitriding
  (iv) The cost of equipment and the ease of operation.

11.10.1 Types of Heat-Treatment Furnaces

Furnaces used in heat treatment can be classified in the following ways:

  (i) According to use
  (ii) According to type of work
  (iii) According to source of heat
  (iv) According to working medium.

ACCORDING TO USE Heat-treatment furnaces can be classified as annealing furnaces, carburising furnaces, tempering furnaces, etc. Some furnaces are suitable for more than one heat treatment operations.

ACCORDING TO TYPE OF WORK Heat-treatment furnaces can be classified as batch-type semi-continuous and continuous type furnaces. In batch-type furnaces, one batch of work after the other is heat treated. The batch consists of a limited amount of parts of small or large size. Box-type furnaces with side loading can be used for small parts which can be manually char-

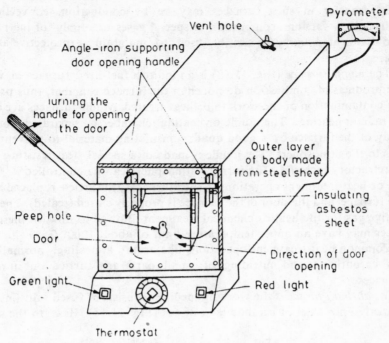

Fig. 11.13 *Heat treatment furnace*

ged, whereas large and heavy parts can be loaded or unloaded on moveable car-bottom furnace. Large parts are also heat treated in pit furnaces. The part is kept in a vertical position; if it is heated in the horizontal position it may warp because of its own weight.

For mass-production work, continuous furnaces are used. Parts are moved mechanically from one end to the other in such furnaces. Movement of the parts in the heating chamber may be effected by conveyors, pushed-and-pulled cars, spiral screws or rollers. Tunnel and rotary furnaces are common continuous heat-treatment furnaces. Heating in such furnaces is done by gas firing, oil firing or electrically.

ACCORDING TO SOURCE OF HEAT Furnaces are classified as coalfired, oil-fired, gas-fired and electric furnaces. Electric furnaces are preferred because of their high thermal efficiency, better temperature control, simple construction and freedom from unwanted waste gases as they have neither combustion chamber nor chimney. Conditions of work are considerably more hygenic than with other furnaces. Electric furnaces may be of the resistance type or electric-induction type.

ACCORDING TO WORKING MEDIUM The parts to be heat treated may be heated in air, a protective gas atmosphere (as in gas carburising or nitriding), salt baths, mineral oils or in the presence of solid powdered materials.

Heat transfer in such furnaces may be by conduction, convection or radiation. A vacuum, combustion of special gases or supply of inert gases and dissociated ammonia can be used to maintain a protective atmosphere.

The *muffle furnace* (Fig. 11.13) is an indirect fuel-fired furnace in which the products of combustion do not enter the furnace chamber, thus preventing contamination of the work to be heat treated. Muffle furnaces are based on radiant heating. The muffle or heating chamber is separated from the body of the furnace by a good quality insulating material to prevent heat loss to the atmosphere. The muffle is made of a special heat-resisting alloy or refractory material. The furnace atmosphere can be controlled.

For heat-treatment operations in such furnaces, the article is placed inside the furnace and the door of the furnace is properly closed (sealed). The protective gas for the heat or chemical treatment allowed to flow in these furnaces may have an upper temperature limit of about 1200 °C.

Common heat-treatment operations like bright annealing, normalising, gas carburising and nitriding of small parts are carried out in muffle furnaces.

In *salt-bath furnaces* the working medium consists of fused salts inside a properly lined steel or ceramic tank of pot (Fig. 11.14). Heat to the salt is

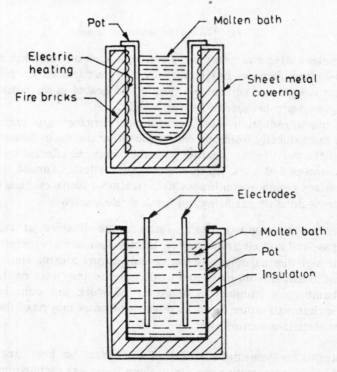

Fig. 11.14 *Thermocouple pyrometer*

supplied either by an external or internal heating arrangement, through oil or gas firing or electrical heating. The heating temperature depends upon the melting point of the salt.

These furnaces are commonly used for isothermal quenching in hardening and tempering processes. Sodium or potassium saltpetres are used for tempering springs and similar articles; sodium chloride, calcined soda and barium chloride salt baths are used for hardening low-carbon and low-alloy steels. For high-speed steel hardening a sodium chloride salt bath is used.

Lead bath furnaces use molten lead for hardening hand files, reamers, drills and other hand tools requiring low hardening temperatures.

Oil baths are used for heating uniformly upto 250 °C, as when quenching and tempering steel parts. Oil bath furnaces use mineral oil electrically heated in a metallic container. Air heating cannot warm the oil so effectively.

A salt-bath furnace is preferred when uniform heating on the centre surface of the work or a part of it is required. This can be done by immersing that portion in the fused salt. Due to proper protection of the salt, the chances of overheating, burning, scaling, oxidation and decarburization are eliminated. Heating of the work is quick and the time of heat treatment operation is small.

In all salt baths furnaces, care should be taken that the articles to be heat treated should be perfectly clean and dry before they are immersed in the bath, otherwise formation of steam can cause dangerous explosions. Another point that needs consideration is the production of dangerous fumes, against which proper protection should be taken. An objection in the use of salt-bath furnaces is that the surface finish or brightness of the surface, as maintained in nitriding operations cannot be maintained.

11.10.2 Heat-Treatment Furnace Atmosphere Control

A controlled atmosphere may be required in the furnace to produce either a chemically active atmosphere or a protective atmosphere in order to prevent undesirable effects (or changes) or to create chemical changes in the surface of the material being treated. This is a sort of control in the composition and distribution of the furnace atmosphere to produce a specific effect.

To ensure the success and quality of heat-treatment processes, furnace atmosphere control is very necessary. Control may be obtained by regulating the air or gas supply into the heating chamber of the furnace. As already discussed, such controlled atmospheres are used to add carbon or nitrogen to the outer case of the steel as in gas carburising and nitriding operations or to prevent decarburisation or scaling of steel.

The furnace atmosphere can be controlled either by infra red (to control the supply of CO, CO_2 and their concentrations) or by gas chromatography in which proper mixing of a few gases can be done to get the desired heat-treatment atmosphere.

11.11 TEMPERATURE MEASUREMENT OR PYROMETRY

Temperature measurement of the heat-treatment furnaces is a very important activity. The soundness of the heat-treatment process depends on the accuracy of temperature recording. There are different methods of measuring temperatures, but the most reliable is the use of an instrument called a pyrometer. The two types of pyrometers in common use are

- (i) thermoelectric pyrometers which may be used for recording temperatures upto 1500 °C
- (ii) optical pyrometers used for temperatures upto 4000 °C.

11.11.1 Thermoelectric Pyrometer

The thermoelectric pyrometer (Fig. 11.15) is based on the principle of the thermocouple. When two dissimilar wires are joined to complete an electric

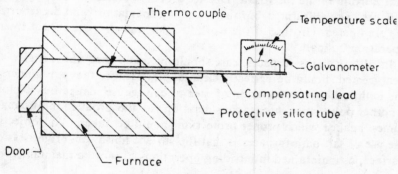

Fig. 11.15 *Thermoelectric pyrometer*

circuit and the two junctions are maintained at different temperatures, it forms a thermocouple and an electric current starts flowing in the circuit. The magnitude of the current produced depends upon the temperature difference between the two junctions and the material of the wire. The hot junction is placed in the furnace the temperature of which is to be recorded. The hot junction is made of welded platinum alloy wires enclosed in a protective silica tube. The leads from these wires are connected to a sensitive galvanometer which forms the cold junction of the thermocouple. Deviations of the galvanometer indicate the current flowing between the two junctions. Actually the galvanometer is so calibrated that it reads temperature in degrees instead of electrical units.

Thermocouple pyrometers are mainly used to measure the temperature of molten metal as the hot junction can be inserted in the metal. Other pyrometers cannot give reliable data under such conditions.

11.11.2 Optical Pyrometer

The optical pyrometer shown in Fig. 11.16 is of the disappearing filament type in which the intensity of a lamp filament is adjusted to match the radiation intensity of the furnace heat so that the former disappears

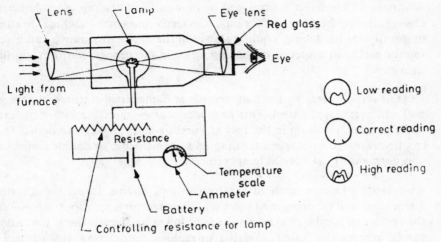

Fig. 11.16 *Optical Pyrometer*

against the radiation background. The current required to produce this effect is varied by resistance adjustments, current intensity being indicated on an ammeter. This instrument is calibrated in degrees of temperature instead of units of electric current. It gives reasonably accurate results and is very suitable for higher temperature ranges (above 1500 °C and up to 4000 °C) where thermocouples do not give reliable results.

11.12 DEFECTS IN THE HEAT TREATMENT OF STEEL

Common defects in steel due to heat treatment are described below:

DECARBURISATION   Heating articles for long periods at high temperatures in oxidizing atmospheres cause the loss of carbon from the surface. Heating in protective atmospheres can avert this defect.

OXIDATION   An oxidizing atmosphere in the furnace will result in a thick layer of scale formed on the surface of the article. This defect can also be avoided by using the right atmosphere or using carburising agents during heat treatment in oxidizing atmospheres.

QUENCHING CRACKS   Due to the critical rate of cooling in the transformation of martensite, as required during hardening, quenching cracks usually appear at the grain boundaries. This defect can be prevented by

(i) tempering immediately after quenching
(ii) heating to the minimum suitable hardening temperature
(iii) avoiding sharp protections and corners in the article so that it is free of stresses.

WARPING This defect is produced by non-uniform heating and volumetric changes during heating and cooling. Corrective measures include heating uniformly for hardening, cooling slowly in the martensitic range and holding the article in a special quenching jig or in the proper position during quenching.

OVERHEATING Heating for long periods at higher temperatures than normal will overheat the steel. This produces coarse-grained micro-structures and fractures, resulting in the loss of ductility and impact strength. This can be prevented by single annealing and normalising, or double annealing and normalising, at normal temperatures.

SOFT SPOTS Due to localised decarburisation, bubble formation (during quenching) and inhomogeneity of the initial structure, soft spots appear on the surface of hardened steel resulting in lower hardness. Their formation can be avoided by using effective quenching, protecting against decarburization during heat treatment and using homogeneous structured steel.

CORROSION AND EROSION These are produced when heating in excessive oxidizing and reducing conditions, as in salt baths or in the flame hardening processes. These defects can be prevented by carefully controlling the flame heating and salt compositions in salt baths. Erosion is produced by the chemical action of salts and heating electrodes. Adding deoxidising salts and properly positioning the article with respect to the electrodes can prevent this defect.

EXCESSIVE OR INSUFFICIENT HARDNESS AFTER TEMPERING Insufficient or excessive holding time while tempering produces this defect. A proper tempering temperature and holding time or subsequent annealing can prevent this defect.

REVIEW QUESTIONS

1. What is the object of heat treatment? List the various heat treatment processes
2. (a) What are the objects of annealing?
   (b) Describe process annealing and full annealing briefly.
3. What is the need for hardening? Explain the various requirements of the hardening reaction
4. Define the term hardenability. What factors affect hardenability? Describe a method for determining the hardenability of steel.

5. Explain the different methods of hardening.
6. Explain the processes of austempering and martempering.
7. (a) Why is hardening followed by tempering?
   (b) Explain briefly the theory of tempering.
   (c) What are temper colours?
8. Explain the subzero treatment of steel.
9. State and explain the differences between case-hardening and surface-hardening. List the various case-hardening and surface-hardening processes.
10. Explain briefly:
    (a) Nitriding
    (b) Carburizing
    (c) Cyaniding
    (d) Flame hardening
    (e) Induction hardening.
11. Describe the various types of heat-treatment furnaces.
12. What do you understand by pyrometry? Describe one type of pyrometer.
13. What are the common defects observed during heat treatment? Describe a few precautions for their prevention.
14. (a) How much carbon must be present in steel before it can be hardened?
    (b) It is not customary to grind a piece of work made of low carbon steel that has been case hardened by cyaniding. Why?
    (c) Suggest two possible ways to harden one part of a piece of work without its hardening all over.   (IAS 1978)
15. Write a brief essay on annealing of non-ferrous metals.   (IAS 1979)
16. Write a short note on age hardening.   (UPSC)
17. (a) Explain hardenability.
    (b) Describe the various case-hardening methods.   (AMIE Winter 1985)
18. (a) Why are metals heat treated?
    (b) Describe the operation and consequent effect on metals of:
        (i) Annealing
        (ii) Normalising
        (iii) Tempering   (AMIE Winter 1985)
19. Explain why steel is heat treatable and discuss
    (a) the different methods of hardening steel
    (b) changes occuring at different stages of tempering
20. Write short notes on the following:
    (a) TTT diagrams   (AMIE Winter 1985)
    (b) Pearlite   (AMIE Summer 1983)
    (c) Flame hardening   (AMIE Summer 1982)
21. Explain briefly the following heat treatment operations:
    (a) Annealing
    (b) Normalising
    (c) Tempering
    (d) Hardening
    (e) Nitriding   (AMIE Summer 1982)

22. (a) Explain the important processes of heat treatment of steel with illustrative examples.
    (b) Discuss the effect of heat treatment on the mechanical properties of steel. (AMIE Winter 1982)
23. (a) Describe martensitic transformation in steel.
    (b) Name the 'quenching media' that are used for heat treatment. Discuss their merits and demerits. (AMIE Winter 1980)

12
ORGANIC MATERIALS

12.1 INTRODUCTION

Organic materials are derived directly from carbon. In most cases their structure is complex. A number of important engineering materials are organic compounds in which carbon is chemically combined with hydrogen, oxygen or other non-metallic substances. The common organic materials include.

(i) natural organics like wood, coal, rubber, animal fibre, petroleum, etc.
(ii) synthetics including plastics, synthetic rubbers and fibres, adhesives, solvents and many other important engineering materials of industrial importance.

It will not be untrue to add that plastics, due to their wide acceptance as engineering materials, are replacing metals. That is why they have been explained extensively in this chapter.

As already said, all organic materials have carbon as one of their constituents. They usually have long molecules or macromolecules. Since carbon shares the valence electron very readily, the covalent or homopolar is the most common type of bond in organic materials, which enables it to cover a very large range of structural arrangements from viscous liquid to rigid solid.

12.2 POLYMERS

The term polymer is derived from two Greek words. The word '*mer*' means a unit and a monomer is a single unit end a polymer consists of many units. The monomers are joined together end to end in a polymerization reaction. A polymer, therefore, consists of thousands of monomers joined together. The polymer molecule is called a macromolecule. A polymeric material consists of a large number of these long-chain molecules. Polymers are also made from inorganic chemicals such as silicates and silicons (including cement and concrete). The naturally occuring polymers include proteins, shellac, starch, resins and cellulose (commonly observed in leather, wool, cotton, rubber, etc.). There are also superior types of polymers which are obtained synthetically. Such polymers are sub-divided into plastics, fibres and elastomers. Common synthetic polymers are polyethylene, polystyrene, teflon, nylon, terylene and rayon.

12.3 MECHANISMS OF POLYMERIZATION

The process of linking together of monomers is called polymerization. It is the process of growing large molecules from small ones.

There are three mechanisms of polymerization:
(i) Addition polymerization
(ii) Co-polymerization
(iii) Condensation polymerization

12.3.1 Addition Polymerization

This is a reaction which results in the bonding of additional monomer molecules to the growing chain without the elimination of any by-product molecules. The classic example of this type of reaction is the bonding of many ethylene molecules (C_2H_4) to form polyethylene ($C_2H_4)_n$, as shown schematically in Fig. 12.1 (a).

There must be some agency, called an *initiator* to initially break the carbon-carbon double bond (covalent bond). Such bonds tend to separate spatially from each other as much as possible. Typical initiators can be produced by thermal or photochemical decomposition. Once initiated, free radicals or charged ions are formed which advance the polymerization process rapidly. Each added monomer produces yet another free radical until the monomer molecules are gradually consumed and the probability of chain termination increases. This termination may be caused by the joining of two chains or by interaction between the end free radical and another molecular species which does not produce another free redical to continue the reaction.

12.3.2 Co-Polymerization

Co-polymerization is the addition polymerization of two or more different monomers. Many monomers will not polymerize with themselves but will co-polymerize with other compounds (Fig. 12.1 (c)).

$$\begin{array}{c} H\ H \\ |\ \ | \\ C=C \\ |\ \ | \\ H\ H \end{array} \rightarrow \begin{array}{c} H\ H \\ |\ \ | \\ -C-C- \\ |\ \ | \\ H\ H \end{array} \rightarrow \begin{array}{c} H\ H\ H\ H\ H \\ |\ \ |\ \ |\ \ |\ \ | \\ -C-C-C-C-C- \\ |\ \ |\ \ |\ \ |\ \ | \\ H\ H\ H\ H\ H \end{array}$$

CO-POLYMERS WITH DIFFERENT MONOMERS A copolymer may have properties quite different from those of either component member. Thus, a

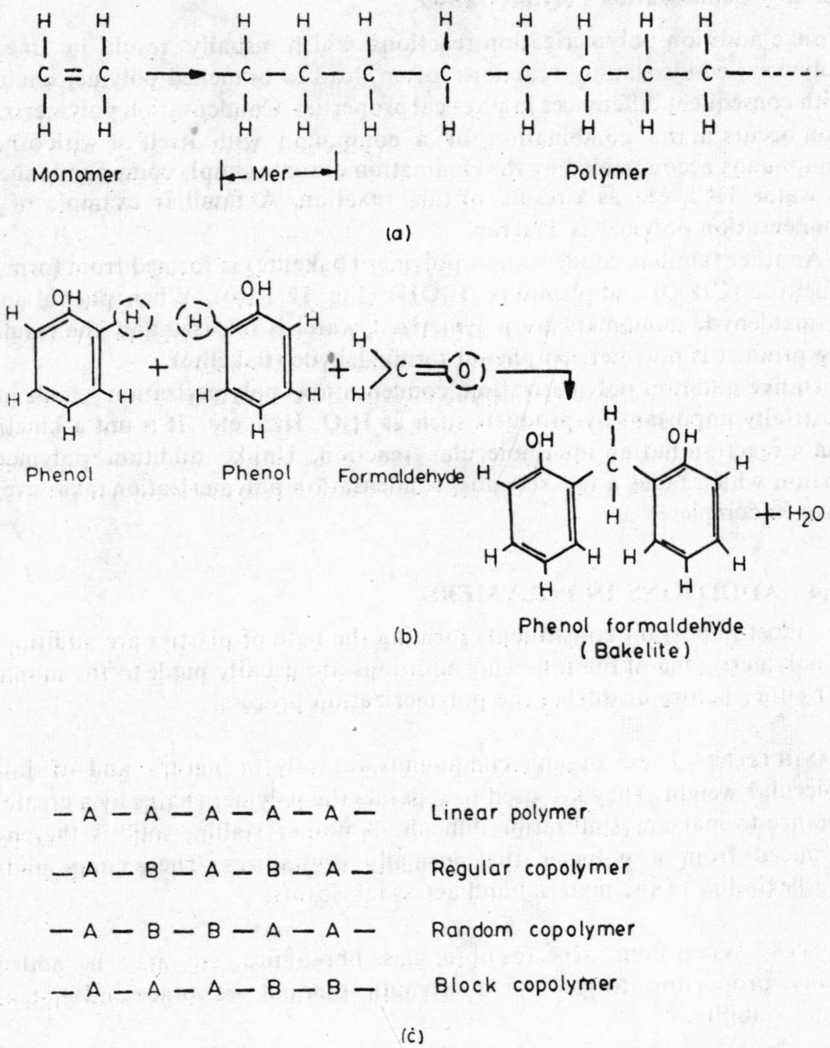

Fig. 12.1 *Polymers: (a) Linking together of monomers to form polymer. Also shows the arrangement of addition polymerization*
(b) Condensation polymerization
(c) Types of polymers and copolymers

wide variety of plastics may be obtained by this process. Vinyl chloride acetate and butadiene-styrene copolymers are examples of this type of polymerization. The buna-s rubbers are artificial rubbers mainly required for the tyre industries.

12.3.3 Condensation Polymerization

Unlike addition polymerization reactions which usually result in linear polymers, condensation reactions often lead to branched polymer chains with consequent differences in physical properties. Condensation polymerization occurs in the combination of a compound with itself or with other compounds accompanied by the elimination of some simple compounds such as water, HCl, etc. as a result of this reaction. A familiar example of a condensation polymer is Dacron.

Another familiar condensation polymer (Bakelite) is formed from formaldehyde (CH_2O) and phenol (C_6H_5OH) (Fig. 12.1 (b)). When phenol and formaldehyde monomers are polymerized, water is released and the resulting product is polymerised phenol formaldehyde (Bakelite).

Unlike addition polymerization, condensation polymerization yields industrially important by-products such as H_2O, HCl, etc. It is not a kinetic chain reaction but an intermolecular reaction. Unlike addition polymerization which takes a few seconds, condensation polymerization takes even days to complete.

12.4 ADDITIONS IN POLYMERS

The most important constituents forming the base of plastics are additions to polymers. One of the following additions are usually made to the monomer either before or during the polymerization process:

PLASTICIZERS These organic compounds are oily in nature and of low molecular weight. They are used to separate the polymer chains by a greater distance to make crystallization difficult. A non-crystalline solid is thereby produced from a polymer that normally crystallizes. These compounds give flexibility to the material and act as lubricants.

FILLERS Wood flour, asbestos fibre, glass fibre, mica, etc. may be added in any proportion to give good strength, thermal resistance and dimensional stability.

CATALYSTS Catalysts are used for quick and complete polymerization.

INITIATORS These are used to start polymerization. H_2O_2 is a familiar initiator.

DYES AND PIGMENTS In order to give good materials of different shades, dyes and pigments are used. Their effect on hardness and adhesiveness should be taken into consideration along with their own properties like translucency, resistance to chemicals etc.

12.5 POLYMER STRUCTURES

Since the process of polymerization is random, there is a little co-relation between the lengths of individual chains. So far, it has been assumed that a linear polymer molecule consists of a straight chain or backbone of carbon atoms, with various possible side units attached. This is, however, far from the case in practice, because the tetrahedral nature of the covalent bond will cause the carbon backbone to zig-zag (Fig. 12.2).

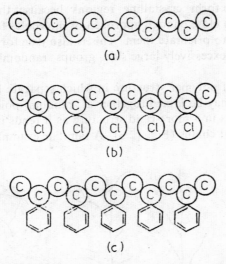

Fig. 12.2 *Schematic representation of molecular chains of different polymers*

Furthermore, in the absence of any external force which would keep the molecule extended in this way, the large freedom of rotation about the single bonds in the chain means that there is a high degree of flexibility and extensive bending and twisting of the chain results. Thus, the general configuration of a linear polymer molecule tends to be that of an irregular coil.

These long-chain molecule are generally mixed together in a solid polymer in a random manner just like a tangled bundle of coiled fibres. Whereas the molecules in any individual chain are held together by strong covalent bonds, the bonds between the various molecules are only weak secondary bonds (commonly of the Van der Waals type). The basic structure, therefore will be amorphous. Occasionally there will be some regions in the material where the linear chains are aligned close enough to one another so that the weak intermolecular forces are sufficient to keep them parallel. Over these regions the material is said to be crystalline. These crystalline regions are usually quite small and, in fact, one individual chain may form a crystalline region on its own due to regular arrangement of folds. The individual molecule may pass through several crystallites. The crystallinity of a polymer has a considerable influence on its physical properties. A polymer with a predominantly crystalline structure is appeciably stronger, more rigid and less permeable to fluids than the corresponding amorphous polymer. Usually the crystalline regions are oriented in a completely random manner, but application of an external force, as for example in the drawing of fibres, may orientate the crystalline regions leading to anisotropy of the physical properties. Branched chain molecules are less likely to form crystalline regions because the presence of the branches prevents close enough approach of molecules for the weak intermolecular forces to orientate them. This is also true for those linear molecules which have excessively large side groups randomly arranged along the chain.

Another type of polymer structure in which separate linear or branched chain molecules may be joined together at random points is called cross-linking and results in a more rigid structure as it tends to prevent slipping between individual chains (Fig. 12.3). A familiar example of cross-linking

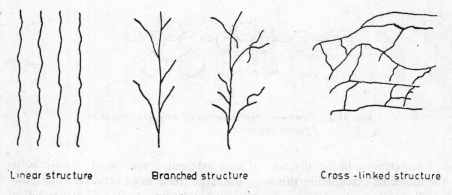

Linear structure   Branched structure   Cross-linked structure

Fig. 12.3 *Polymers structures*

occurs in the vulcanization of rubber in which the raw material is heated in the presence of 1-2% by weight of sulphur, resulting in a random array of sulphur cross-linkages. If the percentage of sulphur is increased to about 30, the rigid material ebonite is produced. Cross-linking is one of the major effects produced by irradiating polymers with high-energy radiation such as X-rays. In addition, it may be caused accidently by oxygen or sulphur atoms, catalysed by light, leading to a decrease in elastic properties. Various Organic Structures are shown in Fig. 12.4.

12.6 PLASTICS

Plastics are mouldable organic resins. They are macromolecular materials or large molecules. These organic resins are either natural or synthetic

SILICON RUBBER
1. A recent development in synthetic rubbers
2. Good resistance to chemicals and used as gaskets and seals
3. Very suitable for high temperature (250 °C) service because of thermal stability
4. A high-cost rubber
5. Outstanding at low temperatures (to -80 °C)
6. Valuable in surgical devices

12.8 FIBRES AND FILAMENTS

The characteristics of large molecule polymers are well observed in fibres which are composed of long linear molecules oriented more or less parallel to the fibre axis. This alignment gives them directional properties; their strength can be increased manifold in a direction parallel to the fibre length. The fibres embody both the strength of the constitutive molecular units and the flexibility of their conformation. All synthetic fibre-forming polymers belong to a large group of thermoplastics which show a high tendency to crystallize. When crystallized, the result is a fibre of high tenacity and a high softening point. In a few cases materials should be left in the amorphous state to attain adequate flexibility of the fibres which otherwise are brittle and rigid. A filament, on the other hand, is an individual strand of continuous length.

There are, in general, three methods of producing synthetic fibres. These are melt extrusion, dry spinning and wet spinning. Products of all three methods exhibit polymer characteristics such as heat stability, melting point and solubility in solvents.

In melt extrusion, a polymer is forced through a series of orifices, through a spinneret, and solidified into thin filaments by a stream of cool air. Polyamide (Nylon), Dacron and PVC are produced by melt extrusion.

Dry spinning is accomplished by dissolving the polymer in a suitable solvent which is then extruded through a spinneret. As warm air is circulated in the system, the solvent evaporates and the filament solidifies. This method is used to produce the commercially known copolymers Orlon (cellulose acetate) and Acrilon (polyacrylonitrile).

The wet spinning method is used to produce rayon fibres. An alkaline viscose solution is forced through a spinneret into an acid bath where coagulation of fibres takes place resulting in solid filaments.

After the production of fibres the next treatment is improving the chain molecules by stretching the filament. To increase crease resistance, waterproofing and fireproofing qualities, a number of chemical treatments are required. Like polymers, metal and glass fibres are also produced and a familiar example is of fibre glass which has many industrial applications because of its high strength (safety helmets, racing cars, bullet proof bodies of automobiles and durable furniture items).

12.8.1 Wood

Wood is a composite of cellulose and lignin. Cellulose fibres are strong in tension and flexible. Lignin binds these fibres together to give them stiffness. The differences in structure and composition have a large effect on properties. Wood is organic in origin and the oldest of all structural materials. The growth rings in a cross-section of a tree trunk can be distinguished as a result of colour and density variations. The high affinity of wood for moisture is an important factor in controlling its dimensional stability

Types of wood are:

(i) Pine
(ii) Teak
(iii) Shisham
(iv) Mahogany
(v) Deodar
(vi) Kail
(vii) Oak
(viii) Cypress
(ix) Cherry
(x) Hickory

Advantages of wood:

(i) Cheapest and strongest cellular material
(ii) Easy to join and work
(iii) Light weight
(iv) Easy to obtain a good surface
(v) Easy to repair

Defects of wood:

(i) Twisted fibres
(ii) Distortion
(iii) Knots
(iv) Dry rot
(v) Honey-combing
(vi) Cavities
(vii) Rind galls

12.9 COMPOSITE MATERIALS

In order to develop certain specific properties which do not occur naturally in individual materials, composite materials or a combination of materials is used. Composite materials retain the (Fig. 12.5) characteristics of each material for particular applications. They may be either natural such as wood, or artificially made. Composite materials may be classified as:

(i) Laminates or layered composites
(ii) Reinforced composites

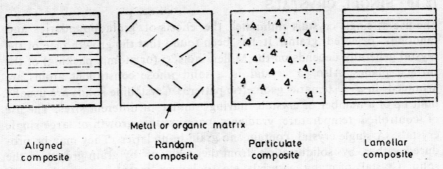

Fig. 12.5 *Different types of composites showing dispersion of materials*

(iii) Agglomerated materials
(iv) Surface-coated materials

12.9.1 Laminated Composites

Layered composites are very commonly used. Plywood is a laminated composite of thin layers of wood, successive layers having different orientations of grain or fibre. Laminated plastic sheeting, laminated glass, metal-to-metal laminates such as bimetallic thermostats, contact switches, etc. are the other uses of laminated composite materials.

Tufnol is another composite material consisting of layers of woven textile impregnated with a thermosetting resin. The woven textile provides strength while the resin gives rigidity to the composite.

12.9.2 Reinforced Materials

These materials form the largest and most important group of composite materials. The use of strong fibres, thread, rods and multiphases may improve the strength of the composite. Metallic fibres such as steel or tungsten wires are used in the metal matrix of aluminium to improve the toughness of the composite. Reinforcement with glass, carbon, ceramics and polymer fibres are used to produce a variety of composites. Glass reinforced plastic is a very familiar example of reinforced composite materials. This composite is used for helmets, boat hulls, automobile bodies, lightweight furniture and other applications where high strength and ease of manufacture are the main considerations.

Reinforced concrete is another example of this type of composite very widely used as construction material. The low strength of concrete in tension is compensated for by the high strength of steel in tension when steel rods or wires are used in concrete to reinforce it. Similarly the high compression strength of concrete compensates for the weak steel in the composite in order to improve the strength of the resulting structure. RCC is very economical, durable, fire resistant, not affected by termites or fungus, requires little maintenance and is a very strong composite material.

12.10 SINGLE CRYSTALS

The use of single crystals eliminates the effects of grain boundaries and randomly oriented grains. It has been found that the growth rate of the crystal generally depends on the Miller indices (or atomic packing) of the surface planes. Also, a crystal is a solid whose constituent atoms are arranged in a systematic geometric pattern. Crystalline solids are usually built up of a number of crystals having similar or different sizes. The use of controlled temperature gradients permits the growth of large single crystals. A single crystal contains no grain boundaries. They may be produced either by solidification from the melt or by grain growth in the solid. Ctystals of many materials, e g., polymers, metals, quartz, diamond, etc. are being grown for solid-state electronic components (such as transistors, photo diodes), lasers and masers, as optical materials, piezoelectric materials, in research work for studying the mechanical properties of single crystals and as (crystal of tungsten) filaments in electric lamps. High-purity materials are generally used when growing single crystals. Single crystals are commercially available from a number of manufacturers.

12.10.1 Whiskers

Whiskers are very thin filaments, hair-like single crystals of about 13 mm length and perhaps 10^{-4} cm diameter. Whiskers are far stronger than polycrystals of the same material because they are produced as dislocations of free crystals. Whiskers are used as reinforcements in materials to increase strength by embedding fibres of one material in a matrix of another. The properties of these fibre or whisker-reinforced composites can often be tailored for a specific application. The major disadvantage of the method is the cost of whiskers and the expensive fabrication.

Whiskers are currently the most defect-free crystalline solids available. Probably the best-known composite, fibre glass, which consists of glass-reinforcing fibres in a matrix of either an epoxy polymer or polyester whiskers can bear considerably high stresses both at low and relatively elevated temperatures. The diameter of the whiskers affects its strength as the larger diameters contain more defects and adversely affect the strength of the material. By means of special techniques single crystals of Al_2O_3, SiC, graphite, Boron, S-Glass, iron, silver, copper and tin can be produced.

12.11 AGGLOMERATED STRUCTURES

Particles, when bonded together into a mass, i.e. a macrostructure or coarse structure, are known as an agglomerate. By this process a number of engineering materials are formed from small particles to a usable product. For example:

  (i) Concrete is an agglomerated material. It is obtained by mixing sand, gravel, portland cement and water.

(ii) An agglomeration of asphalt and stone is used for paving (roads) highway surfaces.
(iii) An agglomeration of abrasives for the manufacture of grinding wheels is commonly obtained by the bonding of abrasive grains either by glass or resin.
(iv) An agglomeration of moulding sand is used in the shell moulding process.
(v) An agglomeration of small particles is formed into bulk materials using the sintering process. The technique of powder metallurgy is used to make hard cemented carbide tools, magnetic and dielectric materials and many commercially important cements.

12.12 PROTECTIVE COATINGS

Protective coatings provide a continuous physical barrier between the surface to be protected and the environment. Such a barrier must be chemically inert to the environment under the given conditions of temperature and pressure and should not permit the environment to penetrate to the material to be protected. Protective coatings can be metallic, ceramic (inorganic) or organic. Coatings are provided in order to protect the surfaces against corrosion, improve the surface finish, decorate the surface and impart special properties to the surface.

The protective value of the coating depends upon the thickness of the coating, type of material, its impermeability to the corroding environment and the required degree of protection. Under extreme conditions, mastics and linings are used for protection of the surface to be coated.

Metallic coatings are:

  (i) Anodic coatings
 (ii) Cathodic coatings
(iii) Metal spraying
 (iv) Hot dipping
  (v) Cladding
 (vi) Vapour plating
(vii) Cementation

Organic coatings are:
 (i) Chemical or electrochemical coatings such as anodizing, phosphating, chromating, oxide coating, etc.
(ii) Vitreous coatings

12.12.1 Organic Coatings

Organic coatings consist of organic materials applied as a thin protective barrier on the surface of metals or other materials to decorate and protect against corrosion. Usually, the maximum thickness of these barriers is upto 15 mils. A perfect organic coating is impermeable to water, salt and gases and has good surface adhesion. It should also be chemically inert to the environment.

12.12.2 Types of Organic Coatings

1. Paints
2. Varnishes
3. Lacquers
4. Enamels
5. Dispersions ⎫
6. Emulsions ⎬ Some other types of coatings
7. Latex   ⎭

PAINTS A paint is a mechanical mixture of an oil and a pigment. The oil acts as the vehicle and its function is to bind together the pigment particles in a paint. The oil may be linseed, cotton seed or soyabean. A solvent thinner such as turpentine oil is added to produce a thermosetting finish and to facilitate the paint to be easily brushed or sprayed over the surface. The pigments provide colour to the finish and covering power. When the paint is applied on a material, the thinner evaporates and the linseed oil, on drying, leaves behind a dry protective pigment film. To prevent the cracking of this film, plasticizers such as Mn, Pb, Co, Zn and fatty acids are added. For maximum protection, three coats or coverings of paint should be applied.

VARNISH It is a mixture of a resin and a drying oil and does not contain pigments. Sometimes a thinner is also mixed. When applied on a surface, it gives a transparent film on evaporation of the oil-resin mixture, oxidation and polymerization of its particles.

LACQUERS A lacquer is a film-forming organic finish that dries quickly by evaporation of solvents and forms a thin film of its non-volatile constituents. A lacquer dries in an extremely short period. It is a solution or dispersion of cellulose nitrate or other cellulose derivatives, including resins and plasticizers, in solvents and diluents. Lacquers give a decorative as well as protective coating. The lacquer film is based on a thermoplastic resin which does not polymerize during heating. In certain cases, lacquers may be heated at elevated temperature. Unlike varnishes or enamels, the lacquer film is not hard. Today lacquers are modified or made with vinyl, acrylic or rubber derivatives and asphalt for better abrasion and moisture resistance.

ENAMELS An enamel is a dispersion of pigments in varnish or a resin vehicle. It dries into a very lustrous and glossy finish. Depending upon the nature of the varnish vehicle and resin, the enamel varies widely in its properties. In oil dry enamel, drying occurs at room temperature by oxidation or polymerization. In baking enamel, the drying occurs at elevated temperatures. The resins used are alkyl, phenolic, acrylic, epoxy, etc.

12.12.3 Some other Types of Coatings

Besides the above-mentioned types of treatment, there are
  (i) dispersion coatings
  (ii) emulsion coatings and
  (iii) latex coatings. All are organic in nature and used for the purpose of protection of metal surfaces. These coatings are essentially suspensions of finely divided olioresinous liquids or resinous-solid particles in water or inorganic media, stabilized by various emulsifying agents. In a dispersion coating, the resin particles coagulate and the organic thinners evaporate, leaving a resin film on the surface. The dispersants commonly used are esters, ketones and glycol ethers. An emulsion is a suspension of particles of drying oil and pigment in water. The other consituents may be a varnish, an alkyl resin and solvents with or without pigments. After application, the water evaporates, leaving the mixture of oil and pigment as an organic film. This may be dried in air or heated to obtain the desired properties of coatings. Emulsion coatings may be used as decorative finishes on walls (i.e. internal building finishes). Like emulsion coatings, latex coatings also form films of closed-packed mass of resinous solid in water. Latex films are porous but more resistant to re-emulsification films.

12.12.4 Comparison of Protective Coatings

1. Organic coatings like enamels and paints are cheap, flexible and easy to apply. They are soft and have temperature limitations and oxidizing properties.
2. Metal coatings, for example noble metals and electroplating are deformable, thermally conductive and insoluble in organic solutions.
3. Ceramic coatings, for example oxide coatings and vitrous enamelled coatings, are temperature resistant, and harder, but brittle and are thermal insulators.

REVIEW QUESTIONS

1. (a) List the principal additives used in the manufacture of plastic articles.
   (b) What are the special properties of plastics that make them useful engineering materials?
2. What is meant by thermoplastics and thermosetting plastics? Distinguish between these clearly. State their properties and industrial applications.
   (AMIE, Summer 1982) (AMIE, Nov 1969, May 72)

3. Write short notes on the following:
   (a) Fillers for plastics
   (b) Plasticizers
   (c) Bakelite
   (d) Polyvinyl chloride   (IETE, June 1974)
4. (a) List the principal materials used in manufacture of plastic articles.
   (b) Distinguish between thermosetting and thermoplastic materials.
   (c) What is meant by polymerization?
   (d) Write a short note on plastic as insulating material.
                                    (AMIE, May 1966, 1969, 1971)
5. What is the difference between compression moulding and transfer moulding of plastic parts? For what type of work is the latter used?   (AMIE, Nov. 1979)
6. What are linear polymers? Explain the difference between addition and condensation polymerisation.   (AMIE, Summer 1982)
7. What are thermoplastic and thermosetting polymers? Explain addition and condensation polymerisation. Give suitable examples illustrating the structure and properties of these materials.   (AMIE, 1982)
8. Write short notes on the following:
   (a) Wood as a natural polymer   (AMIE, Summer 1984)
   (b) The difference between thermosetting and thermoplastic resins.
                                    (AMIE, Winter 1982)
   (c) The difference between resins and rubber.   (AMIE, Summer 1983)
9. Write short notes on the following:
   (a) Protective coatings   (AMIE, Winter 1985)
   (b) Electroplating   (AMIE, Winter 1982)
10. Explain the different methods used to protect steel against corrosion.
                                    (AMIE, Summer 1984)

13
ELECTRICAL AND MAGNETIC PROPERTIES OF MATERIALS

13.1 INTRODUCTION

The position of electrons in the outermost shell of the atoms controls its electrical and magnetic behaviour. Recent developments in the field of electrons have accelerated the growth of a number of solid-state devices. In this chapter we present a brief description of electrical and magnetic properties of materials.

13.2 RESISTIVITY

The electrical resistivity of solids is probably the most important of all physical properties. Metals with high electrical resistance are required for certain applications such as the heating element of any electrical appliance, while low electrical resistance is recommended for long-distance, transmission lines. The actual resistance depends on the shape of the product as well as on the resistivity of the metal. Values of resistivities of a number of materials which are important from the engineering point of view are given in Table 13.1. For a conductor of constant cross-section, the following equation is valid:

$$R = \rho \frac{l}{A}$$

Where R is the resistance in ohms,

l is the length

in centimetres, A is the area in square centimetres, and ρ is the resistivity in ohm-centimetres. Many factors influence the value of ρ for a given metal. For electrical properties, materials may be divided into three broad categories:

(i) Conductors
(ii) Semiconductors
(iii) Insulators

TABLE 13.1 *Electrical Properties of Solids*

Material Metals and Alloys	Resistivity at 20°C (ohm cm)	Temperature Coefficient $\alpha/°C$
Copper, annealed	1.67×10^{-6}	4.29×10^{-3} (0–100 °C)
Copper, reduced 75% by cold drawing	1.71×10^{-6}	—
Cartridge brass annealed, 70% copper 30% Zn	6.20×10^{-6}	1.48×10^{-3} (20 °C)
Aluminium, annealed	2.65×10^{-6}	4.29×10^{-3} (20 °C)
Iron, annealed	9.71×10^{-6}	6.57×10^{-3} (0–100 °C)
Constantan 55% Cu, 45% Ni	49×10^{-3}	0.02×10^{-3} (25 °C)
Manganin 84% Cu, 12% Mn, 4% Ni	44×10^{-6}	0.009×10^{-3} (25 °C) -0.42×10^{-3} (100°C)
Nichrome, 80% Ni, 20% Cr	108×10^{-3}	0.14×10^{-3} (0–500°C)
Semiconductors		
Germanium	10^{-3} to 60	—
Silicon	10^{-3} to 2.5×10^5	—
Insulators		
Alumina	10^{13}	—
Diamond	10^{14}	—

Conductors are metals and alloys. Gold, silver, copper are among the best conductors of electricity, followed by aluminium, iron and nickel. Some semimetals like graphite also fall in this group. By the free electron theory, when the outer orbit of an atom has less than one half of the maximum eight electrons, the material is usually a metal and a conductor. The electrical resistivity of conductors range from 10^{-7} to 10^{-2} ohm cm.

Semiconductors are materials which behave as insulators at 0 °K but develop significant conductivity as the temperature rises. At room temperature their electrical conductivity falls between that of a conductor and an insulator. They have resistivities between 10^{-1} and 10^5 ohm-cm. These form the base material for a number of electronic devices. Germanium and silicon are the common materials used as semiconductors. By the free electron theory, when the outer orbit of an atom has exactly one-half the maximum eight electrons, the material has both metal and non-metal properties and is usually a semiconductor. The electrical resistivity of a semiconductor is usually strongly dependent on temperature.

The third category of materials are insulators. Mica, PVC, rubber, porecelain and bakelite are the common insulators. The resistivity range of such materials extends from 10^6 to beyond 10^{19} ohm-cm. By the electron theory, when the outer orbit of an atom has more than one half of the maximum eight electrons, the material is usually a non-metal and a non-conductor.

13.3 CONDUCTIVITY

Electrical conductivity is the reciprocal of electrical resistivity and is expressed in siemens/metre. This term may be defined as the movement of electrical charge from one point to another. It depends on the number of charge carriers, the charge per carrier and the mobility, i.e. ease with which the electron moves through the substance. The relationship is expressed by the following equation:

$$\sigma = ne\mu$$

where σ is the conductivity,

n the number of charge carriers in a material,

e the charge carried by each of them,

μ the mobility of the carriers.

The ideas of Drude, at the beginning of this century, on electric conduction in metals are fundamental and provide a simple view of the behaviour of electrons in all materials. His work was based on the concept of a metal as a lattice of positive ions through which electrons move freely according to the laws of kinetic theory. The electrons move at random thermal speeds but, under the influence of an applied field, gain an additional drift velocity. Collisions with the lattice are the source of electrical resistance.

13.3.1 Expression for Conductivity of Conductors from the First Principle
(AMIE 1975)

Let E be the electric field applied to the conductor, e the charge on the electron, m the mass of the electron and v the velocity corresponding to the temperature equilibrium with the material of any metal. According to the laws of kinetic theory:

$$m \frac{dv}{dt} = - eE$$

$$dv = - \frac{eE}{m} dt$$

$$v = - e \frac{Et}{m} + \text{Constant}$$

If the average time between collisions is 2τ then, during this time, the electron is acted upon by a force $F = - eE$. But at $t = 0, v = 0$ (immediately after each collision) and the integration constant is zero.

Hence the velocity immediately before a collision,

$$v = - \frac{eE \, 2\tau}{m}$$

The mean velocity $= -eE\tau/m$ (1)

τ is called the relaxation time and is the time interval in which there is unit probability of a collision.

When there is collision of lattices (resistance), the current density J from n electrons per unit volume of charge e and drift velocity v is given by

$$J = nev \tag{2}$$

$$\therefore \quad J = ne\frac{eE\tau}{m} = \frac{ne^2 \tau E}{m} \tag{3}$$

$$= \sigma E \text{ (Ohm's law)} \tag{4}$$

So that the conductivity

$$\sigma = \frac{J}{E} = \frac{ne^2 \tau}{m} = \frac{ne(e\tau)}{m} \tag{5}$$

The velocity in a unit electric field V/E defines the mobility of the electron.

Hence $\quad \mu = \dfrac{v}{E} = \dfrac{1}{\underset{v}{E}} \dfrac{(e\tau E)}{m} = \dfrac{e\tau}{m}$

It follows from the definition of mobility and Eqs. (4) and (2) that

$$J = ne\mu E$$

and $\quad \sigma = ne\mu$ (6)

Eq. (6) is of fundamental importance. The electrical conductivity depends on two factors

(i) the number n of carriers per unit volume and
(ii) their mobility μ

The way these two quantities vary with temperature provides the key to the understanding of electrical properties of materials. For example, in a metal n is constant and μ varies with temperature. In semiconductors the exponential dependence of n on temperature is of prime importance while in some insulators it is the exponential dependence of μ on temperature that is significant while n is constant.

13.3.2 Band Model of Conductivity

This is used to study the characteristics of electrons to classify solids in terms of conductivity. Energy bands are certain ranges of energy levels which electrons may have. Although these permitted energy ranges appear continuous, they are actually collections of closely spaced energy levels. An electron of an isolated atom may have only certain energies. In a solid, a large number of electrons exist very close together. This causes their free or valence electrons to interact and resulting into single energy levels to be splitted into a number of lines to establish new energy levels which are dis-

Electrical and Magnetic Properties of Materials 231

crete but infinitesimally different in energy contents. Each band contains as many discrete levels as there are atoms in the crystal.

To move from one location to another, an electron must receive extra energy. An external electric field can supply the small amount of energy required to raise an electron to the next level within the band and this added energy permits movement of electrons in one direction. The adjacent energy bands in materials do not always overlap, so an energy gap may be present. This situation does not affect the conductivity of the material if there are still vacant levels within the energy band. Energy gaps vary in size from material to material.

Figure 13.1 shows electrons in crystals arranged in energy bands, separated by regions in energy for which no electron energy states are allowed. Such forbidden regions are called energy gaps or band gaps. If the

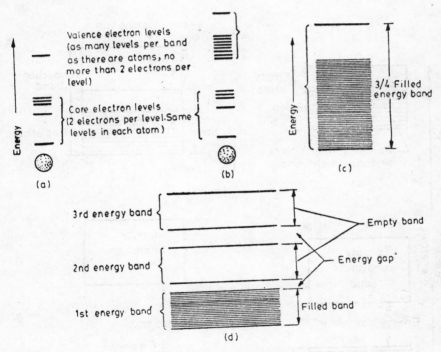

Fig. 13.1 *Energy bands*: (a) *Energy bands in a single atom* (b) *In several atoms* (c) *Three-fourth filled energy band* (d) *Filled and empty energy bands*

number of electrons in a crystal is such that the allowed energy bands are either completely filled or empty, then no electrons can move in an electric field and the crystal will behave as an insulator. If one or more bands are partly filled, say 10-90% filled, the crystal will act as a metal. If the bands are entirely filled, except for one or two bands which are slightly filled or slightly empty, say that the crystal is a semiconductor. From the above we can conclude the following points:

(i) A completely filled band can not contribute to electric conduction because the electrons in it cannot move freely in any direction.
(ii) There is always free movement of electrons in the partly filled bands which contribute to conduction (as in metals).
(iii) An empty band contains no charge and cannot contribute to conduction as in insulators.

The outermost band that is full or partly filled is called the valence band. The band that is above the valence band and empty at 0 K is called the conduction band. As mentioned above, solids can be classified on the basis of their band structure as metals, semiconductors and insulators (Fig. 13.2) The energy of the highest filled state is termed the Fermi level which

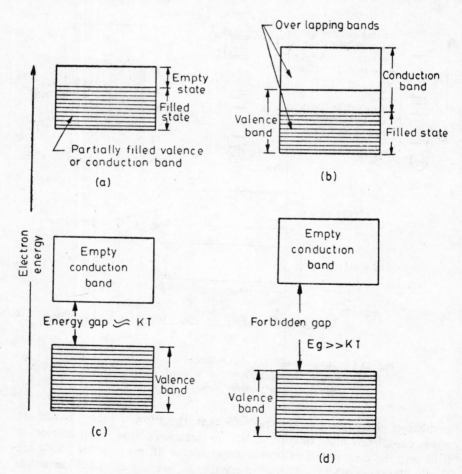

Fig. 13.2 *Band models for metals, semiconductors and insulators*

depends upon the number of electrons per unit volume. This is the level which divides the filled and vacant energy levels. At normal temperature, some electrons from the valence band jump to the conduction band, i.e. slightly above the Fermi level, due to heat energy. Therefore, a few energy levels below the Fermi level become empty and a few energy levels above the Fermi level are filled. There is no forbidden energy gap in the metal, therefore a very small amount of energy is required for excitation from the filled states to the Fermi level. In metals, the valence band is either partly filled or overlapping a higher empty band. In such cases electrons are free to move because of the empty states provided by the extra band.

At room temperature, the number of electrons that can be thermally excited across the gap in diamond turns out to be extremely small. This accounts for the insulating properties of diamond and its high electrical resistivity at room temperature. (10^{14} ohm-cm.) The energy gap of diamond is 5.4 eV, silicon 1.1 eV and germanium 0.7 eV. Those materials which have an energy gap of about 2–3 eV or less are usually called semiconductors, while those with more than 3 eV are known as insulators. When the energy gap is 2 eV or less, an appreciable number of electrons can be excited across it at room temperature. Therefore semiconductors conduct much better than insulators at room temperature. The above energy gap or band model has its importance for the study of effects of temperature, composition and structure on the number of available carriers and their mobility.

13.4 SEMICONDUCTORS

Semiconductors conduct electricity better than insulators but not as well as conductors. Semiconductors are insulators at absolute zero (0K) but develop significant conductivities at room temperature since they exhibit completely filled valence bands and completely empty conduction bands. But, at room temperature, the electronic structure of these elements makes more valence electrons available in the conduction band and they develop useful conductivities. They have conductivities ranging from 10^{-9} to 10^{4} ohm^{-1} cm^{-1}. Carbon, silicon, germanium, zinc sulphide, magnesium oxide, gray tin and selenium are the common materials termed as semiconductors. Silicon is the most widely used semiconductor crystal. Germanium is another useful semiconductor crystal. A number of solid-state devices such as junction rectifiers, transistors, photo cells, solar batteries and thermistors also use intrinsic or extrinsic semiconductor crystals.

13.4.1 Intrinsic Semiconductors

In intrinsic semiconductors, conduction is due to intrinsic processes without the influence of impurities added intentionally or otherwise. A pure crystal of silicon or germanium is an intrinsic semiconductor. The electrons that are excited from the top of the valence band to the bottom of the conduction band by thermal means are responsible for conduction.

Figure 13.3 shows the type of energy band that characterises intrinsic semiconductors. The Fermi energy level for this type of semiconductor lies midway in the forbidden gap. The movement of some electrons across

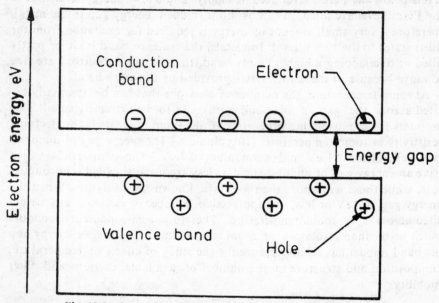

Fig. 13.3 *Mechanism of conduction in intrinsic semiconductors at room temperature*

the gap leaves some vacancies in the lower valence band. These are called holes. Under an externally applied field, the moved electrons accelerate using the vacant energy levels in the conduction band. At the same time the holes in the lower band also move, but in a direction opposite to that of the electrons. Both conduction electrons and holes serve as charge carriers. These are called *electron-hole pairs*. In an intrinsic semiconductor, the number of holes is equal to the number of conduction electrons. The forbidden gap in an intrinsic semiconductor is narrow between the lower valence band and higher conduction band. We can account for the number of electrons moved across the gap (charge carriers) by the following equation:

$$n = Ne^{-E_g/2kT}$$

where N = number of electrons available for movement

E_g = Energy gap

kT = 0.025 eV at 20 °C

13.4.2 Extrinsic Semiconductors

In extrinsic semiconductors, conduction is due to external factors, such as impurities, with which the crystal is doped. The presence of impurities produces a marked effect on the conductivity of the semiconducting materials. There are two types of impurities which, when added to a crystal of germanium or silicon, produce either the n-type or p-type material. The foreign atom is called a donor or an n-type impurity and the pure (intrinsic) semiconductor, on being doped with the foreign atom, becomes, an n-semiconductor. Similarly (aluminium) a foreign atom may provide an excess hole by accepting the valence electron from the intrinsic semiconductor. The foreign atom is called an acceptor or a p-impurity and the intrinsic semiconductor, on being doped with this foreign atom, becomes a p-semiconductor. In extrinsic semi-conductors the electrons are excited to acceptor level (p-type) or from donor level (n-type). At room temperature, the resistivity of extrinsic semiconductors is 10^{-2} ohm-cm. In extrinsic semiconductors, at absolute zero temperature, the valence electrons are tightly bound by covalent bonds. Moreover, the surplus electrons of n-type impurities and surplus holes of p-type impurities are attached to their parent atoms. Thus, at absolute zero (0K) temperature, no current flows. On increase of temperature the covalent bonds are broken and electron-hole pairs created, but the impurity carriers remain constant and show higher current values than intrinsic semiconductors.

13.4.3 n-Type Crystal

Phosphorous and arsenic are used as donor impurities. These have five valence electrons. It is known that germanium and silicon have four valence electrons. In a covalent bond each valence electron of an atom is shared by one of its four nearest neighbours as the electrons are tightly bound to the nucleus. When the impurity is introduced in the preparation of a single crystal, say a silicon crystal which is doped with a fifth column (having five valence electrons) element phosphorous, four of the five electrons in the outermost shell of phosphorus take part in the covalent bonding with four silicon neighbours. The fifth electron cannot take part in covalent bonding and is loosely bound to the parent atom. This electron represents a surplus particle in the regular structure of the crystal and serves as a current carrier. All such elements which provide extra electrons are called donors and a semiconductor with such an impurity is called n-type. The electrons in an n-type material are called the majority carrier and the holes the minority carrier. The mechanism of band formation for n-type materials is illustrated in (Fig. 13.4 (a)).

13.4.4 p-Type Crystal

On the other hand, when a trivalent impurity such as indium, aluminium or boron which have three valence electrons is added to pure silicon or germanium, only three of the electrons pairs or bonds can be filled and an extra electron is needed to complete the fourth bond. The extra electron

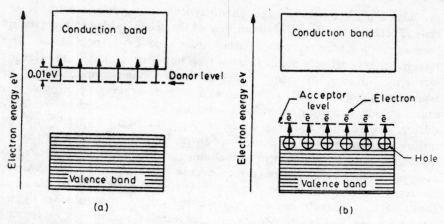

Fig. 13.4 *Mechanism of conduction at room temperature in extrinsic semiconductors (a) n-type semiconductor (b) p-type semiconductor*

can only come from one of the neighbouring silicon atoms, thereby creating a vacancy (hole) on the silicon. The aluminium atom with the extra electron becomes a negative charge and the hole with a positive charge can be considered to carry the current. The trivalent impurities make available positive carriers or holes that can accept electrons and are called acceptors. The energy required to do the conduction is just above the top of the valence band (Fig. 13.4 (b)). As the majority carriers are positive charges (holes), this extrinsic semiconductor is called a *p*-type semiconductor.

13.5 CLASSIFICATION OF SEMICONDUCTORS ON THE BASIS OF FERMI LEVEL AND FERMI ENERGY

Fermi Dirac statistics are applicable to electrons in solid crystals which follow the conditions imposed by Pauli's Exclusion principle. According to Fermi, the state of energy E and probability $P(E)$ is given by the following relation:

$$\frac{P(E) N(E)}{M(E)} = \frac{1}{e^{(E-E_f)/kT} + 1}$$

where $N(E)$ is the density of electrons per unit energy

$M(E)$ is the density of its quantum states per unit energy

E_f the Fermi level energy, is the highest energy of the electron in the valence band of a crystal at $0°K$.

Figure 13.5 (a) (b) and (c) show different types of semiconductors on the basis of their Fermi levels. When all the electrons are present in the valence band then the probability of energy is given by $P(E) = 1$.

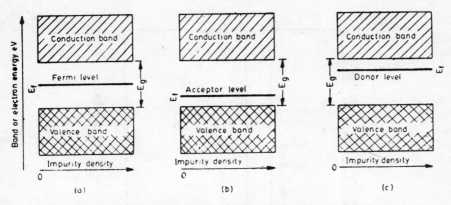

Fig. 13.5 Classification of semiconductors on the basis of Fermi levels at low temperature

When $P(E) = 1/2$ the Fermi level is between the conduction and valence bands in the forbidden energy gap. When we add an impurity to the pure semiconductor, the Fermi level E_f rises above the mean level.

For an acceptor impurity the Fermi level lowers the mean level in the forbidden energy gap. The Fermi level is nearer the valence band in p-type and nearer the conduction band in n-type semiconductors, as shown in the above figs.

13.6 p-n JUNCTION AS RECTIFIER

A piece of semiconductor in which a section of p-type material is joined to one of n-type material is called a p-n junction. It is on the properties of this junction that the semiconductor diode and the transistor work. Such a junction is not usually made by joining two separate pieces of material but by taking a single crystal of silicon and introducing donor and acceptor impurities perhaps by diffusion, into the appropriate parts of the solid. We shall now consider the electrical properties of such a junction (Fig. 13.6).

On one side of the junction is a material containing electrons moving freely and randomly at high velocities, while on the other side is a material with free holes moving at similar velocities.

The Fermi level of n-type shifts up from the middle of the energy gap towards the donor level. Similarly the Fermi level of p-type shifts down towards acceptor level. Therefore, for a p-n junction the Fermi level has to be constant at thermal equilibrium, as shown in Fig. 13.6. At equilibrium there is no net current flow across the p-n junction. The concentration of electrons in the conduction band on the p-side is small, but they can accelerate down the potential hill across the junction to the n-side, resulting

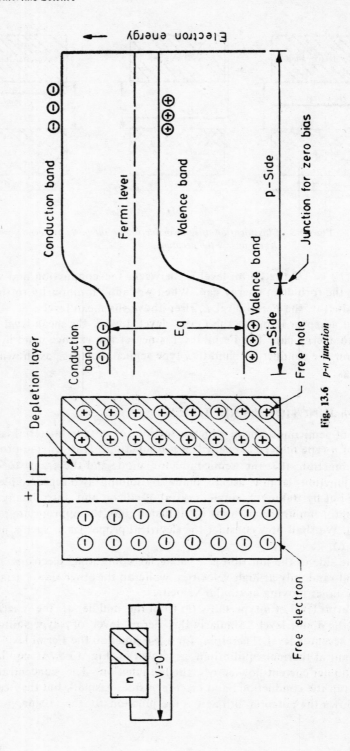

Fig. 13.6 *p-n junction*

in a current I_o that is proportional to their number. The concentration of electrons in the conduction band on the n-side is very large due to donor contribution. However, only a small number of these can flow to the p-side by climbing the potential barrier at the junction. At equilibrium, the forward and reverse currents are the same and equal to I_o. There is equal contribution to I_o from the flow of holes across the junction. The donor atoms in the n-type material are positively charged, while the acceptors on the other side have negative charge. Thus the whole crystal becomes neutral.

13.7 TRANSISTORS

A transistor is a device that can be used to amplify small alternating voltages or currents. It amplifies by converting power injected from a dc source (e.g., a battery) into ac power at the frequency of the input signal. The transistor can be regarded as two p-n junctions connected back-to-back Fig. 13.7. We now have two potential barriers, one between the left-hand

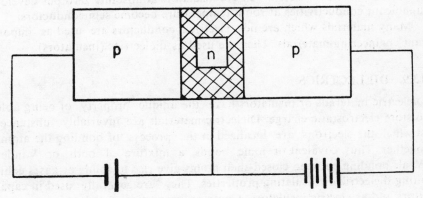

Fig. 13.7 *The junction transistor with its biasing voltage*

(emitter) and centre (base) and one between the base and the right-hand (collector) regions. If we apply a forward bias, reducing the barrier height between emitter and the base regions, holes pass into the base region from the emitter. The width of the base region is kept sufficiently small so that the electrons flowing into the base region are not lost by recombination with the holes that are dominant in this region.

In an ideal transistor, almost all the electrons emitted reach the collector by flowing down the potential hill at the base collector junction. A small change in the input voltage results in a large forward current from the emitter, to the base, and on to collector so that this appears as a large change across the load resistor. Thus, amplification results. For small power requirements, small sizes of semiconductor devices are commonly used now-a-days. The combination of a number of diodes, transistors, resistors and capacitors built into a single crystal is called an integrated circuit (IC).

13.8 INSULATORS

In an ideal insulator, all valence electrons are occupied in bond formation and none are available for conduction. Thus, an insulator has a much wider energy gap and consequently fewer charge carriers. Non-metallic crystals are held together by ionic and covalent bonds and the valence electrons are much more closely associated with their own atoms than in the metallic bond. A good insulator may have a resistivity as high as 10^{16} ohm-cm. Thermal agitation and imperfections adversely affect the insulation properties of the material but, as in real materials, a few electrons are always free to conduct electricity.

The effect of an increase in temperature is to release more and more conducting electrons. As a result, the conductivity of insulators increases with temperature.

However, the materials which are insulators at absolute zero but develop significant conductivities at room temperature become semiconductors.

Many materials which are not electrical conductors are used as important engineering materials. They are used as dielectrics (insulators).

13.9 DIELECTRICS

Dielectric materials or insulators have the unique property of being able to store electrostatic charge. Dielectric materials are invariably substances in which the electrons are localized in the process of bonding the atoms together. Thus, covalent or ionic bonds, a mixture of both, or Vander Waals bonding between closed-shell atoms give rise to solids or gases exhibiting dielectric or insulating properties. They are generally used in capacitors and as electric insulators. Common properties of dielectrics are:
  (i) Dielectric constant
  (ii) Tangent of loss angle which is best explained in terms of a capacitor or phase difference
  (iii) Dielectric strength
  (iv) Loss factor
  (v) Polar and non-polar materials
  (vi) Insulation resistance
  (vii) Surface resistivity

13.9.1 Dielectric Constant

We know that, in the electric field, the charge density D is directly proportional to the applied field E.

$$D = \epsilon E$$

where E is the dielectric constant or permittivity of the material placed between the electrodes.

If the same field is used in vacuum the relationship will be:

$$D = \epsilon_0 E$$

where ϵ_0 is called the absolute permittivity and is equal to 8.854×10^{-12} farad/metre.

The relative permittivity of a dielectric constant may therefore be defined as:

(i) The ratio of the electric field density produced in the medium to that produced in vacuum by the same electric field strength.
(ii) The ratio of the capacitances of a condenser containing a given dielectric to the same condenser with vacuum as dielectric

$$\therefore \quad \epsilon_r = \text{relative dielectric constant} = \frac{\epsilon}{\epsilon_0}$$

$$\text{Charge density} = \epsilon_0 \epsilon_r E = \epsilon_0 E + \epsilon_0 (\epsilon_r + 1) E$$

where $(\epsilon_r - 1)$ is known as electric susceptibility

$$= \chi = \text{Bound charge density/free charge density}$$

Permittivity is influenced in the following manner by various factors such as permeability, stress distribution, frequency and temperature.

The dipoles form long chains with a positive charge at one end and a negative charge at the other. A dipole moment is due to non-coincidence of positive and negative charges.

13.9.2 Phase Difference

The difference $90 - \phi = \delta$ is the loss angle. The perfect dielectric capacitor has a current which leads the voltage exactly by $90°$ (ϕ = phase angle).

13.9.3 Dielectric Strength

If the voltage imposed across a dielectric is large enough, it may cause breakdown of the material as an insulator. This voltage per unit thickness of material is called the dielectric strength and is commonly measured in volts per mil. Dielectric failure involves deterioration and electronic breakdown or cascading. Resistance to electronic breakdown is intrinsic dielectric strength of the material. The other factors involve influences other than the electric field itself.

Intrinsic breakdown is caused by imperfections. When the potential becomes high enough, a few electrons are broken loose at points where their bonds are strained by the presence of imperfections. Upon being freed, these electrons are accelerated rapidly through the material. This process results in complete loss of insulating capacity of the material.

Thus, dielectric strength is a structure-sensitive property. The actual thickness of the material affects the breakdown potential per unit thickness (dielectric strength). Thicker materials usually possess lower dielectric strength than thinner materials.

13.9.4 Power Factor

This is the ratio of power loss in the material to the product of applied voltage and current. The loss angle and power factor depend upon the nature of the material, temperature, applied voltage, frequency and humidity.

13.9.5 Dielectric Loss

Imagine a capacitor charged by the application of a steady voltage. There is an initial rush of current which follows the exponential law but after this, instead of falling to zero, the current persists at a low value for an appreciable time. The current is described as soaked in the dielectric and the phenomenon is known as absorption.

In charging the capacitor, there is molecular movement which is rapid at first and corresponds to the initial charging current; the movement thereafter being much slower. Due to the movement, work has to be done and therefore power expanded when the process of charging and discharging is repeated in rapid cycles. This expended power is the dielectric loss and the energy conveyed is converted into heat which raises the temperature of the dielectric. This is analogous to hysteresis loss in a magnetic field.

13.9.6 Polar and Non-Polar Materials

Materials in which there is free rotation of dipoles at certain frequencies and temperatures are known as polar materials. Such materials are permanently unbalanced in electric charge. The combined effect of all moments per unit volume of a given dielectric is called the dielectric polarisation.

POLARISATION A material placed in an electric field, as between the plates of a condenser, interacts with the field strength of charged particles within the material. If the material is a conductor, some of the free electrons simply move to the side nearest the positive electrode until they completely counteract the applied field. Thus, no field is left within the material. The shift occurs almost instantaneously, bringing about equilibrium.

In a non-conducting material or dielectric, electrons can only be displaced locally because they are bound to the individual atoms. This local displacement, however, is sufficient to polarize the material. In each atom the negative electron cloud is displaced relative to the positive nucleus, creating a small induced dipole whose negative pole is toward the positive side of the field. All dielectrics are subjected to such electronic polarization; (Fig. 13.8).

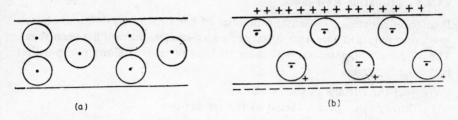

Fig. 13.8 *Electronic polarization* (a) *No field present* (b) *Formation of induced electronic dipoles in the presence of a field*

Another form of induced polarization occurs in materials containing ions. In an ionic crystal, for example the negative ions are attracted towards the positive side of the field and vice versa (Fig. 13.9).

The net result is that the centres of positive and negative charge are displaced relatively and the material becomes polarized.

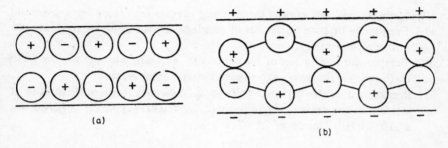

Fig. 13.9 *Ionic polarization* (a) *No field present* (b) *Ions displaced towards opposite charged plates*

Polarization also takes place in certain materials by alignment of permanent dipoles. Some molecules contain permanent charges which produce dipole moments in the individual molecules. Many polymers are also of this type (polar polymers) and glasses contain permanent dipoles. The dipole moment of a single pair of charges is

$$\mu = Qd$$

where Q = One of the charges,
d = distance between the charges.

Most hydrocarbons are non-polar. The best dielectrics are polystyrene, polycarbonate, polyethylene, mineral oil, polymide, pure alumina and pure silica.

13.9.7 Insulation Resistance

Insulation resistance separates a number of conductors at different electrical potentials and prevents a large flow of electric current between them. The difference of potential will cause leakage of current along two parallel paths, i.e.

(i) over the surface of the insulation
(ii) through the solid material of the insulation.

The resistance offered along the two paths is not the same due to the nature of the material. While the former is due to the surface resistivity of the material, the latter is the result of the volume resistivity of the material. The combined effect of the two is known as the insulation resistance of the material.

The surface resistivity may be defined as the resistance between the two opposite edges of a square of unit area of the insulation surface.

Volume resistivity is the resistance presented to the flow of an electric current by a material of unit cross-section and of unit length at 0 °C.

The resistive properties of a real dielectric are:

(i) Mobility with which any conducting species moves under a potential.
(ii) Frequency or time variation of conductivity to analyse the time response of dielectrics.
(iii) Temperature variation of conductivity to state why the conductivity of the dielectric varies exponentially with temperature.
(iv) Breakdown or the highest voltage a dielectric will withstand is of interest in the design of equipment. Insulators typically breakdown at 10^6 to 10^9 V/m dc at 20 °C.

13.9.8 Types of Dielectrics

Dielectrics can be classified as follows:

Subdivision	Effect of field
Simple dielectrics	Creates dipoles
Paraelectrics	Orients dipoles
Ferroelectrics	Orients domains of aligned permanent dipoles

13.10 FERROELECTRICITY

This is the name given to the spontaneous polarization that takes place in polar crystals, which is a function of temperature. This effect is due to the presence of permanent dipoles in the materials. Thus a ferroelectric crystal can be switched by the application of a field and a hysteresis loop is associated with the switching. If a graph of applied field against polarization is drawn, a typical hysteresis loop is obtained similar to the *B-H* loop for a ferromagnetic (Fig. 13.10). Thus the material will have an extremely large dielectric constant.

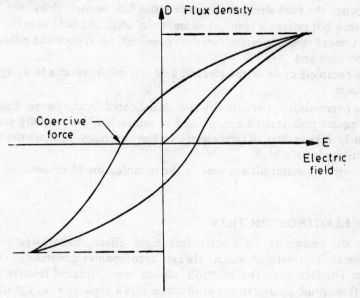

Fig. 13.10 *Hysteresis loop for a ferro electric material*

One of the conditions for spontaneous polarization is a proper temperature range. Like ferromagnetic materials, ferroelectrics have a curie temperature. At temperatures higher than the curie point thermal agitation prevents spontaneous alignment of dipoles, while at lower temperatures the material becomes ferroelectric. When the temperature of a ferroelectric is lowered through its curie point, polarization begins at scattered points throughout the material. If no field is present, the orientation of these points is random. As polarization spreads from these nuclei, however, domains are formed in which the directions of polarization are the same as those of the nuclei. The result is a network of randomly oriented domains analogous to those in unmagnetised ferromagnetic materials.

With the application of an electric field, a ferroelectric having randomly oriented domains aligns the domains in a similar way as a magnetic field aligns magnetic domains. Those oriented in the direction nearest the direction of the field grow by movement of domains walls while other domains have their polar axes rotated by a simultaneous rotation of the dipoles in the domains.

This motion takes energy and consequently polarization is not reversible. In Fig 13.10 we observe a hysteresis loop analogous to that of ferromagnetism. If the dielectric flux density D is plotted against the electric field strength E, it is found that

(i) the variation is not linear,
(ii) a saturation value of polarization is reached when all the domains are aligned
(iii) when the field strength is reduced, the flux density does not become zero but retains a residual value, even when the field is zero.
(iv) A negative or co-ercive force is required to reduce the polarization to zero and
(v) a complete cycle of the electric field strength results in a hysteresis loop

Like permeability, permittivity is also indicated by the figure. Because of the complete polarization approached at saturation, permittivity can reach extremely high values in ferroelectrics. For barium titanate the relative permittivity is very high.

Ferroelectric materials are used in the manufacture of miniatured capacitors.

13.11 ELECTROSTRICTION

This is also known as an electromechanical effect. Electrostriction is a mechanical deformation which always accompanies polarization in a dielectric. The fact that the electron clouds are displaced relative to their nuclei (electronic polarization) produces a small expansion in the direction of the applied field. This expansion varies with the square of the field strength and is the same whether the field is positive or negative. Its magnitude is ordinarily very small. There is no inverse effect because mechanical deformation simply displaces each atom as a whole and does not distort its electrical charge. Therefore no polarization results (Fig. 13.11).

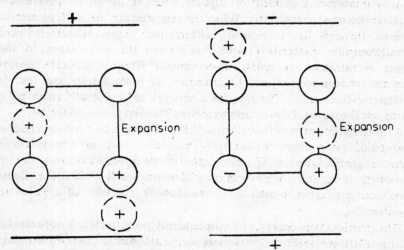

Fig. 13.11 *Electrostriction : Material expands for both directions of field*

Ferroelectric crystals exhibit a special kind of electromechanical effect associated with the presence of permanent dipoles in the crystal structure. The mass realignment of permanent dipoles in domains causes relatively large deformations in the crystal. The strain is directly proportional to the applied field instead of its square. Of greater importance is the inverse effect, the distortion of a ferroelectric crystal that causes it to be polarized. The permanent dipoles are reoriented under strain to create a polar axis in one direction or the other.

The combination of electrical to mechanical and mechanical to electrical actions is called the piezoelectric effect or pressure-electric effect. All ferroelectric materials have piezoelectric properties. Certain other crystals are also piezoelectric because of their permanent dipoles although they are not spontaneously polarized like the ferroelectrics. The basic requirement is that the crystal structure has no centre of symmetry. Many crystals fulfil this requirement. Quartz is one of the commonest examples.

13.12 PIEZOELECTRICITY

Piezo is derived from the Greek word meaning to press and the piezoelectric effect is the production of electricity by pressure. It occurs only in insulating materials and is manifested by the appearance of charges on the surface of a single crystal which is being mechanically deformed. Piezoelectric materials have the property of becoming electrically polarized in response to a applied mechanical stress. This property has an inverse; when an electric stress (a voltage) is applied the material becomes strained. The strain is directly proportional to the applied field E. Natural quartz and many synthetic materials grown by the hydrothermal process are being commonly used as piezoelectric materials (Fig. 13.12).

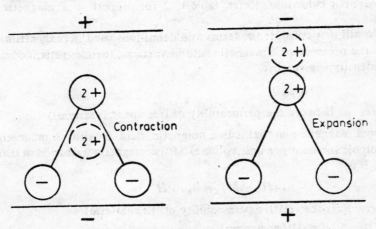

Fig. 13.12 *Piezoelectric effect—Material contracts for one field direction and expands for reversed field*

Piezoelectric materials have the following applications:
(i) Gramophone pickups
(ii) Air transducers (ear phones, hearing aids, microphones, etc.)
(iii) Ultrasonic flaw detectors
(iv) Underwater sonar transducers
(v) Filters
(vi) Frequency resonators

13.13 USES OF DIELECTRICS

1. Dielectrics are most generally used as ordinary insulators in power cables, signal cables, electric motors, etc.
2. In transformers and various forms of switchgear and generators where the dissipation problem of heat is active, and a common way of solving it is to insulate with a transformer oil (mineral oil).
3. In capacitors, resistors and strain gauges
4. Radiation detectors, thermionic valves, electric devices, dielectric amplifier.
5. Piezoelectric and electro-optic devices.

13.14 MAGNETIC PROPERTIES

Materials in which a state of magnetism can be induced are called magnetic materials. The ability of certain materials—notably iron, nickel and cobalt and some of their alloys and compounds to acquire large permanent magnetic moments is of prime importance. The many applications of magnetic properties for transformer cores and for magnetron oscillators of loudspeakers cover a wide range including the use of almost every aspect of magnetic behaviour. Refer table 13.2 for properties of magnetic materials.

We will first consider the terms and definitions used in magnetism.

In the presence of a magnetic field in vacuum, the magnetic induction or flux density is given by:

$$B = \mu_0 H$$

where μ_0 is called the permeability of free space (vacuum).

When a magnetic material is in a magnetic field H with a magnetization (or dipole moment per unit volume) M the magnetic induction in the material, B is

$$B = \mu_0 (H + M) = \mu_0 \mu_r H$$

where μ_r is the relative permeability of the material
(= 1 for vacuum)

Hence $\mu_r = 1 + (M/H) = 1 + \chi$

where χ is the magnetic susceptibility of the material

B, M and H are vectors. The unit of B is the testla or weber m^{-2}. while M and H have the same units of ampere per metre.

Hence χ has no unit.

Magnetism is dipolar, i.e. characterized by having two opposite poles—north and south. The strength of a magnetic dipole is measured by the product of the pole strength and the distance between the poles. This is known as magnetic moment. One source of magnetism in an atom is the orbital motion of electrons. Each electron revolving around the nucleus constitutes a circulating electric charge or current and thus produces a small magnetic field. In addition, each electron, spinning on its axis can be conceived as a circulating charge and also produces a small magnetic field.

13.14.1 Magnetic Moments—Dipoles

The magnetic properties of a substance depend upon the presence of dipole moments. When there is an angular momentum in the charged particle, there is always a dipole moment. Permanent magnetic moments can arise, from three sources:

(i) Orbital magnetic moment of electrons
(ii) Spin magnetic moment of electrons
(iii) Spin magnetic moment of the nucleus

Of the three, spin dipole moments of electrons are important in most magnetic materials. A discussion on this source of magnetism follows:

In the quantum state, an atom or a molecule has two electrons of opposite spins and the net magnetic moment of the filled state is zero. Magnetic properties originate due to the imbalance of spin orientation in atoms. A number of atoms or molecules have paired and unpaired electrons. There is one electron in the outermost s-shell which is unpaired and can align itself in an applied field, giving rise to magnetism. So we can conclude that the magnetic moment of an atom in the solid state is due only to an incomplete inner shell. The electrons in the incomplete inner shells of transition elements have their spins alligned in the same direction, or we may say that the electrons arrange themselves among the energy levels to give the maximum possible total spin angular moment consistent with Pauli's principle. The sequence of filling the energy levels in a subshell is governed by Hund's rule and we may deduce from it that there must be an interaction between the electrons, when they have different functions, which tends to make their spin axes parallel. The net magnetic moment due to electron spin in a sodium atom is one unit, called a Bohr magneton. (One Bohr magneton = 9.27×10^{-24} Am^2).

The atomic moments of the 3rd transition elements in the solid state are obtained by adding the spin magnetic moments of the electrons in the unfilled shell. Table 13.3 shows the spin direction of the 3*d* electrons of the given elements.

13.15 CLASSIFICATION OF MAGNETIC MATERIALS

There are five classes into which magnetic materials may be grouped (Figs 13.13 and 13.14).

(i) Diamagnetic
(ii) Paramagnetic
(iii) Ferromagnetic
(iv) Anti-ferromagnetic
(v) Ferrimagnetic

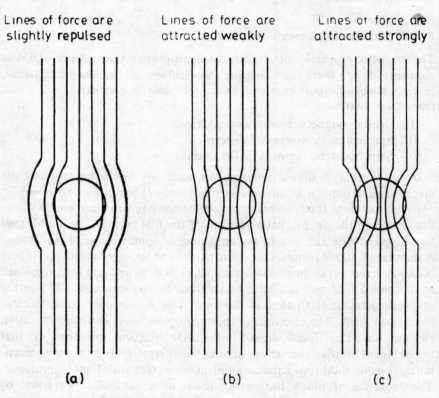

Fig. 13.13 *Classification of magnetic materials*
   (a) Diamagnetic, having relative permeability less than one
   (b) Paramagnetic, having relative permeability slightly greater than one
   (c) Ferromagnetic, having relative permeability much greater than one

Electrical and Magnetic Properties of Materials   251

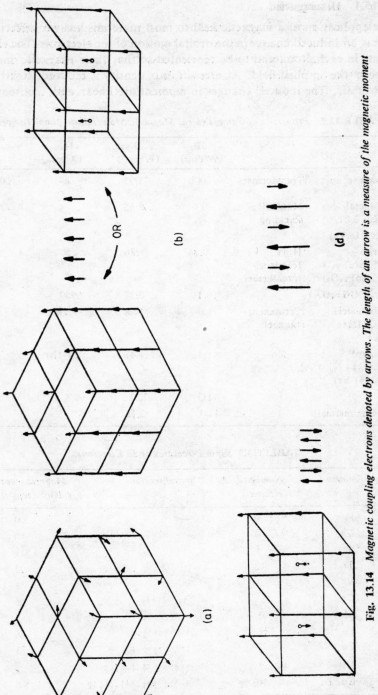

Fig. 13.14 Magnetic coupling electrons denoted by arrows. The length of an arrow is a measure of the magnetic moment

13.15.1 Diamagnetic

The application of a magnetic field to most materials has two effects. The first is an induced change in the orbital motion of the electrons. The electron orbits in each atom tend to be reoriented so that their magnetic moments oppose the applied field. As a result, flux density is reduced slightly from that of air. The induced changes in moment are linear and, consequently,

TABLE 13.2 *Properties of Some Typical Magnetic Materials at Room Temperature*

Material	Use	B_r (Wb/m²)	B sat (Wb/m²)	H_e (A/m)	r_γ at H=0
Silicon iron Fe-3% Si	Transformers	0.8	1.95	24	500-1500
Mu-metal (5 Cu 2 Cr, 77 Ni, 16 Fe)	Magnetic shielding	0.6	0.65	4	2×10^4
Ferrite (48% MnFe₂O₄, 52% ZnFe₂O₄)	High frequency transformers	0.14	.36	50	1400
Ferrite (NiFe₂O₄)		0.11	0.27	950	15
Carbon steel 0.9 C, 1 Mn, 98 Fe	Permanent magnets	1.0	1.98	4×10^3	14
Almico V (24 CO, 14 Ni, 8 Al, 3 Cu, 51 Fe)	—do—	1.31	1.41	5.3×10^1	20-250
Fe		1.3	2.16	0.8	10^1
Fe (Commercial)		1.3	2.16	7	150

TABLE 13.3 *Spin Directions of 3d Electrons*

Element	Number of 3d electrons	Spin directions	Magnetic moment in Bohr magnetons
K	0	—	0
Ca	0	—	0
Sc	1	↑	1
Ti	2	↑ ↑	2
V	3	↑ ↑ ↑	3
Cr	5	↑ ↑ ↑ ↑ ↑	5
Mn	5	↑ ↑ ↑ ↑ ↑	5
Fe	6	↑ ↑ ↑ ↑ ↑ ↓	4
Co	7	↑ ↑ ↑ ↑ ↑ ↓ ↓	3
Ni	8	↑ ↑ ↑ ↑ ↑ ↓ ↓ ↓	2
Cu, Zn	10	↑ ↑ ↑ ↑ ↑ ↓ ↓ ↓ ↓ ↓	0

B remains proportional to H but slightly smaller. This effect is common to all materials and is called diamagnetism. If it is the only effect of a magnetic field on a given material, the material is said to be diamagnetic. Organic materials and the rare gases are examples of diamagnetic materials.

Diamagnetic solids have negative magnetisation or susceptibility. Magnetic lines of force are repelled by such solids. The motion of the charge-carrying electrons in the solid are modified when a field is applied to it. The modified motion produces a local magnetic moment that opposes the externally applied field as per Lenz's law.

13.15.2 Paramagnetism

The second effect of a magnetic field occurs in materials whose orbital and spin moments do not entirely cancel. In materials with a random distribution of direction of magnetization, an applied field can produce a small degree of alignment but this disappears on the removal of the field. This behaviour is called paramagnetism. It is well known that a suspended magnetic needle aligns itself spontaneously with the earth's magnetic field. Paramagnetic materials have relative permeability slightly greater than unity and are magnetised slightly.

In such materials the individual atoms act as permanent magnets and are reoriented by the field so that their moments add to the field. This behaviour increases the flux density.

13.15.3 Ferromagnetism

Ferromagnetic solids are those in which permanent magnetic moments due to spin are already aligned due to bonding forces. Ferromagnets become very strongly magnetised in a weak external field and may possess a spontaneous magnetic moment even in zero field. Ferromagnetism only exists below certain temperature, the curie temperature T_c, above which the substance becomes paramagnetic. The relative permeability of these materials is very large and positive. They strongly attract magnetic lines of force. They are important commercially.

In ferromagnetic materials, the atomic moments are all parallel. The most important of the ferromagnetic materials are all metals and iron cobalt, nickel and gadolinium are the only examples which are ferromagnetic at room temperature. As stated above, Hund's rule manifests the energy levels of the spin electrons within a shell. When the electrons have the same momentum and the same spin, they must be separated from one another by a certain distance in order to be consistent with Pauli's exclusion principle. This tends to increase their energy levels and thus stabilize the electrons filling in shells. It is observed that in the case of Fe, Co and Ni, the $3d$ electrons are able to wander through the lattice just like $4s$ electrons.

Ferromagnetic materials have permanent magnetic moments which result in additional forces that tend to align the atoms parallel to each other, even in the absence of an external field, and against the effects of thermal agitation. This behaviour results in spontaneous magnetization of small regions in the material, called domains (Fig. 13.15).

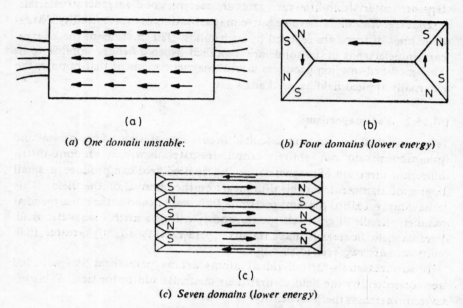

(a) *One domain unstable;*

(b) *Four domains (lower energy)*

(c) *Seven domains (lower energy)*

Fig. 13.15 *Magnetic domains in a single ferromagnetic crystal*

The magnostatic energy of each domain varies with its volume—the third power of a linear dimension. There is, however, a new source of energy in the domain walls. This energy varies with surface area—the second power of its linear dimension. Each of these small regions is magnetised with respect to another.

Another characteristic phenomenon of ferromagnetic materials is that of magnetic hysteresis.

MAGNETIC HYSTERESIS One of the characteristic features of a ferromagnetic below its curie temperature is the relation between magnetization M or magnetic induction B and the applied field H, which shows a hysteresis. Thus, a susceptibility cannot be defined uniquely below a temperature equal to θ without knowing the state of the specimen. A typical hysteresis curve is shown in Fig. 13.16 and refers to a ferromagnetic specimen being initially demagnetised and, in an initial state corresponding to the point O. As H is increased positively, the path OA is followed and, subsequently, as H is reduced and increased negatively, $ABCD$ is traversed. The susceptibility at any point is the slope dM/dH. For the corresponding B-H curve, dB/dH will be the instantaneous permeability.

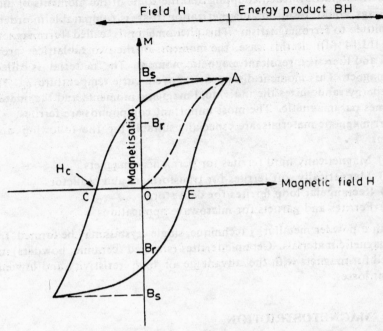

Fig. 13.16 *Magnetic hysteresis B_r is residual magnetism which is large and H_c is a small coersive force*

The slope at the origin is called the initial susceptibility or permeability. The field corresponding to the point C is that required to reduce the magnetisation to zero and is called the coercive force while the magnetic induction appropriate to the point B is the remanence. This is accompanied by a certain dissipation of energy and the work done is given by the area of the loop. It depends upon the quality of the material.

13.15.4 Antiferromagnetic

In some substances, for example manganese oxide, the net magnetic moment of the compound is zero because of the antiparallel arrangement of the moments. The material is said to exhibit anti-ferromagnetism. In such a material the paired moments are parallel but opposite in direction (Fig. 13.14 (c)). The simplest model of an antiferromagnet is based on a structure of two interleaved sublattices. the individual moments on each being ferromagnetically aligned but with antiparallel coupling between the two sublattices. The net moment is thus zero.

13.15.5 Ferrimagnetic

In some compounds, the constituent atoms may be antimagnetically coupled but with different magnetic moments. This would give rise to a net magnetic moment in each coupling and the some of the moments of all the couplings could result in magnetization that is comparable in order of magnitude to ferromagnetism. This phenomenon is called ferrimagnetism (Fig. 13.14 (d)). In this case, the moments of the two sublattices are unequal and there is a resultant magnetic moment. The material is called a ferrimagnet. This moment disappears above a curie temperature T_c. Thermal energy randomizes the individual magnetic moments and the material becomes paramagnetic. The most important compounds are ferrites.

Ferrimagnetic materials are especially suitable for the following applications:

    (i) Magnetically hard ferrites for permanent magnets
   (ii) Magnetically soft ferrites for transformers and inductors
  (iii) Rectangular loop ferrites for data stores
  (iv) Ferrites and garnets for microwave applications

With a powder metallurgy technique, single crystals may be formed from ferrimagnetic materials. Ceramic ferrites (sintered ceramic powders) may be used for magnets with the advantage of high resistivity and low eddy current losses.

13.16 MAGNETOSTRICTION

The spontaneous magnetization of a domain in a ferromagnetic crystal is accompanied by an elongation or contraction in the direction of magnetization, called magnetostriction. It is associated with the inherent anisotropy of the crystal structure and the fact that the preferred directions of magnetization are always definite crystallographic directions.

The magnetostriction coefficient λ is the fractional change (extension or contraction) in length $\delta l/l$ associated with a change in magnetisation from zero to saturation.

When the material is saturated, on the other hand, all domains are oriented in the direction of magnetization. Then the expansion in the direction of magnetization is the same at saturation for a single domain or for a number of domains. This expansion λ_s is called the saturation magnetostriction. It is positive in some materials and negative in others. A negative value of λ_s means that there is a contraction in the direction of magnetization and vice versa. Any elastic deformation, whether expansion or contraction, causes an increase in strain energy.

Magnetostriction versus applied field curves for nickel, cobalt, iron, and 45 permalloy are shown in Fig. 13.17. For 45 permalloy and iron, the dimensions tend to increase in the direction of H and perpendicular to H. For cobalt and nickel the opposite is true. Nickel has the highest negative magnitude and permalloy (45% nickel) has one of the highest positive magnetostrictions.

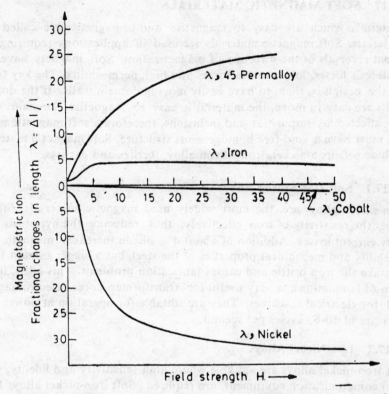

Fig. 13.17 *Magnetostriction of some common materials showing expansion and contraction*

Magnetostriction is an important engineering property. In transducers, it transforms electrical energy into mechanical energy. It can also cause demage by setting up serious vibrations in magnetic parts of electrical machinery.

Magnetostriction or mechanical deformation of magnetic domains has other important effects. When a ferromagnetic material is strained, the domains tend to realign themselves into positions of lower energy. The permeability of the material is thereby changed and it becomes easier or more difficult to magnetise. In materials having positive magnetostriction, realignment puts the domains in a better position to be polarized and permeability is increased; the opposite is true for materials having negative mechanical deformation.

At higher temperatures, the domains are hindered in their reorientation as the atoms are in constant motion. Thus, due to thermal agitation, magnetostriction is decreased.

13.17 SOFT MAGNETIC MATERIALS

Materials which are easy to magnetise and demagnetise are called soft materials. Soft magnetic materials are used in applications requiring frequent reversals of the direction of magnetization. Soft magnets have low hysteresis losses, low coercive forces and high permeability. The key factor in the design is, then, to have easily moving domain walls. If the domain walls are easy to move, the material is easy to magnetise. Domain walls are affected by impurities and inclusions, therefore a soft magnetic material must have a void-free homogeneous structure. Soft magnetic materials include soft iron-nickel, iron-silicon alloy, ferrites and garnets.

13.17.1 Soft Iron-Silicon Alloys

Iron-silicon alloys are the most widely used magnetic materials. Silicon raises the resistivity of iron effectively, thus reducing the hysteresis and eddy current losses. Addition of about 4% silicon increases magnetic permeability and mechanical properties of the steel, but a bigger amount tends to make the iron brittle and causes fabrication problems. This alloy, in the form of laminations, is very useful for transformer cores, electromagnets and for electrical machines. They are suitable for operation at power frequencies of 50-60 cycles per second.

13.17.2 Iron-Nickel Alloys

Soft iron-nickel alloys are suitable where high sensitivity and fidelity, such as in communication equipment, are required. Soft iron-nickel alloys have high initial permeability, as found in instrument transformers. These alloys are sensitive to heat and mechanical treatment. Heating and cooling affects the permeabilities of the alloys. Permalloy, supermalloy and perminvar are examples of various iron-nickel magnetic alloys. Iron-nickel alloys tend to reduce the area under the hysteresis loop making them suitable for high frequencies.

13.17.3 Ferrites

In some materials with characteristics crystals, the ions are so arranged that although the magnetic moments point in opposite directions, those in one direction are larger than those in the other (Fig. 13.14 (c)). This is a form of antiferromagnetism but produces typical ferromagnetic properties, although on a smaller scale than in ferromagnetic materials. This is the type of magnetism found in ferrites. Ferrites are oxides similar to magnetite. Their crystal structure is cubic and corresponds to that of the mineral spinel ($MgAl_2O_4$). They are manufactured in the form of a ceramic, usually polycrystalline. Their magnetic permeability is small compared to ferromagnetic materials, a typical value being $\mu_{max} = 1000$. For permalloys, $\mu_{max} = 100,000$ is common.

Ferrites are another group of soft magnetic materials suitable for high-frequency fields. Metals and alloys are unsuitable as soft magnets due to the very high eddy currents produced at higher frequencies. Ferrimagnetic oxides such as ferrites and garnets are most suitable for these applications due to their much higher electrical resistivity than alloys. This reduces eddy current losses to a negligible value.

The selection of ferrites depends on the application. For audio and TV transformers, nickel-zinc ferrites are used. For microwave work such as isolators and gyrators, magnesium (50%)-manganese (50%) ferrites are used. With a higher manganese to magnesium ratio, the ferrite exhibits a square hysteresis loop and is used for memory cores in computers.

13.18 HARD MAGNETIC MATERIALS

Materials which retain their magnetism and are difficult to demagnetise are called hard magnetic materials. Many industrial applications call for a constant magnetic field independent of electric field. For permanent magnets, a hard material is needed with a high saturation magnetisation, high remanence and high coercivity. Such materials are characterized by fat hysteresis loops, the shape of the loop in the second quadrant often being used as a measure of the effectiveness of the material as a permanent magnet.

The product BH is plotted as a function of H and the maximum value of this product, which is a measure of the maximum amount of stored energy, is used as index. Figure 13.18 shows a demagnetization curve in the second quadrant of a B-H graph, together with the corresponding B-BH curve.

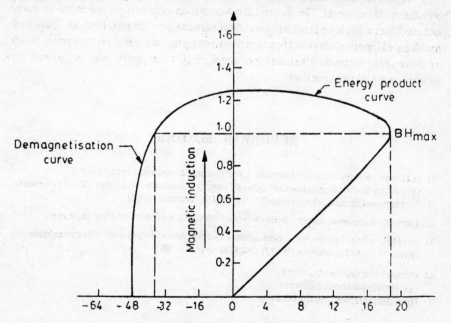

Fig. 13.18 *Demagnetization curve and a B-BH curve*

Hard magnets are commonly used in magnetos, electrical power meters, electrical clocks, in refrigerator doors, etc. Hard magnetic materials are classified below.

13.18.1 Hardened Steels

In order to have a good martensite transformation upon cooling, they include carbon, chromium, tungsten and several cobalt steels. These materials have anistropic properties, leading to a large magnetization in one particular direction. They possess high hardness as well as high magnetic energy.

13.18.2 Alnico Alloys

Alnico is an alloy made in such a way as to precipitate very small particles of magnetic material (an iron-rich phase) into an essentially non-magnetic matrix based on aluminium. Among the alloys used for this purpose, the best known are aluminium-nickel-cobalt. These alloys are directionally solidified and subsequently given a suitable heat treatment in a magnetic field that enhances the desired properties. Alnico-5 alloy contains 8% Al, 14% Ni, 24% Co, 3% Cu and the balance is iron. Alnico alloys may be machineable or non-machineable magnetic alloys. They are commercially important magnetic materials.

13.18.3 Sintered Magnets

Very recently, permanent magnets of cobalt alloyed with rare-earth elements have been developed. These are also known as oxide-type permanent magnets and have high coercivity and high saturation magnetization but low mechanical properties. Such permanent magnets are used in magnetic tapes to store information. Barium oxide and Fe_2O_3 magnets are produced by powder metallurgy method.

REVIEW QUESTIONS

1. (a) Explain electrical conductivity by showing a band energy diagram.
   (b) What is the importance of a band energy diagram and how does it compare the conductor and insulator?   (AMIE, Summer 1984)

2. Derive an expression for conductivity of conductors from the first principle.

3. Explain what insulators, conductors and semiconductors are. Give examples of each.   (AMIE, Summer 1982) (AMIE, Winter 1985)

4. Explain the following terms:
   (a) Intrinsic semiconductors
   (b) Extrinsic semiconductors

5. Define the terms
   (a) Electron hole pair
   (b) Donor
   (c) Acceptor

   Give the examples of each with suitable materials.

6. Describe n-type and p-type semiconductors.

7. Describe a p-n junction and transistor with neat sketches.

8. Define a dielectric. Explain the common properties and uses of a dielectric.

9. Write short notes on
   (a) Ferroelectricity
   (b) Piezoelectricity
   (c) Electrostriction

10. Explain the following terms
    (a) Magnetization
    (b) Magnetic domains
    (c) Curie point
    (d) Magnetic moment

11. Classify the magnetic materials.

12. Explain
    (a) Paramagnetic
    (b) Diamagnetic
    (c) Ferromagnetic
    (d) Antiferromagnetic
    (e) Ferrimagnetic materials.   (AMIE, Summer 1982) (AMIE, Summer 1984)

13. Explain the principle of ferrimagnetism with a diagram.

14. Describe the effects of magnetostriction on the properties of some common materials.

15. Classify soft and hard magnetic materials.

16. Write short notes on the following:
    (a) Ferromagnetism   (AMIE, Winter 1982)
    (b) Hard magnetic materials   (AMIE, 1983)
    (c) Soft ferrites.   (AMIE, 1983)

14
CORROSION

14.1 INTRODUCTION

Corrosion is defined as a gradual attack on a metal, as a result of a chemical or electrochemical reaction, by its surroundings such that the metal is converted into an oxide, salt or some other compound. In short, it is the deterioration and loss of material due to chemical attack. The engineer must understand the mechanisms of corrosion if its effects are to be minimised. He will then be better able to

(i) avoid severely corrosive environments
(ii) provide protection against corrosion.

There are two types of attack on metals causing corrosion. In the first instant, corrosion phenomena are essentially of an electrochemical nature and due to the presence of an electrolyte in contact with metal. This can be called wet corrosion if the electrolyte is an aqueous solution of acid, salt or alkali. In the second case, the chemical reaction between metal and gas or liquid in the absence of electrolytes is known as dry corrosion.

If the corrosion is localized, severe pitting may produce holes. The result may be disastrous in case of liquid vessels as extensive corrosion reduces mechanical strength of the section. Corrosion has been a serious problem for many years but with the increased use of materials and increasing occurance of corrosive conditions it has become a much more serious problem in recent years. Although corrosion is an undesirable process, there are commercial processes which negate corrosion e.g. electroplating, painting, electrochemical machining, stainless steels, porcelain coatings on metals, etc.

14.2 FACTORS INFLUENCING CORROSION

1. Chemical nature of the metal
2. Internal structure
3. Working conditions or environment, e.g., stresses, temperature, or concentration

4. Presence of dust, dirt or foreign matter
5. Surface film
6. Blow holes, inclusions and trapped gases
7. Distribution of secondary phases
8. Nature of engineering application
9. Eddy electric currents

14.3 TYPES OF CORROSION

The following types of corrosion are described in this chapter
  (i) Direct corrosion
  (ii) Electrochemical corrosion
  (iii) Galvanic corrosion
  (iv) High-temperature oxidation

14.3.1 Direct Corrosion

Direct conversion is the simplest corrosion produced by means of chemical solution. This is ordinarily a chemical attack that occurs as a result of chemical reaction between materials, particularly metals and corrosive solutions. This includes all kinds of corrosion in which there is no appreciable flow of current through the metal or intervention of electrolysis. It includes such things as oxidation, in which the oxygen of the atmosphere combines with all or a part of the surface of material, and other chemical reactions in which the whole or a part of the surface is attacked. The best known form of this type of corrosion is oxidation. Generally speaking, oxidation at normal temperatures is not a serious problem unless associated with electrochemical corrosion. If the temperature is elevated, then oxidation can become very serious. The development of the process depends mainly on the physical and chemical nature of the oxide film which is formed.

So far we have considered chemical corrosion as it occurs in a gaseous atmosphere, but chemical reaction can also take place at the interface between a liquid and a metal. Liquid attack may take one of the three forms. First the liquid atmosphere may dissolve the surface of the parent metal by a simple solution action. Secondly the liquid may react with the parent metal surface to form a compound similar to oxide films. The third possibility is that the atoms from the liquid may diffuse into the surface of the solid metal and cause serious changes in mechanical properties, perhaps causing embrittlement or some similar weakness.

The answer to such problems lies in the selection of a parent material which is not prone to attack or, alternatively, using a protective layer of some corrosion resistant metal on the surface of the parent material.

The rate of direct corrosion tends to be relatively high as compared with other mechanisms. The direct corroded surface has an etched appearance. Common examples of direct corrosion are:

(i) Rusting of iron and steel
(ii) Corrosion of copper flashing in the atmosphere
(iii) Tarnishing of silverware
(iv) Acid pickling used to clean metal surfaces
(v) Reactions of dry chlorine, hydrogen, etc.

Some examples of direct chemical corrosion are cited below:

RUSTING OF IRON AND STEEL Iron and steel do not rust under ordinary atmospheric conditions according to the following equation.

$$2Fe + O_2 + 2H_2O \rightarrow 2Fe(OH)_2$$

In the presence of moisture in the air, the iron hydroxide [$Fe(OH)_2$] formed frequently decomposes to the appropriate oxide according to the equation

$$Fe(OH)_2 \rightarrow FeO + H_2O$$

The rusting of iron in water involves the following reaction

$$Fe + 2H_2O \rightarrow Fe(OH)_2 + H_2$$

High-temperature oxidation of iron involves the following reaction

$$2Fe + O_2 \rightarrow 2FeO$$

$$\text{Metal} + O_2 \rightarrow \text{Metal oxide}$$

or $\quad Fe^{2+} + O^{2-} \rightarrow FeO$

TARNISHING OF SILVERWARE In the presence of moisture in the atmosphere, silverware is tarnished by hydrogen sulphide by the following reaction

$$Ag + H_2S \rightarrow AgS + H_2$$

These reactions are common to all metals and the scales formed are uniform throughout the surface i.e. direct attack proceeds uniformly over the entire surface of the metal.

14.3.2 Electrochemical Corrosion

The type of corrosion most frequently encountered is electrochemical corrosion. It is probably the most serious form of corrosion, taking place in the form of chemical reaction in conjunction with electrolysis. It takes place at or near room temperature as a result of the reaction of metals with water or with aqueous solutions of salts, acids or bases and the flow of electric current. For electrochemical corrosion to take place there should be an electrolyte, current through the circuit, a difference of potential between a metal and its surrounding or between different parts of the same metal and a closed circuit.

In electrochemical corrosion, anodic and cathodic regions on the metal surface are involved. The term is used to represent that portion of the metal surface which is corroded and dissolved as ions and from which the current goes direct to the cathode through the medium of the electrolyte. The metal surface from which current leaves the electrolyte and returns to the metal is called the cathode. The cathodic area does not corrode. The cathodes and anodes may be separate and independent units. Dry battery cells with zinc anode and graphite cathodes are a common example. Different areas on the same piece of metal may also represent the anode and cathode. Numerous tiny anodes and cathodes get formed on the surface of iron due to surface-imperfection grain orientations and localized stresses. During the electrochemical reaction, electrons are generated and pass through from anodes to cathodes immersed in the electrolyte. The rate of reaction depends on the current intensity between anodic and cathodic sites and the nature of the electrolyte. Multiphase metals and alloys corrode at higher rates than pure metals.

When a piece of metal is partly dipped in an electrolyte a potential difference is spontaneously set up between the metal surface and the solution. Its magnitude depends upon the temperature, concentration of electrolyte and nature of the metal. The production of an electrode potential is due to the formation of ions and electrons to form metal atoms, keeping the solution in equilibrium. When the ions leave the surface of the metal, an equivalent number of electrons must accumulate on its surface. To measure the electrode potential of any material, we must first determine the voltage difference between the metal and a standard hydrogen electrode. With hydrogen, equilibrium occurs through the following reaction:

$$H_2 \rightleftharpoons 2H^+ + 2\bar{e}$$

Hence all electrode potentials are determined under standard conditions with reference to a standard hydrogn electrode whose electrode potential is assumed to be zero. When metals are arranged in order of their standard electrode potentials, an electrochemical series of metals is obtained.

The sign of an electrode potential varies according to convention (Table 14.1)

The ability of a metal to resist corrosion is to some extent dependent upon their position in the electrochemical series. A greater negative potential will produce more positive ions which pass into solution. Metals with high negative potentials are chemically very active.

A number of factors other than the presence or absence of protective layers influence the course and rate of electrochemical corrosion under given conditions. One of the most important is the relative chemical nature of the anodic and cathodic regions of the corroding material. When these two regions are represented by two different metals, the electromotive series (Table 14.1) determines which of the metals, will be anodic under standard conditions. However, the actual conditions under which corrosion is occuring may result in different relative positions of the two metals in a corresponding series. Therefore the galvanic series (Table 14.1) have been deter-

TABLE 14.1 *The electrochemical series and galvanic series in sea water*

Electrode potential of metals or electrochemical series		Galvanic series in sea water
Anodic (corroded end)		*Anodic (corroded) end*
Element	Electrode potential at 25°C	Magnesium
		Zinc
		Aluminium
Li	+3.02	Cadmium
K	+2.92	
Na	+2.72	Steel
Mg	+2.34	Wrought iron
Al	+1.67	Cast iron
Zn	+0.76	Stainless steel (18-8)
Fe	+0.44	
Pb	+0.13	Lead
H_2	0.00	Tin
		Nickel
Cu	−0.34	Brass
Ag	−0.80	Copper
Au	−1.7	Silver
Cathodic (protected) end		Cathodic (protected) end

mined for practical corroding environments. The galvanic series for sea water can be used to predict that iron will corrode in contact with copper and that the copper will be protected. Thus the relation of two metals in the galvanic series plays an important part in determining the course of corrosion when the metals are in contact in a corroding environment.

MECHANISM OF ELECTROCHEMICAL CORROSION   As described in the earlier part of this chapter, the electrode supplying electrons to the external circuit is called the anode and the electrode receiving electrons from the external circuit is called the cathode. In the electrochemical mechanism of corrosion, the flow of electrons between anode and cathode is a big factor. This causes the anodic action, formation of corresponding ions and dissolution of metal. On the other hand, the cathodic reaction depends upon the nature of the corrosive conditions, for example absorption of oxygen and evolution of hydrogen bubbles. Thus electrochemical corrosion mechanisms are:

(i) Hydrogen evolution
(ii) Oxygen absorption

Hydrogen Evolution   The hydrogen evolution type of corrosion occurs in acid environments such as industrial waters and solutions of mono-

oxidising acids where the concentration of dissolved oxygen is low. The anodic reaction is one in which the ions of metals pass into solution. The tendency of hydrogen to move out increases with the acidity of the solution. It is removed as gas bubbles.

Anodic reaction $\qquad M \rightleftharpoons M^{2+} + 2e$

The cathodic reaction consists of the elimination of hydrogen according to the equation

$$2H^+ + 2e \rightleftharpoons H_2$$

Figure 14.1 shows a steel tank containing sea water and a small copper piece in contact with a steel plate. In this corrosion cell copper becomes cathodic and steel anodic.

Dissolution of steel in contact with copper takes place by the reaction:

$$Fe \rightleftharpoons Fe^2 + 2\bar{e}$$

It is evident that all metals above hydrogen in the electrochemical series will have a tendency to dissolve in acid solution.

Oxygen Absorption Mechanism The mechanism of oxygen absorption is complex. The two possibilities are the setting up first of an oxygen cathode,

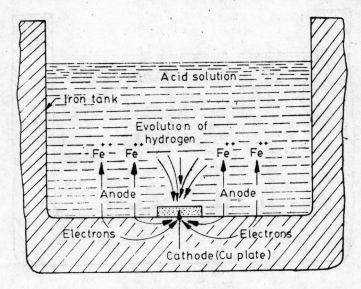

Fig. 14.1 *Mechanism of electrochemical corrosion showing hydrogen evolution*

and second a hydrogen cathode with oxygen as depolariser. The combination of dissolved oxygen with hydrogen ions or atomic hydrogen to form water is called depolarisation and this permits the continuation of corrosion. This corrosion mechanism is most likely to occur when dissolved oxygen is present in the electrolyte. Corrosion of iron in the presence of oxygen is very commonly observed. Drops of water on the surface of a steel plate will result in the formation of an oxide layer. Water acts as a neutral aqueous solution, any small crack in the plate as anode and the oxide film as cathode.

The dissolution of steel is given by the reaction:

$$Fe \rightleftharpoons Fe^{2+} + 2Z\bar{e} \tag{1}$$

and the reaction of oxygen as depolariser by

$$H_2O + \tfrac{1}{2}O_2 + e \rightarrow 2OH \tag{2}$$

The corrosion rate of iron is thus controlled not by the primary reaction with hydrogen ions but the depolarisation reaction involving oxygen. That is why the rate of corrosion is determined by the rate at which oxygen can diffuse to the reacting metal. An opposite characteristic of oxygen and oxidizing media is their ability to make certain metals and alloys passive. A metal is less reactive in the passive condition than in the normal or active condition. The passivity is believed to be caused by the presence of complex oxide films at the metal surface. Figure 14.2 shows the oxygen absorption mechanism of cracks.

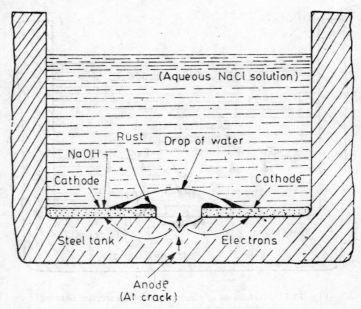

Fig. 14.2 *Oxygen absorption mechanism of corrosion*

14.3.3 Galvanic Corrosion

Galvanic corrosion is the mechanism by which most metals are corroded by liquids, in which the anode metal is made to dissolve or corrode continuously. Such corrosion occurs when two dissimilar metals are in electrical contact with each other and are exposed to an electrolyte. A less noble metal, say zinc, will dissolve and form the anode, whereas the more noble metal, such as copper, will act as the cathode (Fig 14.3)

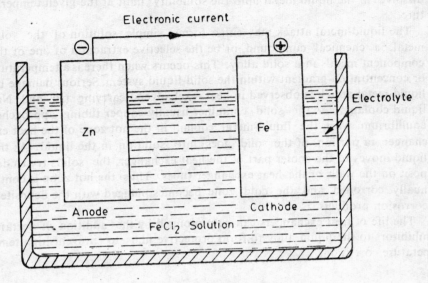

Fig. 14.3 *Galvanic corrosion*

The potential difference and the rate of corrosion of the anodic metal will depend upon the relative position of the two metals in the galvanic series. The cathodic action may occur either by the hydrogen evolution process or the oxygen absorption process, depending on the corrosive conditions. The combination of electrodes (Cu and Zn) and electrolyte is called a galvanic cell. Corrosion of dissimilar metals in contact with each other like the galvanic cell is rapid. If, however, direct contact between dissimilar metals can be avoided, the corrosion of anodic metal can also be avoided. That is why, in fabrication work, the composition of electrodes in case of welding and of rivets in rivetting should be of the same chemical composition as the base metal. Galvanic corrosion also occurs on the unprotected steel of ships because of the selective action of steel on the brass screws.

14.3.4 High-Temperature Oxidation or Corrosion

High-temperature oxidation, as distinguished from wet corrosion, is known as high-temperature dry corrosion. The rusting of ferrous alloys at high temperatures forms scales and oxides. This corrosion can be prevented by adding alloying elements which are resistant to oxidation in some metals and have good properties of high temperature service.

Another form of high-temperature corrosion occurs when liquid metals flow past other metals. The corrosion reaction is essentially a process of mass transfer and is not dependent upon local cell potentials for its driving force. Actually the corrosion is due to the tendency of the solid to dissolve in the liquid metal upto the solubility limit at the given temperature.

The liquid-metal attack may either form a simple solution of the solid metal, a chemical compound, or be the selective extraction of one of the component metals in a solid alloy. This occurs when there is a temperature or concentration gradient within the solid-liquid system. Serious damage by liquid-metal attack is observed in heat exchangers carrying (Bi and Na) liquid coolants As the solid container usually copper tubing, approaches equilibrium with the liquid-metal coolant in the hot zone of the heat exchanger, a portion of the solid goes into solution in the liquid. As the liquid moves to the cooler part of the heat exchanger, the solids try to deposit on the walls of the heat exchanger tubes. Thus the hot zone is continually corroded and the cold zone becomes plugged with the deposited corrosion products.

The life of heat exchangers can be prolonged by the addition of certain inhibitors to the liquid alloy to form protective films to prevent high temperature corrosion.

14.4 SPECIFIC TYPES OF CORROSION

The above mentioned types of corrosion represent convenient divisions of the subject for some purposes. More specific descriptions are widely used for certain types of industrially important corrosions. In this section, the corrosion problems of most metals and alloys in service, depending upon corroding conditions and mechanisms of attack, are discussed.

14.4.1 Uniform Corrosion

This type of corrosion occurs when the whole surface of the metal is corroded to the same degree and when the metal surface and environment are uniform. Under these conditions the useful life of a given material is easily estimated and unexpected failure need not be feared. Uniform corrosive attack is relatively easily controlled by selecting suitable materials. In practice, uniform corrosion is rare. It is observed in metals like zinc, lead and aluminium.

14.4.2 Pitting Corrosion

Pitting is an especially serious type of localized corrosion and is recognised by the presence of pits or holes. This is non-uniform type of corrosion and is initiated by inhomogeneties in the metal. It may be accelerated by such additional factors as breakdown in the protective layer due to local straining, cut edges and surface roughness. Pitting corrosion results from an electrochemical reaction between the broken film as anode and the surrounding unbroken film as cathode which form a small galvanic cell and produce pits. Pitting is observed in aluminium, steel, copper and nickel alloys (Fig. 14.4).

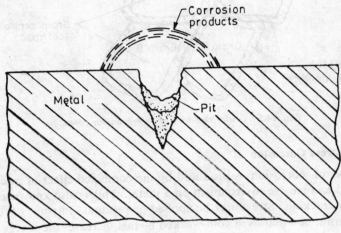

Fig. 14.4 *Pitting corrosion*

Pits are randomly formed on the surface. Corrosion attack is limited to small areas. While the shapes of pits vary widely, they are usually hemispherical with electropolished inner surfaces. Pits have considerable effect on mechanical properties, for example fatigue.

14.4.3 Intergranular Corrosion

Intergranular corrosion occurs when a pronounced difference in reactivity exists between grain boundaries and the remainder of the alloy. When the grain boundaries contain material which has more solution potential than the grain centre, in the particular environment, intergranular attack takes place and produces serious damage. Such corrosion is observed in defective welding and heat treatment of stainless steels, copper and aluminium alloys. Microscopic examination can show intergranular corrosion. It also has adverse effects on the mechanical properties of alloys (Fig. 14.5)

14.4.4 Atmospheric Corrosion

Atmospheric corrosion is very frequent on ferrous materials. Rain water and humid air act as electrolytes and are mainly responsible for this type of corrosion. It is primarily due to the formation and breakdown of films by the moisture and electrochemical attack on the exposed metal. This follows the oxygen-absorption mechanism. Cracks and discontinuities in the film produced expose fresh areas to the formation of corrosion cells.

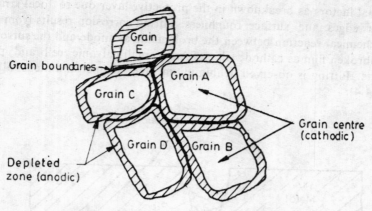

Fig. 14.5 *Grain-boundary or intergranular corrosion*

14.4.5 Stress Corrosion

This type of corrosion is produced by the combined effect of mechanical stress and a corrosive environment on a metal. This is also known as stress-corrosion cracking. Stress may be applied or residual. Pile-up of dislocations at grain boundaries in cold-worked metals increases the energy in such regions and makes anodes in the structure in certain environments. This gives rise to the formation of cracks which propagate progressively with the continuation of corrosion. Well-known examples of stress corrosion are season cracking in brass, especially in the presence of moisture and traces of ammonia, and caustic embrittlement of steel exposed to solutions containing sodium hydroxide. The most effective control is the elimination of tensile stress from the component part. Pure metals are relatively immune to stress corrosion. The stress corrosion rate is dependent upon the corrosive agent, time and temperature of exposure, behaviour of protective films and stress magnitude. The absorption of hydrogen in metal under high tensile stress causes embrittleness. The phenomenon is known as hydrogen embrittleness. This can be observed in mild steel exposed to alkaline solutions at high temperatures and stresses (Fig. 14.6).

14.4.6 Corrosion Fatigue

This is the combined action of corrosion and repeated stresses and is far more serious than the sum of these two factors acting individually. It is also the reduction of fatigue strength of materials due to to the presence of a corrosive medium. It is most common to environments which cause pitting on the surface of the material. The effect of corrosion fatigue is clearly observed in heat exchangers where variable stresses are produced because of thermal expansion and contraction.

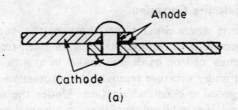

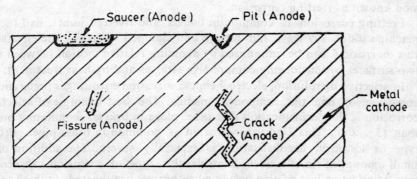

Fig. 14.6 *Stress corrosion*

The influence of corrosion fatigue strength is expressed by the **damage ratio**:

$$\text{Damage ratio} = \frac{\text{Corrosion fatigue strength}}{\text{Normal fatigue strength}}$$

This ratio for salt as a corroding medium is about 0.2 for carbon steels, 0.5 for stainless steels, and 1.0 for copper. Suitable protective measures against corrosion fatigue include treatment of the corroding medium and surface protection of the metal. Nitriding of steels is often useful for this process.

14.4.7 Erosion Corrosion

Erosion corrosion refers to the combined effect of the basic corrosion mechanism on a metallic surface and mechanical abrasion. Mechanical abrasion is produced by the turbulent flow of liquids or the impingement of trapped gases. Such corrosion causes accelerated attack because it mechanically removes the protective layer that normally builds up on the corroding surface. Erosion action contributes to the formation of localised pitting cells. This type of corrosion is associated with the formation of cavities in the metal by fast-moving liquids. This type of corrosion is encountered in pump impellers, turbines, condenser tubes, piping, ship propellers, diesel engine cylinders and vessels in which streams of liquid strike on the walls of the vessel with rapid movement past the metal surface.

14.4.8 Fretting and Selective Corrosion

Fretting corrosion occurs where contact areas between the surfaces of any two materials are subjected to vibrational stresses. In steel, fretting corrosion appears as patches of iron oxide and makes deep pits. The relative movement of one part under pressure removes the protective surface film, leaving the surface exposed to chemical reaction. Under the action of vibration stress and a corrosive environment there is always localised corrosion known as fretting corrosion.

Fretting corrosion is common in bolted and rivetted joints, ball races, machine slides, key ways, hub fittings and clamped surfaces. The effects of this corrosion can be minimised by avoiding relative motion between the two surfaces by introducing compressive stresses and by heat treatment.

In selective dissolution, electrochemical corrosion encourages preferential corrosion of one the component metals. The best example of selective corrosion is the dezincification that takes place in brasses containing more than 15% zinc. This kind of corrosion is common in brass pipes. Plug-type or localized dezincification is especially severe. The addition of a small amount of tin effectively controls this type of corrosion. A second type of selective loss of zinc occurs when brasses are heated to high temperatures at which the zinc vapourizes rapidly. Selective dissolution of iron from cast irons is known as *graphitization*.

14.5 CONTROL AND PREVENTION OF CORROSION

Only under ideal service conditions, corrosion can be completely avoided. As many factors control the numerous types of corrosion, so to minimise corrosion, many methods of diverse natures are used. The use of a particular method will depend upon its successful implementation and on economic considerations. In order to control corrosion, the materials would have to be absolutely uniform with no heterogeneities in either composition or structure, and the environment would also have to be entirely uniform.

The various methods by which corrosion can be controlled and prevented are:

(i) Use of high-purity metal and alloy additions
(ii) Design against corrosion
(iii) Modification of corrosive environment
(iv) Use of protective surface coatings
(v) Use of inhibitors
(vi) Cathodic protection.

14.5.1 High-Purity Metal and Alloy Additions

The corrosion resistance of a given metal may be improved by avoiding galvanic couples, i.e., limiting designs to only one metal, but this is not always feasible. In special circumstances, the cells may be avoided by selecting suitable electrically insulating metals of different composition or by alloy additions. The corrosion resistance of most metals can be increased by alloying them with suitable alloying elements. Alloying may increase corrosion resistance by increasing the energy level of the solution, reducing the driving emf in the galvanic cell, protective coatings and decreasing the mobility of the corrosive ions. Alloying also helps in the following ways:

(i) In heat treatment to avoid carbon precipitation.
(ii) Making steels with high chromium content (chromium corrodes less readily)
(iii) Making steels containing strong carbide formers. Such elements include titanium and tantalum. These elements do not allow carbon precipitation at the grain boundary.

This technique is used in stainless steel which must be fabricated by welding. The above methods reduce intergranular corrosion in metals.

14.5.2 Design Against Corrosion

The corrosion rate can be reduced by proper design and fabrication. The following points should be ensured to prevent corrosion.

(i) Avoidance of dissimilar metal contacts
(ii) Avoidance of recesses and sharp corners
(iii) Cathodic protection
(iv) Proper fabrication to avoid excessive stress concentration
(v) Air-tight joints to prevent the retention of liquid
(vi) Clean surfaces free from foreign matter like dust, dirt and soot
(vii) Two different metals should be as close as possible
(viii) Selection of corrosion-resistant materials wherever possible.

14.5.3 Modification of Corrosive Environment

There is large saving in materials if these are used in less corrosive environments. Corrosive environments play a major role in corroding the parts in service. The effects are reduced by preventing the attack of harmful constituents in the environment and removing the harmful components in the environment. For this, the environment can be modified in the following ways:

(i) Decreasing the temperature, pressure, concentration, velocity, etc.
(ii) Altering the chemical composition of the environment by changing humidity, oxygen content, oxidizing agents and solid impurities. A purified and dehumidified atmosphere around the structure definitely reduces corrosion.
(iii) Introducing inhibitors. Alkaline neutralizers are used to prevent the corrosive effect of the environment by neutralizing the acidic character of corrosive conditions. Vacuum and inert gases are also used to prevent high-temperature corrosion of ferrous metals and alloys.

14.5.4 Protective Surface Coatings

When an easily corroded steel is used, some provision is made to prolong its life or improve its appearance by protecting it from its surroundings. The most important protective measure is some kind of coating. Protective coatings are widely used instead of alloying elements to minimise corrosion of metals. The applied coating may be,

(i) Metallic
(ii) Non-metallic

The coating should have a good corrosion resistance when in use, perfect adherence to the underlying metal and the surface should be completely covered.

METALLIC COATINGS Two factors are involved in the protection of the underlying metal by a metallic coating. One is mechanical isolation of the metal from the corroding environment, the other is the galvanic relation of coating metal and the base metal. If the coating metal is higher in the galvanic series than the base metal, discontinuities in the coating are not a serious problem since the base metal is cathodic and is protected. Zinc and cadmium are examples of metals that afford galvanic protection to steel. Nickel and chromium plating give an attractive appearance and the excellent corrosion and fabrication qualities of tin plate are made use of in food containers. A number of methods are used to apply metallic coatings:

(i) Hot dippings
(ii) Electroplating
(iii) Metal cladding
(iv) High temperature diffusion
(v) Metal spraying

Zinc coatings on steel (galvanized coating) and tin plate are usually produced by dipping clean sheet steel into molten zinc or a tin bath. The lives of galvanized coatings depend on the thickness of the zinc layer and the conditions of the environment.

Thinner, uniform coatings of zinc and tin can be obtained by electroplating. This process is becoming increasingly popular for these metals. Cadmium, nickel and (Fig. 14.7) chromium platings are almost always produced by electroplating. In this process the coating metal is deposited on the base metal by passing a direct current through the electrolyte solution containing a salt of the coating metal. The coating material is made the anode and the base metal the cathode. Coating properties depend on the current intensity, agitation, temperature of the solution and the chemical composition of the plating solution. These coatings are relatively thick and are usually applied to cast or machined parts.

In metal cladding a comparatively thick lining is placed upon the metal surface forming a strong alloy bond between plates of the two metals. This

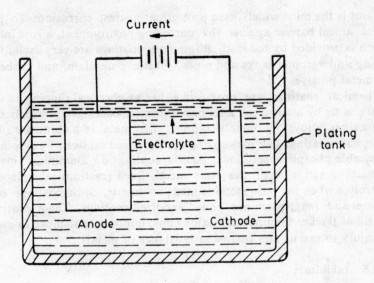

Fig. 14.7 *Electroplating*

bond is formed either by casting or hot rolling. The production of a corrosion resistant surface is the principal objective for cladding. Metal cladding is also used to make bimetallic strips for thermostats.

The high-temperature diffusion process makes use of the powder metallurgy technique. The base metal is heated with powder of the metal which forms the coating. When diffusion coatings are formed on steels by high-temperature treatment, the innermost core is rich in the base metal followed by successive richer coating surfaces. This treatment in the presence of powdered aluminium is known as calorizing, with powdered zinc as shearadizing, and with chromium as chromising. These coating are also thick.

By metal spraying cr metallized coatings, thick coatings of any metal can be produced. Coatings are obtained by impinging heated metal particles against the base metal surface. In this process a special gun melts and atomizes the coating metal and then uses compressed air to drive the small particles against the prepared surface. Since the metal particles solidify in air, the bond they form is primarily mechanical and the coating is somewhat porous. However, the coating is sufficient to resist corrosion. Worn and damaged parts can be repaired by metal spraying. Metal spraying is very advantageous when a very large surface is to be treated, greater speed of working is required and the coating is to be applied to hidden areas. This process is more widely used.

NON-METALLIC COATINGS The most common non-metallic coatings are paints, porcelain enamels and chemical coatings. Coatings of plastics, pitch or bitumen are also frequently used these days.

Paint is the most widely used protection against corrosion. To provide a mechanical barrier against the corroding environment, a rust inhibiting action is provided by red lead. Bituminous coatings are very useful for protecting underground tanks and pipes. Plastic, porecelain and rubber make the metal passive.

Chemical coatings are produced either by chemical dipping in suitable solutions or by a chemical reaction. Chromate coatings are widely used on magnesium alloys by merely dipping the metal in a chromate solution. Phosphate coatings like parkerizing are produced on steels by dipping them in suitable phosphate solutions. Anodic coatings on aluminium involve a chemical reaction with the base metal. Such reactions take place in the electrolyte when the base metal is made an anode. Such coatings on aluminium and magnesium are called anodized coatings. These coatings are 100 times thicker than natural oxide films, improve resistance to mechanical injury to metals and corrosion resistance of metals.

14.5.5 Inhibitors

Inhibitors are chemical compounds added to electrolytes, antifreeze mixtures and corrosive solutions to retard or slow down a chemical reaction. Corrosion inhibitors are added in cooling waters, steam and acids. Inhibitors, when added in small quantities in corrosive conditions decrease the corrosion rate. These are used in liquids, solids, semi-solids and packing materials. They may be classified as acidic, neutral or basic or as per corrosion reactions such as anodic, cathodic or mixed. Inhibitors are commonly used to produce an oxide film on manganese and aluminium, antifreeze mixtures for refrigerators and automobile radiators, in water treatment, for surface cleaning processes like pickling, etc. and for protective greases.

Anodic inhibitors suppress the anodic reaction or metal dissolution. Oxidizing substances such as chromates, nitrates, phosphates, etc., are used as anodic inhibitors for the protection of iron and steel. Here oxygen works as the anodic inhibitor and forms a protective film by forming ferric hydroxide by the oxidation of ferrous hydroxide.

Cathodic inhibitors prevent oxygen absorption at the cathodic area and also the evolution of hydrogen. Such organic inhibitors try to shield the cathodic areas. Common among these are calcium bicarbonates, zinc, manganese, chromium, nickel, nitrogen compounds and amines, etc.

14.5.6 Cathode Protection

In electrochemical corrosion there is always a flow of current between the anode and cathode. The anode is corroded and at the same time the cathode is protected. Thus corrosion is reduced or eliminated by making the whole surface cathodic to an extraneous surface which acts as anode. This can be done by impressed voltage or by the attachment of a sacrificial anode. This phenomenon of galvanic protection is known as cathodic protection and is the most effective method of corrosion control. In fact it is the only method that can completely prevent corrosion. The impressed voltage can be externally applied for the protection of structures like oil and water pipes, steam boilers and all types of underground pipes. Galvanized iron is a common example of cathodic protection.

The sacrificial anode is so called as the anodic metal is reduced by the dissolution solution and has to be replaced from time to time. For uniform

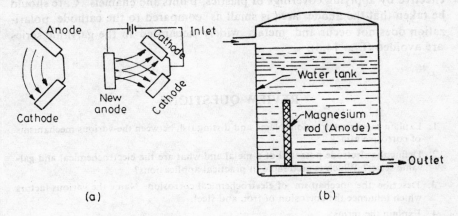
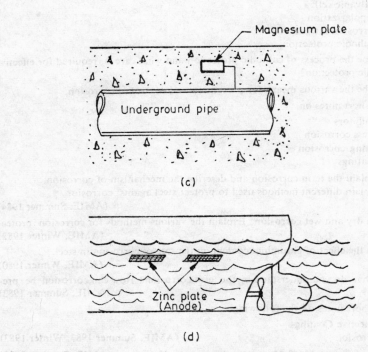

Fig. 14.8 *Cathodic Protection with galvanic or sacrificial anode*

protection of large surfaces, it is necessary to insert a number of zinc, aluminium or magnesium anodes distributed at different points. The additional anodes will establish a potential with the metal to be protected and will prevent its corrosion. This protecting metal (Zn) is known as the sacrificial anode.

For cathodic protection of domestic and industrial water heaters, magnesium anodes have been mostly used. Cathodic protection is made very effective by applying coverings of plastics, paints and enamels. Care should be taken that the anode area is small as compared to the cathode, polarization does not occur and metals widely separated in the galvanic series are avoided (Fig. 14.8).

REVIEW QUESTIONS

1. Explain the term corrosion. Name and distinguish between the various mechanisms of corrosive attack.
2. Define the electrode potential of a metal and what are the electrochemical and galvanic series? Which is preferable in practical applications?
3. Describe the mechanism of electrochemical corrosion. Name the various factors which influence the corrosion of iron and steel.
4. Explain the terms:
    (a) Galvanic cell
    (b) Depolarization
    (c) Corrosion fatigue
    (d) Cathodic protection
5. Describe the process of cathodic protection and what care is required for effective cathodic protection?
6. Describe the various means for prevention and control of corrosion.
7. Write short notes on:
    (a) Inhibitors
    (b) Stress corrosion
    (c) Pitting corrosion
    (d) Coatings
8. (a) Explain the term corrosion and describe the mechanism of corrosion.
    (b) Explain different methods used to protect steel against corrosion.
    (AMIE, Summer 1984)
9. Explain dry and wet corrosion. Explain the various methods for corrosion protection. (AMIE, Winter 1982)
10. Discuss the effect of chromium content on the rate of corrosion in steel.
    (AMIE, Winter 1980)
11. (a) What do you understand by the term corrosion? How can corrosion be prevented? (AMIE, Summer 1982)
12. Write short notes on the following:
    (a) Protective Coatings
    (b) Corrosion   (AMIE, Summer 1983, Winter 1983)
13. What is stress cracking? How can it be prevented?   (AMIE, Summer 1982)

15
MECHANICAL PROPERTIES

15.1 INTRODUCTION

The mechanical properties of materials, their strength, rigidity and ductility, are of vital importance in determining their fabrication and possible practical applications. Materials show a wide range of mechanical properties ranging, for example, from the ductility of pure copper to the hardness of diamond and the surprising elastic behaviour of rubber. Many materials behave quite differently when stressed in different ways, for example steel and wood are stronger in tension whereas cast iron, cement and bricks are much stronger in compression. From this we conclude that stresses can produce a shape change and may also cause a material to break or fracture. Both fracture and the different phenomena responsible for shape changes will be discussed in the subsequent sections of this chapter. Initially, however, a brief review is given of some definitions of common mechanical properties.

15.2 DEFINITIONS OF COMMON TERMS USED IN MECHANICAL PROPERTIES

ISOTROPY A polycrystalline material in which the grains or crystals are randomly oriented behaves isotropically, i.e. its properties are independent of direction. Steels, cast irons and aluminium have random distribution of crystals. The behaviour of a solid material when it is subjected to a force, is of fundamental interest to engineers since most things we design must withstand applied forces. Many solid materials behave elastically when subjected to a force or load. By this it is meant that the solid deforms when it is loaded but returns to its original condition when the force is removed. Robert Hookes studied the elastic properties of metals and devised a mathematical expression called Hooke's law. As shown in Figs 15.1 (a, b, c), when a specimen of length l and cross-sectional area A is loaded with a tensile load P, the length increases by an amount δl. The increase in length is called the elongation. As the load increases, the elongation increases.

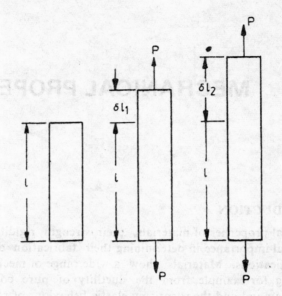

Fig. 15.1 *A specimen subjected to axial tensile load*

Amorphous materials, except when produced in such a manner as to cause some alignment of the molecular structure (a preferred orientation), are also isotropic.

ANISOTROPY This is a state of having different properties in different directions. Various composite materials, wood and laminated plastics are highly anisotropic.

STRESS AND STRAIN This is the resistance of the body to deformation due to the application of external forces. Stress is the force per unit area. If P is the total load acting on the original cross-sectional area A_0, then stress $\sigma = P/A_0$.

Strain is the deformation produced per unit length of a body. It is the ratio of the change in length of the specimen to its original length. If δ_l is the change in length and l_0 is the original length of the sample, then strain,

$$\epsilon = \frac{\delta_l}{l_0} = \text{longitudinal strain}$$

Strain can be lateral strain or shear strain depending upon the type of loading. It is simply a ratio and has no unit.

HOOKE'S LAW This states that, within elastic limits, the relationship between stress and strain is constant and is represented by E (Young's modulus of elasticity).

Thus,   Stress ∝ Strain

∴ $E = \dfrac{\sigma}{\epsilon} = $ Constant

As this proportionality constant is characteristics of a material and is different for different materials and for different nature of stresses, it is called the modulus of the material. When tensile or compressive stresses are used it is called the Young's modulus of elasticity (E). When shear stress and strain are used, this constant is called the modulus of rigidity (G). For volumetric stresses and strains, the constant is called the bulk modulus (K).

POISSON'S RATIO   When a test sample is stressed by a uniaxial force, it is strained in the direction of the force and also in a direction perpendicular to the direction of the force. The strain in the direction of force is known longitudinal strain and that perpendicular to it as lateral strain. The relationship between the lateral strain and axial strain is called Poisson's ratio:

$$\text{Poisson's ratio} = \frac{\text{Lateral strain}}{\text{Longitudinal strain}} = \frac{l}{m} = \text{Constant}$$

It is an important elastic constant and its value for most common engineering materials is between 0.3 to 0.6.

15.2.1 Stress and Strain Relation

The stress and strain relation can be shown better by drawing a graph or curve from the data obtained in a tensile test, in which an increasing tensile stress is applied to a specimen (Fig. 15.1). There are resulting changes in length which can be observed or recorded by strain-measuring devices.

In ductile materials, at the beginning of the test, the material extends elastically. The strain, both longitudinal and lateral, at first increases proportionally to the stress and the specimen returns to its original length on removal of the stress. The limit of proportionality is the stage upto which the material obeys Hooke's law perfectly.

Beyond the elastic limit the applied stress produces plastic deformation so that a permanent extension remains even after the removal of the applied load. In this stage the resultant strain begins to increase more quickly than the corresponding stress and continues to increase till the yield point is reached. At the yield point the material suddenly stretches.

The ratio of applied load to original cross-sectional area is termed the nominal stress. This continues to increase with elongation, due to strain hardening or work hardening, until the tensile stress is maximum. This is the value of stress at maximum load and can be calculated by dividing the maximum load by the original cross-sectional area. This stress is termed the ultimate tensile stress.

At a certain value of load the strain continues at a slow rate without any further stress or loading. This phenomenon of slow extension increasing with time, at a constant stress, is called creep. At this point a neck begins to develop along the length of the specimen and further plastic deformation is localized within the neck. The cross-sectional area decreases in proportion to the increasing length during elastic elongation. The volume of the test bar remains constant.

After necking has begun the nominal stress decreases until the material fractures at the point of minimum cross-sectional area within the neck.

Figure 15.2 is a stress–strain diagram for mild steel showing the limit of proportionality, elastic limit, yield point, ultimate tensile stress and fracture stress at the breaking point. The diagram shows a well-defined yield point.

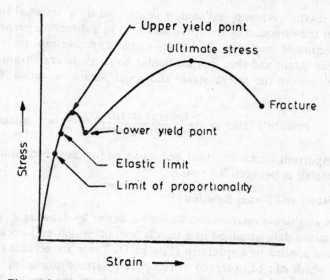

Fig. 15.2 *Stress–strain curve for a ductile material (mild steel)*

Figure 15.3 shows a poorly defined yield point (as in brittle materials). To determine the yield strength in such materials, it is general practice to draw a straight line parallel to the elastic portion of the curve at a predetermined strain ordinate value (say 0.1%). The point at which this line intersects the stress–strain curve is called the yield strength at 0.1% or 0.2% of set strain.

For hard steels and non-ferrous metals stress is specified corresponding to a definite amount of permanent elongation. This stress is known as proof stress. For aircraft materials, the stress corresponding to 0.1% of strain is the proof stress. The proof stress is applied for 15 seconds and, when removed, the specimen should not lengthen permanently beyond 0.1%. In Fig. 15.3 the method of finding the proof stress from the stress–strain curve is shown.

The properties of ductile metals can also be explained with the help of stress–strain curves. The greater the angle of inclination of the line of pro-

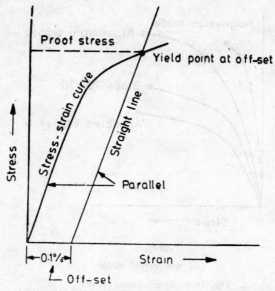

Fig. 15.3 *Stress–strain curve for a brittle material*

portionality to the ordinates, the more elastic is that metal. A higher yield point will mean greater hardness of the metal. A higher value of the maximum stress point will represent a stronger metal. Similarly, the distance from the ordinates of the breaking stress or load point will indicate the toughness and brittleness of the metal. The shorter the distance the more brittle the metal. Figure 15.4 shows stress-strain curves for ferrous and non-ferrous materials. Brittle materials show little or no permanent defor-

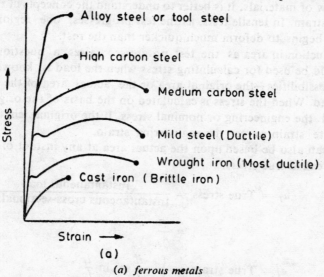

(a) *ferrous metals*

Fig. 15.4 *Stress-strain curves*

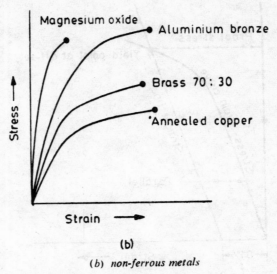

(b) *non-ferrous metals*

Fig. 15.4 *Stress-strain curves*

mation prior to fracture. Brittle behaviour is exhibited by some metals and ceramics like magnesium oxide (Fig. 15.4). The small elongation prior to fracture means that the material gives no indication of impending fracture and brittle fracture usually occurs rapidly. It is often accompanied by loud noise.

ENGINEERING AND TRUE STRESS-STRAIN DIAGRAMS In order to study the plastic flow of materials, it is better to understand the concepts of true stress and true strain. In tensile tests, as the test progresses, one region of the specimen begins to deform much quicker than the rest.

The reduction in area as the test progresses raises a question—which area should be used for calculating stress when the load is known? There are two possibilities—the original area or the actual area of the specimen can be used. When the stress is calculated on the basis of the original area it is called the engineering or nominal stress. If the original length is used to calculate strain we get the engineering strain.

Stress can also be based upon the actual area at any instant of the load. It is then called true stress.

$$\therefore \quad \sigma_T = \text{True stress} = \frac{\text{Instantaneous load}}{\text{Instantaneous cross-sectional area}}$$

$$= \frac{P}{A_i}$$

Similarly $\quad \epsilon_T = \text{True strain} = \int_{l_0}^{l_i} \frac{dl}{l} = \ln \frac{l_i}{l_0}$

where *dl* is the infinitesimal elongation.

The true stress–true strain curve always has a continually rising characteristic as would be expected when consideration is given to the mechanism of work hardening which is the interaction of mobile dislocations with each other. Both the engineering and true stress-strain diagrams are shown for copper in Fig. 15.5. The ultimate strength of the engineering curve corresponds to the beginning of necking. The rapid area reduction that

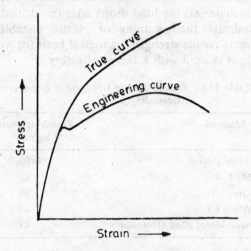

Fig. 15.5 *True and engineering stress–strain curves*

accompanies necking causes the curve to fall. This does not happen with the true stress-strain curve. As strain is not homogeneous after necking begins, when one part of the specimen is elongating more than the rest, it is not meaningful to use the entire specimen length for finding deformation. True strain can be expressed in terms of area by substituting in the equation:

$$\epsilon_{True} = \ln \frac{A_o}{A_i} = \frac{\text{Original area}}{\text{Instantaneous area}}$$

Most of the diagrams used for practical purposes are based on engineering stress versus engineering strain, while true stress-true strain diagrams are required occasionally.

15.3 FUNDAMENTAL MECHANICAL PROPERTIES

   (i) Tensile strength
  (ii) Hardness
 (iii) Impact strength
 (iv) Fatigue
  (v) Creep

15.3.1 Tensile Strength

This is the maximum conventional stress that can be sustained by the material. It is the ultimate strength in tension and corresponds to the maximum load in a tension test. It is measured by the highest point on the conventional stress-strain curve. In engineering tension tests this strength provides the basic design information on the material's acceptance. The significant points on the stress-strain curve have already been discussed in Sec 15.2. In ductile materials the load drops after the ultimate load because of necking. This indicates the beginning of plastic instability. In brittle materials, the ultimate tensile strength is a logical basis for working stresses. Like yield strength, it is used with a factor of safety.

Table 15.1 *Typical Tensile Strengths of Engineering Materials*

Material	Tensile strength kg/mm^2
Alloy steel	60-70
Mild steel	42
Grey CI	19
White CI	47
Aluminium alloy (Duralumin)	47

15.3.2 Hardness

Hardness is the resistance of a material to permanent deformation of the surface. In other words, it can be defined as the resistance of the metal to penetration by an indentor. The hardness of a surface is, of course, a direct result of interatomic forces acting on the surface of the material. It is not a fundamental property, however, but rather a combined effect of compressive, elastic and plastic properties relative to the mode of penetration, shape of penetrator, etc. The main usefulness of hardness lies in the fact that it seems to bear a fairly constant relationship to the tensile strength of a given material and so can be used as a practical non-destructive test for an approximate idea of the value of that property and the state of the metal near the surface.

Several methods of hardness testing have been devised, depending either on the direct thrust of some form of penetrator into the metal surface, or on the ploughing of the surface as a stylus is drawn across it under a controlled load, or on the measurement of elastic rebound of an impacting hammer possessing a known energy. Hardness measurements are the easiest to make and are extensively used for industrial design and in research work. As compared to other mechanical tests where the bulk of the material is involved in testing, all hardness tests are made on the surface or close to it.

In all indentation tests, the mechanism of indentation is that when the indentor is pressed into the surface under a static load, a large amount of plastic deformation takes place. The material thus deformed flows out in all directions. As a result of plastic flow, sometimes the material in contact with the indentor produces a ridge around the impression. Large amounts of plastic deformation are accompanied by large amounts of transient creep which vary with the material and time of testing. As explained in Sec. 15.3.5, transient creep takes place rapidly at first and more slowly as it approaches its maximum. For harder materials, the time required for reaching maximum deformation is short (few seconds) and for soft materials the time required to produce the desired indentation is unreasonably long, up to a few minutes.

Hardness is sometimes stated in terms of macro-hardness and micro-hardness. The macro-hardness of a material relates to its resistance to larger volume displacements in plastic deformation Micro-hardness is the hardness of materials in microscopically small volumes such as in grain boundaries.

15.3.3 Impact Strength

This is a complex characteristic which takes into account both the toughness and strength of a material. Course grain structures and precipitation of brittle layers at the grain boundaries do not appreciably change the mechanical properties in static tension, but substantially reduce the impact strength. Impact strength is determined by using the notch-bar impact tests on a pendulum-type impact testing machine. This further helps to study the effect of stress concentration and high velocity load application.

Impact strength varies with certain factors:

(i) If the dimensions of the specimen are increased, impact strength also increases.
(ii) When the sharpness of the notch increases, the impact strength required to cause failure decreases.
(iii) The temperature of the specimen under test gives an indication about the type of fracture that is likely to occur, i.e., ductile, brittle or ductile to brittle transition.
(iv) The angle of the notch also improves impact strength after certain values.
(v) The velocity of impact also affects impact strength to some extent.

15.3.4 Fatigue

Materials subjected to a repetitive or fluctuating stress will fail at a stress much lower than that required to cause fracture under steady loads. This behaviour is called fatigue and is distinguished by three main features:

(i) Loss of strength
(ii) Loss of ductility
(iii) Increased uncertainty in strength and service life

Fatigue results in brittle fracture with no gross deformation at the fracture due to the effect of inhomogeneities under repeated loading.

Fatigue is an important form of behaviour in all materials including metals, plastics, rubber and concrete. All rotating machine parts are subjected to alternating stresses; aircraft wings are subjected to repeated gust loads, oil and gas pipes are often subjected to static loads but the dynamic effect of temperature variations will cause fatigue. There are many other situations where fatigue failure will be very harmful. Because of the difficulty of recognizing fatigue conditions, fatigue failures comprise a large percentage of the failures occuring in engineering. The point at which the curve flattens out is called the fatigue limit and is well below the normal yield stress. The fatigue strength can be defined as the stress that produces failure in a given number of cycles, usually 10^7.

Fatigue specimens must be carefully made to avoid stress concentrations, rough surfaces and tensile residual stresses.

STRESS CYCLES   Figure 15.6 shows the different arrangements of fatigue loadings. The simplest type of load is the alternating stress shown in Fig. 15.6 (a) where the stress amplitude is equal to the maximum stress and the mean or average stress is zero. The bending stress in a shaft varies in this way.

The second type of loading (Fig. 15.6(b)) produces alternating stresses, superimposed on a steady stress; the maximum stress is equal to the sum of the mean stress and stress amplitude.

FATIGUE FRACTURE   Fatigue fracture results from the presence of fatigue cracks, usually initiated by cyclic stresses, at surface imperfections such as machine markings and slip steps. Although the initial stress concentration associated with these cracks are too low to cause brittle fracture they may be sufficient to cause slow growth of the cracks into the interior. Eventually the cracks may become sufficiently deep so that the stress concentration exceeds the fracture strength and sudden failure occurs.

The extent of the crack propagation process depends upon the brittleness of the material under test. In brittle materials the crack grows to a critical size from which it propagates right through the structure in a fast manner, whereas with ductile materials the crack keeps growing until the remaining area cannot support the load and an almost ductile fracture suddenly occurs.

Mechanical Properties 291

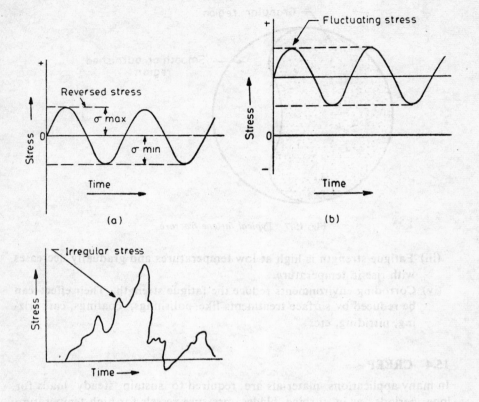

Fig. 15.6 *Stress cycles for fatigue testing*

Fatigue failures can be recognized by the appearance of fracture: a typical fracture is shown in Fig. 15.7. Two distinct zones can be distinguished — a smooth zone near the fatigue crack itself which has been smoothened by the continual rubbing together of the cracked surfaces, and a rough crystalline-looking zone which is the final fracture. Occasionally fatigue cracks show rough concentric rings which correspond to successive positions of the crack.

FACTORS EFFECTING FATIGUE FAILURES   Fatigue failures take place due to the following reasons:

(i) Surface imperfections like machining marks and surface irregularities.
(ii) Stress concentration points like notches, key ways, screw threads and machining under-cuts.

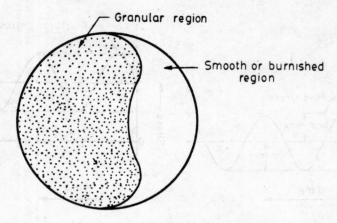

Fig. 15.7 *Typical fatigue fracture*

(iii) Fatigue strength is high at low temperatures and gradually decreases with rise in temperature.
(iv) Corroding environments reduce the fatigue strength. Their effect can be reduced by surface treatments like polishings, coatings, carburizing, nitriding, etc.

15.4 CREEP

In many applications materials are required to sustain steady loads for long periods, as in turbine blades, pressure vessels for high-temperature chemical processes, aircraft, lead coverings (sheaths) on telephone cables, furnace parts and timber beams in roofs of buildings. Under such conditions the material may continue to deform until its usefulness is seriously impaired. Such time-dependent deformations of a structure can grow large and even result in final fracture without any increase in load.

If, under any conditions, deformation continues when the load is constant, this additional deformation is known as creep. The time-dependent strain occuring under stress is also known as kreep. Creep is also the viscous flow in metals involving applied stress, time and temperature. Most materials creep to a certain extent at all temperatures, although engineering metals such as steel, aluminium and copper creep very little at room temperature. High temperatures lead to rapid creep which is often accompanied by microstructural changes.

A characteristic creep curve is shown in Fig. 15.8. There are four main areas of interest:

(i) *OA*, an initial instantaneous strain which can be partly elastic and partly plastic
(ii) *AB*, a primary region of creep of decreasing creep rate

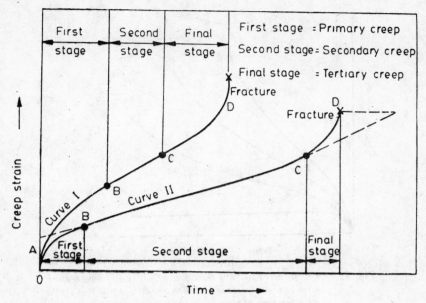

Fig. 15.8 *Creep curve at constant temperature and stress*

(iii) *BC*, a period of dynamic steady state where the creep rate is constant and at the minimum. It is a steady-state creep or secondary creep.

(iv) *CD*, a tertiary creep region of accelerating strain rate which ultimately ends in fracture.

A physical explanation of the three stages of creep is given below.

PRIMARY CREEP   Creep in this stage is mainly due to dislocation movement. The creep rate decreases with time and the effect of work hardening is more than that of the recovery process.

SECONDARY CREEP   During this stage, the rates of work hardening and recovery are equal, so the material creeps at a steady rate (minimum creep rate). Steady-state creep may be essentially viscous or plastic in character, depending upon the state level and temperature. Structural observations show that polygonization is an important recovery process during secondary creep.

TERTIARY CREEP   In this stage, creep rate increases with time until fracture occurs. At high temperatures, tertiary creep can occur due to necking of the specimen or grain boundary sliding and this continues until the specimen fractures.

The effect of temperature and stress on creep is illustrated in Fig. 15.9 which shows a series of tests at constant stress, i.e., $\sigma_1 = \sigma_2 = \sigma_3$, with

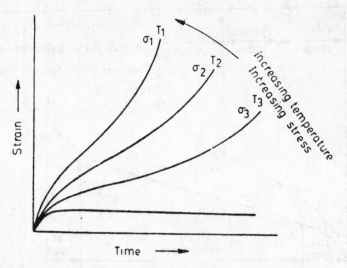

Fig. 15.9 *Creep curves showing the effect of stress and temperature on creep. An increase in either one causes a higher strain as a function of time*

temperatures in the order of $T_1 > T_2 > T_3$, or the effect of increasing stress while holding temperature constant. The maximum operating temperatures and stresses can be obtained from creep data plotted in a suitable form.

Precipitation-hardened alloys are generally used in high-temperature creep resistant applications because in addition to acting as barriers to slip and grain boundary sliding, the fine dispersion of particles apparently restricts the rate of recovery. The phenomenon of creep is important in polymers at room temperature, in alloys of aluminium at 100 °C, and in steels above 300 °C.

15.4.1 Mechanism of Creep

Figure 15.10 shows several mechanisms of creep in crystalline materials. As discussed earlier, creep is a thermally activated process. Some mechanisms that play significant roles during the creep process are discussed below:

(i) Dislocation climb
(ii) Vacancy diffusion
(iii) Grain boundary sliding

At high temperatures, an appreciable atomic movement causes the dislocations to climb up or down. By a simple climb of edge dislocation, the diffusion rate of vacancies may produce a motion in response to the applied stress. Thus edge dislocations are piled up by the obstacles in the glide plane and the rate of creep is governed by the rate of escape of dislocations past obstacles.

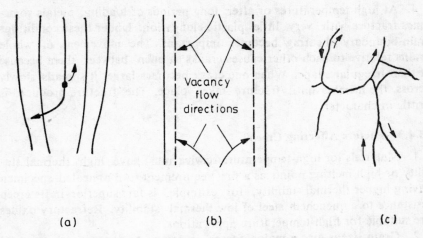

Fig. 15.10 *Mechanism of creep (a) Dislocation climb (b) Vacancy diffusion (c) Grain boundary sliding*

Another mechanism of creep is called diffusion of vacancies. In this mechanism, the diffusion of vacancies controls the creep rate but does not involve the climb of edge dislocations. It depends on the migration of vacancies from one side of a grain to another. Vacancies move in response to the applied stress in the directions shown in Fig. 15.10 (b). Over a period of time this movement would result in creep.

The third mechanism of creep is the sliding of grain boundaries. Grain boundaries become soft at lower temperatures as compared to individual grains. Grain boundaries play a major role in the creep of polycrystals at high temperatures as they slide past each other or create vacancies. At higher temperatures, ductile metals begin to lose their ability to strain-harden and become viscous to facilitate the sliding of grain boundaries. As the temperature is increased, the role of grain boundaries is reversed from one of resisting deformations at low temperatures to aiding deformation at higher temperatures.

15.4.2 Creep Fracture

1. Creep fracture may take place at higher temperatures or longer stress times, as in case of ductile metals. Fracture may take place after a nearly uniform deformation without forming a neck. This behaviour is known as thermal action.

2. Under high stresses and moderate temperature it takes shorter times. The nature of the fracture is exactly the same as that of a ductile material forming a neck before fracture.

3. At low stresses for long time durations fracture is similar to brittle fracture because of the negligible deformation produced in the process.

4. At high temperatures or after long periods of loading, metals sometimes fracture with very little plastic elongation. Under these conditions grain-boundary shearing becomes important. The movement of whole grains relative to each other causes cracks to open between them because of their irregular shape. When one crack becomes larger it spreads slowly across the member until fracture takes place. The fracture produced is brittle in character.

15.4.3 Factors Affecting Creep

1. Materials for high-temperature service must have high thermal stability or high melting points as a first requirement. An annealed specimen having higher thermal stability, for example, is far superior in its creep resistance to a quenched steel of low thermal stability. Refractory oxides are suitable for high-temperature applications.

2. Grain size is also a major factor in creep. A fine-grained material has improved mechanical properties in low temperature applications. On the other hand, a fine-grained material is to be avoided for high-temperature use. Coarse-grained materials exhibit better creep resistance. The grain-boundary sliding problem is greater in fine-grained materials due to the greater amount of grain-boundary materials and their higher tendency towards viscous flow at elevated temperatures.

3. Chemical reactions in materials seriously affect the creep rate. In steel-making furnaces, the lining of the furnace reacts with the stag, so care should be taken in the selection of lining materials.

15.5 MECHANICAL TESTS

To determine mechanical properties, various tests are performed on materials. Standardised test procedures are followed so that results may be compared. Most of the tests are done by a few national organizations set up to improve the use of materials in industries and engineering.

The Indian Standards Institute, the National Physical Laboratory and the National Test House are a few pioneering institutes in India which provide standard test methods in addition to standard specifications for materials and standard definitions of terms.

Mechanical tests may be grouped into two classes, destructive and non-destructive tests.

15.5.1 Destructive Testing

In this type of testing, the material or the component is destroyed and cannot be reused. The tests are conducted under the same conditions and on similar specimens.

Examples of destructive tests are tensile test, hardness test, **impact test**, fatigue test, creep test, etc.

TENSILE TEST This is one of the most widely used mechanical tests. It helps to determine tensile properties such as the limit of proportionality, elastic limit, yield point or yield strength, maximum tensile strength, breaking strength, % elongation, % reduction in area and modulus of elasticity.

The tensile test on a mild-steel test piece is described below.

The specimen to be tested is in the shape of a circular bar or a flat bar Fig. 15.11). One end of it is fastened to the frame of the machine by means

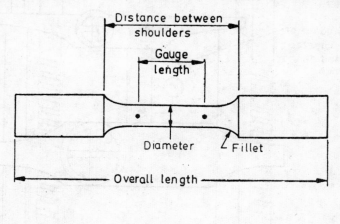

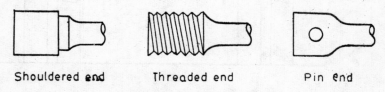

Fig. 15.11 *Standard tension specimen*

of grips or clamps, while the other end is similarly fixed to the movable crosshead (Fig. 15.12). The specimen is gradually loaded by the loading unit of the machine. The power to move the cross-head comes from a mechanical or hydraulic drive system. The magnitude of the load is measured by the load measuring unit. Elongation is measured by attaching an extensiometer or gauge to the specimen. As the load increases, the distance between the marked points on the gauge length increases. The test thus gives elongation as a function of load. Stress and strain can be calculated from load elongation data and a graph of stress versus strain can be made of the material. Many testing machines can automatically record stress-strain curves for materials under test.

During the tensile test, the different values of load and elongation at different intervals are recorded. Just before the pointer on the load scale in the load-measuring unit of the testing machine stops moving in the for-

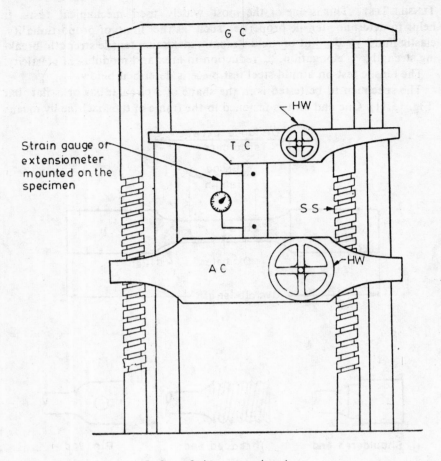

G C — Guide cross head
H W — Hand wheel
A C — Adjustable cross head
T C — Tension cross head
S S — Square screw

Fig. 15.12 *A schematic drawing of a testing mechine showing the specimen in tension*

ward direction, the extensiometer should be removed from the gauge length of the specimen. This is done to prevent any damage to the extensiometer. This is the point where the internal structure of the specimen starts yielding (indicative of yield point). The load is further applied and the maximum travel of the load-scale pointer will indicate the ultimate load. After this the pointer moves in the reverse direction to indicate necking. Finally the pointer stops at a point with a noise to indicate fracture.

After fracture the two pieces of the broken specimen are placed (Fig. 15.13) as if fixed together and the distance between the two gauge marks

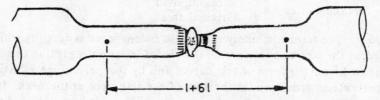

Fig. 15.13 *Increased length and reduced diameter after fracture*

is measured as the final length of the specimen. Similarly, the average diameter at the place of fracture is calculated and the area of fracture calculated. From these measurements and other observations obtained from the tensile test data, the various tensile properties can be calculated as follows:

Elastic limit This is the greatest stress that the metal can withstand without experiencing a permanent strain on removal of the load. It is given by

$$\text{Elastic limit} = \frac{\text{Load within elastic limit}}{\text{Original area of specimen}}$$

Proportional limit This is the stress at which the stress vs strain graph deviates from the straight line law to a curve.

Yield strength or yield stress This is the stress at which the specimen starts yielding without any increase in load

$$\text{Yield strength} = \frac{\text{Load at yield point}}{\text{Original area of the specimen at gauge length}}$$

Ultimate tensile stress or strength This is the maximum travel of the pointer on the dial of the load-measure range divided by the original area of the specimen

$$\text{Ultimate tensile stress} = \frac{\text{Ultimate load}}{\text{Original area of specimen}}$$

Percentage elongation

$$\text{Percentage elongation} = \frac{\text{Final length} - \text{Original length}}{\text{Original length}} \times 100$$

Percentage reduction in Area

$$\text{\% Reduction in area} = \frac{\text{Original area} - \text{Final area}}{\text{Original area}} \times 100$$

Young's Modulus of Elasticity (E)

$$E = \frac{\text{Stress at any point within elastic limit}}{\text{Strain at that point}}$$

Breaking strength This is the stress at the breaking point of the specimen in the tensile test,

$$\text{Breaking strength} = \frac{\text{Breaking load}}{\text{Original area}}$$

Ductility The term commonly referred to in tensile test is ductility which is defined as the strain at fracture. Valuable information regarding this term is obtained from the form of test curves and by the percentage elongation and percentage reduction in the area of the test piece at the neck. It is a measure of the amount of permanent deformation that has occurred when the material reaches its breaking point.

It can be seen that brittle materials such as cast irons show little or no plastic deformation before fracture, i.e. they are not ductile. Ductile materials such as copper show considerable plastic flow due to high ductility before fracture. Due to this property, wires can be drawn.

Toughness Tenacity or toughness is the strength with which the material opposes rupture. It is due to the attraction between molecules which, gives them power to resist tearing apart. It is a measure of the amount of energy that a material can absorb before fracturing. It becomes an important engineering consideration in the ability of a material to sustain an impact load without fracturing.

Toughness is also represented by the total area under the stress-strain curve (Fig. 15.14). It is often measured by impact testing rather than load deformation curves.

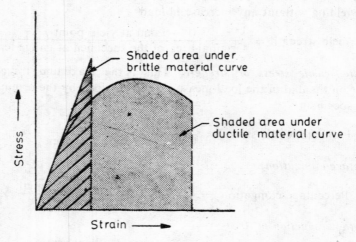

Fig. 15.14 *Areas under stress-strain curves representing the toughness of materials*

HARDNESS TESTS The common hardness tests are (Fig. 15.15):

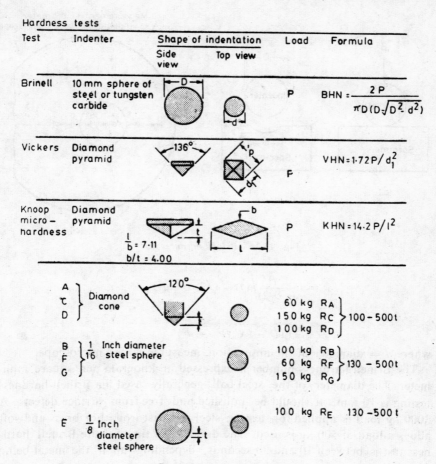

Fig. 15.15 *Hardness tests*

(i) Brinell hardness test
(ii) Vickers hardness test
(iii) Rockwell hardness test
(iv) Rebound hardness test
(v) Scratch test.

Brinell hardness test In this test a standard hardened steel ball of diameter D is pressed into the surface of the specimen by a gradually applied load P which is maintained on the specimen for a definite period of time (Fig. 15.16). The impression of the steel ball (indentor) so obtained is measured by a microscope and the Brinell hardness number (BHN) is found by the following relation:

$$\text{BHN} = \frac{\text{Load in kg}}{\text{Area of impression or indentation of steel ball in m}^2}$$

302 *Materials Science*

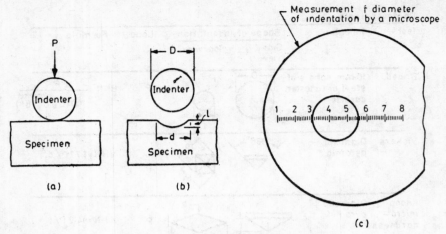

Fig. 15.16 *Brinell hardness test*

$$= \frac{2P}{(\pi D/2)} \bigg/ (D - \sqrt{D^2 - d^2})$$

$$= \frac{2P}{D(D - \sqrt{D^2 - d^2})}$$

where d = diameter of the impression, measured with a microscope.

The Brinell hardness number is expressed in kilogram per square millimeter. The diameter of the steel ball generally used in Brinell hardness testing is 10 mm. It should be polished and free from surface defects. A 3000 kg load is applied for testing steel and cast iron. For brass and soft alloys a load of 500 kg is used. The duration of time for the Brinell hardness test is between 10 and 30 seconds, depending upon the metal being tested.

A Brinell hardness tester is shown in Fig. 15.17. The test procedure is explained below:

(i) The test sample is placed on the top of the test table which can be raised by the elevating screw so that the test sample just touches the ball.
(ii) The desired load is applied either mechanically or by oil pressure.
(iii) During this period the steel ball moves to the position of the sample and makes an impression or indentation.
(iv) The indentation diameter is measured at two places, either on the screen provided with the machine or by coinciding the two points of a reading microscope (Fig 15.16 (e)).
(v) The Brinell hardness is determined from the above relation after substituting the values of P, D and d.

The Brinell hardness test is widely used to examine the effects of heat treatment and cold working in order to control the heat treatment process.

Mechanical Properties 303

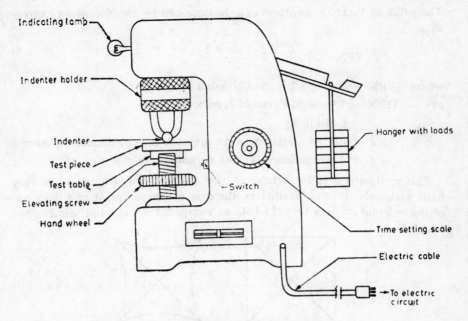

Fig. 15.17 *Brinell hardness testing machine*

The finished parts can be tested without destroying them completely as during tensile testing.

The Brinell hardness number, to some extent, is an indication of the tensile strength of the metal. They may be correlated by multiplying the BHN by a constant which depends upon the character of the metal. The test does not require much preparation of the surface of the metal as is necessary for rebound and Rockwell tests.

The limitations of the Brinell hardness test are:

(i) The test equipment is very heavy
(ii) The area of indentation is so great that it affects the surface quality. Thus it is sometimes considered as a destructive test.
(iii) The thickness of the test piece also limits its use. For example, thin sheets will bulge or be destroyed during the test.
(iv) The test results for very hard materials are unreliable and the ball gets flattened on hard surfaces.
(v) There is difficulty in measuring the indentation diameter accurately.

Vickers' hardness test The drawback of the flattening of the steel ball in testing harder materials is eliminated. In this method of hardness testing a square-based diamond pyramid is used as indentor. Like the Brinell and Rockwell hardness measuring methods, this method also uses the indentation produced by the indentor (diamond pyramid). The indentor gives geometrically similar impressions under different loads from 5 to 120 kg.

The value of Vicker's hardness can be obtained by the following expression:

$$VPN = DPN = \frac{2P \sin \theta/2}{d^2}$$

where VPN = Vicker's Pyramid Number
or DPN = Diamond Pyramid Number
P = Load in kg
$\theta = 136°$ = Angle between opposite face of diamond pyramid
d = Average length of two diagonals, in mm

This method is used to determine the hardness of very thin and very hard materials. It also facilitates the ease of measurement of a diagonal of the indentation area (Fig. 15.18), as compared to circular dimensions

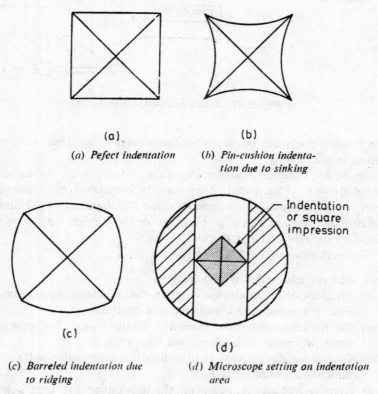

(a) *Pefect indentation*
(b) *Pin-cushion indentation due to sinking*
(c) *Barreled indentation due to ridging*
(d) *Microscope setting on indentation area*

Fig. 15.18 *Diamond pyramid indentations*

which are difficult to measure exactly. The values of Brinell and Vickers hardness are practically the same upto 300. Vicker's test can be carried out accurately on polished surfaces but does not give accurate results when used for rough surfaces.

Another method using the elongated pyramid of Fig. 15.15 is the Knoop indentor, developed especially for studying micro-hardness, i.e., to find the hardness of individual microstructures (metal grains) or check the hardness of a delicate watch gear. The Knoop indentor is a diamond ground to a pyramidal form that produced a diamond-shaped indentation with the long and short diagonals in the approximate ratio of 7 : 1 (shallow indentation). In this test, the indentor is mounted in the Tunken tester which applies a known weight through a balance beam. The load, down to 25 g, is applied for 20 s. The Knoop Hardness Number KHN is calculated from the following relation:

$$KHN = \frac{\text{Applied load}}{\text{Unrecovered projected area}}$$

$$= \frac{P}{A} = \frac{P}{L^2 C}$$

where P = Applied load, kg

A = Unrecovered projected area of indentation, mm²

L = Length of long diagonal, mm

C = a constant related to length of projected area for each indentor supplied by manufacturer

Knoop hardness is calculated from the projected area of the impression or indentation rather than the area of contact as in the case of other indentation tests.

Micro-hardness tests require extra care in all stages of testing. The surface of the specimen must be carefully prepared (good polishing required).

Rockwell Test The Rockwell hardness test is very widely used because of its speed and freedom from personal errors. This test requires much smaller penetrators (steel balls as well as cone-shaped diamond penetrators) and loads than used on Brinell hardness testers.

There are two scales on a Rockwell testing machine—*B*-scale and *C*-scale. The *B*-scale is used to record the hardness of soft metals and its range is $R_B 0$ to $R_B 100$. For the *B*-scale steel-ball indentors are commonly used. The *C*-scale is used to test materials of hardness greater than $R_B 100$ (e.g., cast irons) A diamond-cone indentor is used for measuring hardness. Its usual range is $R_C 20$ to $R_C 70$.

While performing the Rockwell hardness test the following procedure is used:

(i) The test piece is placed on the machine's test table.
(ii) The test piece is raised by turning the hand wheel, till it just touches the indentor and the needle on the machine dial reads zero. This indicates that a minor load of 10 kg has been applied on the test sample by the machine and ensures that the specimen is seated properly.

(iii) After that a major load of 100 kg (for *B*-scale) or 150 kg (for *C*-scale) is applied by pressing the crank provided for this purpose.
(iv) The depth of the indentation is recorded automatically by means of a dial scale. The reading is correlated with arbitrary hardness numbers.

Test Requirements
1. The test piece should be clean, free from dust, oil and scale, etc. It should be flat and preferably machine polished.
2. The thickness of the test piece should be such that a mark is not produced on the other side of the piece. Hence the thickness must be at least ten times the depth of indentation.
3. Successive impressions should not be superimposed on each other and care should be taken that they are not too close to each other (at least 4 times the diameter of the indentation apart).
4. The testing machine should be placed on a rigid support and covered suitably to protect it from vibration and dust.
5. The indentor, work table and test piece should be very clean.

Superficial Rockwell Testers This is a machine or tester available in a special form, called a superficial tester, in which small loads can be used. The value of the hardness obtained is rated the superficial Rockwell hardness number.

Rebound test The rebound hardness testing apparatus is sometimes also known as shore-scleroscope. In this, hardness measurements are made by dropping a hard object, a small diamond pointed hammer, on the surface and observing the height of rebound. As it falls its potential energy is transformed into kinetic energy which is passed on the surface when it strikes. Due to the elastic strain energy of the surface and damping capacity stiffness of the material, the diamond pointed hammer rebounds. The height of the rebound is taken as the index of hardness. In actual tests a 1/12 oz hammer is allowed to fall from a height of 250 mm in a glass tube which is graduated to measure the height of rebound. This apparatus is very handy and is used to test the hardness of sheet rolls.

Scratch test This is a measure of hardness which rates various materials on their ability to scratch one another. Hardness is measured on the Mohs scale, originated by a German mineralogist, Friedrich Mohs, in 1832 (Fig. 15.19). This scale consists of 10 standard minerals arranged in the order of their ability to be scratched. Diamond, the hardest known material, is assigned the hardness number of 10 Mohs, while the softest mineral in the scale is talc with a scratch hardness of one Mohs. Most hard metals fall in the Mohs hardness range of 4 to 8. For most engineering applications, the Mohs scale is not quantitative enough to distinguish clearly between metals of similar hardness. Moreover, the measurement of hardness by scratching

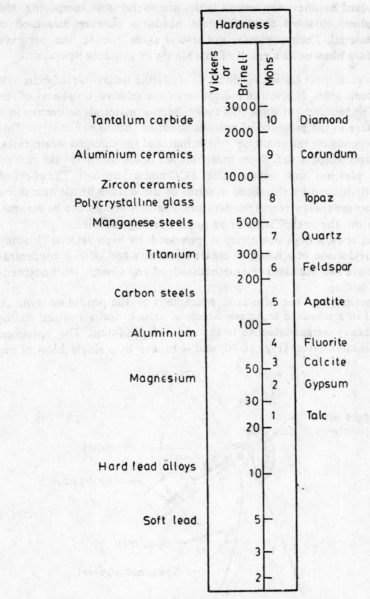

Fig. 15.19 *Approximate relative hardness of certain metals and ceramics*

is difficult to interpret and standardise. Consequently, new hardness scales based on the other methods (indentation hardness and dynamic or rebound hardness) of measuring hardness are of major engineering interest for metals.

The standard hardness conversion tables are useful for comparing the hardness values obtained from different hardness numbers measured on the same material. The results may not always agree due to the presence of thin surface films or as a result of machining or grinding operations.

Impact Tests Impact testing is done to study the behaviour of materials under dynamic load. It gives an indication of the relative toughness of the material. The tendency of some normally ductile materials to behave in a brittle manner in the presence of notches is called notch sensitivity. This property depends on the response of the material to changed strain rates, triaxiality and temperature. Some materials are notch sensitive (like carbon steels and plastics) and others (like FCC metals) are not. The effect of notch sensitivity can be visualized in terms of the ductile-brittle transition curve of fracture energy versus temperature. Fracture energy can be measured in kg-m on the energy scale of an impact testing machine.

In Impact tests, a high strain rate is provided by high-velocity loading and the introduction of a notch to create triaxiality and stress concentration. There are two standard test methods—Izod and Charpy—for notched-bar impact testing.

For impact testing, the standard machine is of the pendulum type. A notch is cut in a standard test piece which is struck under impact conditions by a heavy weight attached to the end of a pendulum. The specimen is held in an anvil (vice) (Fig. 15.20) and is broken by a single blow of the

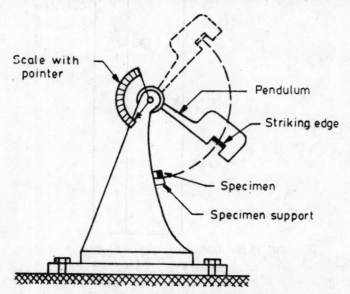

Fig. 15.20 *Impact testing machine*

pendulum weight or hammer which falls from a fixed starting point of a known height. After breaking the specimen the pendulum swings on and the height to which it rises on the other side is measured. Thus, the energy absorbed in breaking the specimen may be determined and if this is low the specimen is brittle. The impact energy required to break the specimen can also be noted from the scale provided on the impact testing machine. When the pendulum is released from the position of maximum height or maximum energy, the pointer on the scale also moves alongwith the pendulum and stops at a particular position to indicate the energy absorbed in breaking and the energy still remaining unutilised. Most machines are constructed so that both types of tests can be used with only minor adjustments.

Izod test This test uses a cantilever test piece of 10 mm × 10 mm section and 75 mm length (Fig. 15.21). The *V*-notch angle is 45° and the depth of the notch 2 mm. The pendulum hammer strikes near its end.

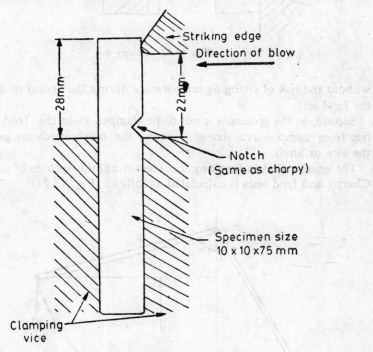

Fig. 15.21 *Izod test*

Charpy test This test is most common. It uses a test specimen of 55 mm length with cross-section 10 mm × 10 mm, with a *V*-notch 2 mm deep of 45° included angle and a root radius 0.25 mm. The specimen is placed on the supports or anvil as a simply supported beam (Fig. 15.22). The blow of the pendulum hammer is from a direction opposite to the notch section. The Charpy test has two advantages over the Izod. First, the case of placing the specimen on the machine facilitates even low-temperature tests

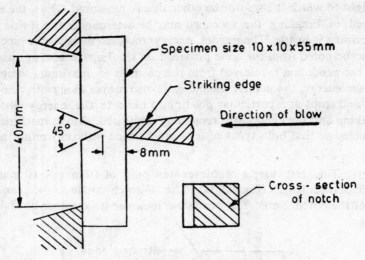

Fig. 15.22 *Charpy test*

without the risk of changing temperature during the period of clamping in the Izod test.

Second, as the specimen is not to be clamped as in the Izod test, it is free from compressive stresses around the notch which are produced by the vice or anvil.

The energy used in breaking or fracturing the specimen in both the Charpy and Izod tests is calculated as follows (Fig. 15.23):

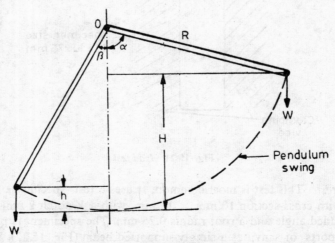

Fig. 15.23 *Impact test*

Initial energy = WH
= Potential energy
= Kinetic energy of striking specimen

Remaining energy after striking specimen = KE of carrying the pendulum to the lowest position = Wh

$\therefore\quad\dfrac{\text{Energy delivered}}{\text{to the specimen}} = \dfrac{\text{Impact value of energy in}}{\text{fracturing specimen}}$

$= WH - Wh$
$= W(H - h)$

where W = Weight of hammer or pendulum used for striking specimen
H = Height of fall of centre of gravity of pendulum or striker
h = Height of rise of centre of gravity of pendulum after striking specimen

FATIGUE TEST Fatigue tests are performed in the laboratory to determine the capacity of a material to withstand repeated applied stresses. Many different machines are used to study the fatigue properties of materials, but the most important classifications are with respect to the type of load and the manner in which it is applied.

The basic types of loading are simple axial loading, bending, torsion, and a combination of these three. Specimens are commonly loaded in constant load machines, the deflection usually increasing as the specimen becomes weaker.

Figure 15.24 shows a constant load machine used for fatigue testing. The test is simple and straightforward. The test specimen, in the form of a

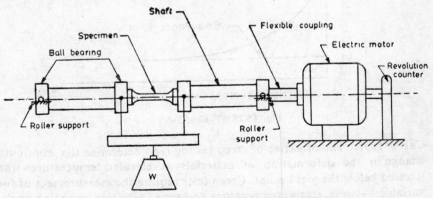

Fig. 15.24 *R.R. Moore fatigue testing machine*

cantilever which forms the extension of a shaft, is placed in the machine. The shaft is driven by an electric motor. A dead load is applied to the machine by means of ball bearings. The bearing relieves the motor of the large bending moment which is applied to the specimen. While the specimen is rotating, any given position on the surface of the specimen alternates between a state of maximum tensile stress and a state of maximum compressive stress.

There is sinusoidal variation of stress as the specimen rotates. The stress is greatest at the surface and zero at the centre. The number of cycles will depend upon the value of the applied stress so that fatigue failure may occur. The number of cycles will be less where the value of the stress is high. On the other hand, if the applied stress is less, the number of cycles will have to be increased. Ultimately a stress is reached below which failure will not take place within the limits of the standard test. This value of stress is known as the endurance limit. For good quality steels the endurance limit value is about 0.46 of the ultimate strength. If fracture does not take place within the specified cycles for the material, then it is assumed that fracture will not take place.

A number of specimens of the same material are fatigue-tested under different stress levels and the results are plotted on a semi-logarithmic scale with stress S on the y-axis and the number of cycles N required to cause failure of the specimen on the x-axis. The result is a S-N graph or curve. Figure 15.25 shows the S-N curve obtained on testing steel and copper specimens.

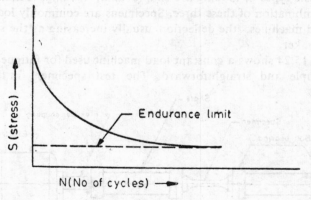

Fig. 15.25 *Typical S-N curve*

CREEP TEST The objective of creep testing is to determine the continuing change in the deformation of materials at elevated temperatures when stressed below the yield point. Creep tests require the measurement of four variables—stress, strain, temperature and time. A testing machine applies the constant load or stress. The test is carried out at constant temperature and the strain of the test piece is noted as a function of time.

The test pieces for creep testing are normally the same as for conventional tensile tests. For a constant load, all that is required is a dead weight and a system of levers to multiply it to the required load.

Figure 15.26 shows the installation for creep testing. The specimen to be ested is placed in an electric furnace (creep testing at high temperature)

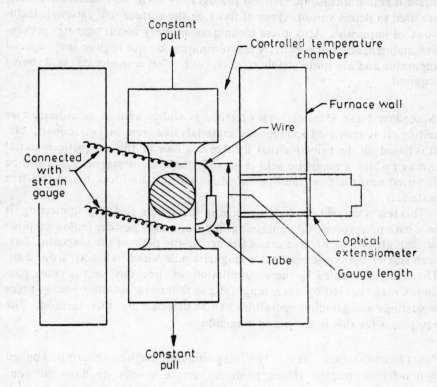

Fig. 15.26 *Creep test*

where it is heated to a given temperature and constantly subjected to a load. The strain in the specimen is measured by a strain gauge or an optical extensiometer. Four or five specimens are tested at each temperature under different loads. Elongation versus time curves are plotted for each specimen.

A platinum alloy wire is spot-welded to the specimen at one end of the gauge length and a platinum alloy tube at the other. The wire slides inside the tube and the reference marks on both are observed through the optical extensiometer.

15.5 Non-Destructive Testing

We have noted the effects of cracks and flaws in various mechanical tests. These should be detected at the early stages and the component replaced or disaster will result. All microscopic flaws can be detected by NDT. In this type of testing the material is not destroyed but will be used after the test. Examples of non-destructive tests are the magnetic dust method, penetrating liquid method, ultrasonic test and radiography. All non-destructive tests are used to detect various types of flaws on the surface or internal inclusions of impurities. Also, these techniques are very useful during preventive maintenance and repair. A few techniques do not require any special apparatus and are quite simple to carry out, only a moderate skill being required.

MAGNETIC DUST METHOD This method is widely used in non-destructive testing. It is restricted to magnetic materials like iron, nickel, cobalt, etc. It is based on the principle that if there is a flaw in the magnetic material through which a magnetic field is passing, the lines of magnetic flux will be distorted near the flaw; the lines of flux will be uniform for defect-free materials.

This test is carried out by magnetising the article and then immersing it in a bath of kerosene oil containing fine iron oxide powder (coloured powder can also be used). If a crack lies across the path of the magnetic flux, each side of the crack becomes a magnetic pole which attracts iron dust. The crack is revealed by the accumulation of iron dust on the crack portion. Cracks caused by quenching, fatigue failure in welding, blow holes in castings and grinding operations can be detected by this method. The equipment for this is known as magnaflux.

PENETRATING-LIQUID TEST The magnetic dust method cannot be applied to non-ferrous metals. Hence, to detect surface cracks or flaws in non-ferrous metals, the penetrating liquid test is used. It is a non-destructive testing process employing a visible colour-contrast dye penetrant technique for the detection of open surface flaws in metallic and non-metallic objects. This technique reveals flaws such as shrinkage cracks, porosity, fatigue cracks, grinding cracks, seams, forging cracks, heat treatment cracks and leaks etc., on weldings, castings, machined parts, cutting tools, pipes and tubes.

The technique essentially consists of spraying the colour penetrant on the surface to be inspected. After allowing a sufficient penetration time, the excess penetrant is removed. A developer is then applied to reveal flaws by absorbing the penetrant from these flaws, thus giving a visible indication of the defects which still retain the coloured penetrant (A developer which contains an absorbant white powder is also sprayed on the surface).

Care should be taken to clean the surface so that it is free from dust, dirt, scale or rust as otherwise the test will give poor results. Precautions should be taken both during use and storage of penetrants as these are highly toxic and flamable.

This test can also be performed by dipping the article in a bath of a penetrating fluorescent substance such as anthracene. The solution enters the cracks, if any, and remains there. The article is then dried and examined under a quartz tube vapour lamp. The cracks can be detected by the fluorescence produced by the ultraviolet light.

ULTRASONIC TEST   In this method, high-frequency sound waves (ultrasonic waves) are applied to the article under test by a piezoelectric crystal. If the article is flawless or free from cracks, it reflects sound waves without distortion. If there are any flaws in the article, the time taken by the sound waves will be less as the reflection will be from the flaw points and not from the bottom of the article. The sound signals are received on a cathode ray tube which has a time base connected to it. The depth of the crack is calculated from the time interval between the transmission of the sound pulse and the reception of the echo signal. This test is predominately used to detect internal cracks like shrinkage cavities, hot tears, zones of corrosion and non-metallic inclusions. It is a very fast method of inspection and is used to test automobiles and aerospace components.

RADIOGRAPHY   X-rays and gamma rays are widely used to detect internal defects in components of large size. This technique is also known as radiography. To investigate internal faults in forgings, castings, press workings, distributions of impurities and the process of self diffusion and the mechanism of friction and wear in metals. Radiography techniques are finding more extensive applications in the field of physical metallurgy and in the treatment of various diseases.

Common radiography techniques are based on the use of X-rays, gamma rays and radioactive isotopes. The components to be tested are exposed to X-rays or gamma rays which can penetrate all substances to varying degrees. The rays, after passing through the components, show a picture on a fluorescent screen or on a photographic plate. The cracks, blow holes and cavities appear lighter, whereas inclusions of impurities appear darker than the component metal. The photographic film, after development, will show lighter and darker areas to represent the radiograph of defects in the component.

In X-rays radiography, the X-rays are emitted from X-ray tubes as discussed in Ch. 6. X-rays after passing through the blow hole in a casting, will be absorbed to a lesser extent than X-rays which pass through sound metal. Fault can be detected by measuring the X-ray absorption. The penetrating power of X-rays is lesser than that of gamma rays, therefore X-rays are used for small thicknesses and gamma rays for greater thicknesses of work.

Gamma-ray radiography is based on the use of gamma rays which have shorter wavelengths and are more penetrating than X-rays. The source of gamma radiation is usually the radioactive isotope of cobalt enclosed in a special sealed container or capsule. Gamma rays give better results for thicker sections and, unlike X-rays, a number of components can be inspected at a time. Gamma rays are undeflected by magnetic fields and do not ionise gases. This method is now becoming more popular because cheap isotopes are available and the test can be performed in a very short time. The limitations of this method are the difficulties in handling isotopes and the precautions required.

The radioactive-isotope method or neutron radiography has become possible after the discovery of artificial radioactivity by Madam Curie and her husband in 1934. They discovered that a radioactive substance has a complex nucleus and such materials continuously and spontaneously emit considerable amounts of energy in the form of radiation that cannot be controlled. The use of this method for studying metals and alloys depends on the characteristic of radioactive decay in unit time. Due to its high sensitivity and accuracy, this method of inspection can be used to find voids, inclusions and concentrations in explosives, lubricants papers, polymers and various defence and space equipments.

15.6 MODELS OF FRACTURE

Fracture is the separation of a specimen into two or more parts by an applied stress and can be classified as either brittle or ductile. Brittle fracture occurs after little or no plastic deformation, whereas ductile fracture occurs after extensive plastic deformation.

Fracture is usually undesirable in engineering applications. Flaws such as surface cracks, lower the stress for brittle fracture whilst line defects (dislocations) are responsible for initiating ductile fractures. Figure 15.27 shows fractures observed in metals subjected to uniaxial tension.

PROCESS OF FRACTURE It consists of two components:
  (i) Crack initiation
  (ii) Crack propagation

15.6.1 Types of Fracture

Figure 15.28 shows different types of fracture and these depend upon:
  (i) Rate of loading
  (ii) Type of stress applied
  (iii) Temperature
  (iv) Nature of material

The different types of fracture are:
  (i) Brittle fracture
  (ii) Ductile fracture
  (iii) Creep fracture
  (iv) Fatigue fracture

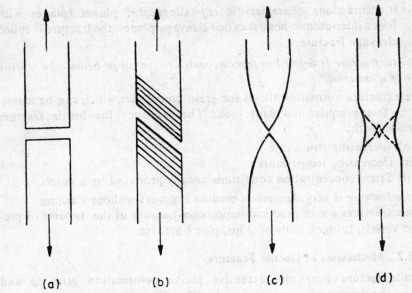

Fig. 15.27 *Fractures observed in metals subjected to uniaxial tension (a) Brittle fracture of single crystals and polycrystals (b) Shearing fracture in ductile single crystals (c) Completely ductile (d) Ductile fracture in polycrystals*

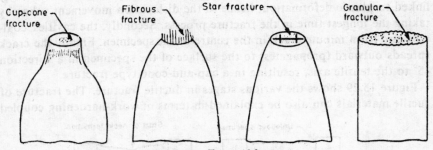

Fig. 15.28 *Types of fracture*

BRITTLE FRACTURE

1. Occurs when a small crack in a material grows. Growth continues until fracture occurs (Crack propagation).
2. The atoms at the surfaces do not have as many neighbours as those in the interior of a solid; consequently they form fewer bonds. As a result, surface atoms are at a higher energy than a plane of interior atoms.
3. In brittle fracture, adjacent parts of the metal are separated by stresses normal to the fracture surface.
4. Because brittle fracture does not produce plastic deformation, it requires less energy than a ductile failure where energy is introduced in the process of forming dislocations and other imperfections within the crystals.

5. It occurs along characteristic crystallographic planes (planes with fewer inter-atomic bonds) called cleavage planes; the fracture is called cleavage fracture.

Brittle fracture is defined as fracture which occurs at or below the elastic limit of a material.

Brittle fracture normally follows the grain boundaries which can be identified by their granular and shiny look. The tendency for brittle fracture increases with:

(i) Increasing strain rate
(ii) Decreasing temperature
(iii) Stress concentration conditions usually produced by a notch.

Brittle fracture is very dangerous because it occurs without warning.

Brittle fractures are of practical importance because of the failures of pressure vessels, bridges, hulls of ships, pipe lines, etc.

15.6.2 Mechanism of Ductile Fracture

Ductile fracture occurs after extensive plastic deformation prior to and during the propagation of the crack. There are three successive events involved in a ductile fracture. First the specimen begins necking and minute cavities form in the necked region. This is the region in which plastic deformation is concentrated. It indicates that the formation of cavities is closely linked to plastic deformation, hence to the dislocation movement, thereby taking the longest time in the fracture process. Secondly, the cavities coalesce and form minute crack in the centre of the specimen. Finally the crack spreads outward (propagates) to the surface of the specimen in a direction 45° to the tensile axis, resulting in a cup-and-cone type fracture.

Figure 15.29 shows the various stages in ductile fracture. The fracture of ductile materials can also be explained in terms of work-hardening coupled

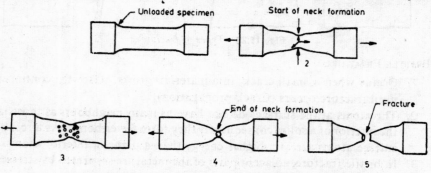

Fig. 15.29 *Successive stages in the ductile fracture of a tensile test specimen*

with crack-nucleation and growth. The initial cavities are often observed to form at foreign inclusions where gliding dislocations can pile up and produce sufficient stress to form a void or micro-crack. Consider a specimen subjected to a slowly increasing tensile load. When the elastic limit is exceeded, the material begins to work-harden. Increasing the load, increases the permanent elongation and simultaneously decreases the cross-sectional area. The associated decrease in area leads to the formation of a neck in the specimen, as illustrated earlier. The necked region has a high dislocation density and the material is subjected to a complex stress, no longer a simple tensile stress. The dislocations are separated from each other because of the repulsive interatomic forces. As the resolved shear stress on the slip plane increases, the dislocations come closer together. The cracks form due to high shear stress and the presence of low-angle grain boundaries. Once a crack is formed, it can grow or elongate by means of dislocations which slip. Crack propagation is along the slip plane for this mechanism. These tend to coalesce. Consequently, one crack grows at the expense of others. Crack growth finally results in failure.

Figure 15.30(a) shows crack nucleation at a slip-plane obstacle due to dislocation pile-up. Figure 15.30(b) shows crack nucleation at low-angle grain boundaries.

15.6.3 Mechanism of Brittle Fracture or Griffith Theory of Brittle Fracture

Griffith has postulated a criterion for the propagation of pre-existing cracks in a brittle material. The crack could come from a number of sources such as a collection of dislocations, a flaw which occurred during solidification, or a surface scratch.

When a solid is stressed in tension, elastic energy is stored when it elongates. As the stress on the solid is increased, the elastic energy stored per unit volume increases. At a sufficiently high stress, the crack becomes larger and the material fractures.

Lengthening of the crack causes an increase in the surface area of the crack and, therefore, the surface energy of the specimen is increased. There is also compensating release of energy. As the crack becomes longer, the material on both sides of the crack can no longer store elastic energy because tensile stress cannot be transmitted across the crack. Thus, an increase in crack length causes the release of elastic energy. Griffith postulated that when the elastic energy relased by extending a crack was equal to the surface energy required for crack extension, the crack would grow.

For an elliptical crack (Fig. 15.31) the maximum stress at the tips is given by:

$$\sigma_{max} = 2\sigma\sqrt{C/P} \qquad (1)$$

where p is the radius of curvature at the tip, σ = applied stress perpendicular to length $2C$.

If the crack is to propagate through the material, the release of stored elastic energy must be sufficient to provide the energy associated with the two surfaces produced.

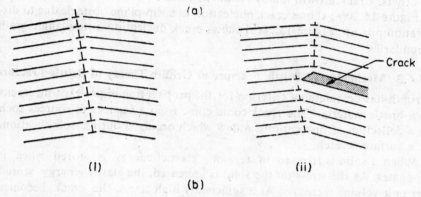

(a) Crack nucleation at a slip-plane when there is obstacle or impurity inclusion. Dislocations pile up at an obstacle and, as shear stress increases, the dislocations move close to each other and form a crack. It is caused by three edge dislocations coming together

(b) Shows crack nucleation at a low-angle grain boundary. Formation of a crack resulting from an active slip plane intersecting the boundary

Fig. 15.30 *Ductile fracture mechanisms*

The elastic strain energy for unit volume is given by,

$$U_E = \sigma^2/2E \times \text{Area} \times \text{Width}$$
$$= \frac{\sigma^2}{2E} \times \pi C^2$$

If the surface energy per unit area is γ then the surface energy for a crack of length $2C$ and unit width will be

$$U_S = (2\gamma C) \times 2 = 4\gamma C \qquad (3)$$

We multiply by two because there are two faces. Applying Griffith's criterion, if the crack is to propagate, the change in surface energy with crack length must just equal the change in elastic strain energy.

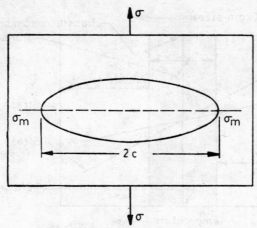

Fig. 15.31 *Model for Griffith's fracture theory*

$$\frac{dU_E}{dC} = \frac{dU_S}{dC}$$

$$\frac{d}{dC}\left(\frac{\pi C^2 \sigma^2}{E}\right) = \frac{d}{dC}(4\gamma C)$$

i.e. $\quad \sigma = \sqrt{\left(\frac{2\gamma E}{\pi C}\right)}$

This result shows that the stress necessary to cause fracture varies inversely as the square root of the crack length and so the tensile stress of a completely brittle material will depend on the length of the largest existing crack in the material.

In the case of crystalline materials, the Griffith theory has to be modified somewhat to take into account the small amount of plastic deformation that occurs. Basically, the fracture mechanism is the same, although the critical crack depth is somewhat larger. In polycrystalline materials, brittle fracture can either be along the cleavage planes or along grain boundaries due to presence of impurities which have a weakening effect.

15.6.4 Ductile to Brittle Transition

The increase in yield stress associated with low temperature or high strain rates can result in a material changing its mode of fracture from ductile to brittle and this is very important when selecting materials for engineering purposes. The transition can be explained with the help of Fig. 15.32. The brittle fracture stress (σ_f) and the yield stress (σ_y) are plotted as a function of temperature or strain rate. The curve for brittle fracture stress rises slightly to the left because surface energy increases as temperature decreases. The yield stress curve shows a strong temperature dependence as in BCC metals and metal oxide ceramics. The two curves intersect and a ver-

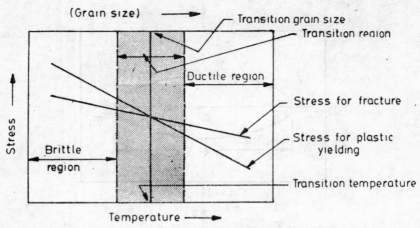

Fig. 15.32 *The Ductile-brittle transitions as a function of grain size and temperature*

tical line is drawn at the point of intersection. This is called the ductile-brittle transition temperature. If a material is stressed at a temperature or strain rate which is to the right side of this line, it will reach its yield point before it reaches the brittle fracture stress and will undergo some plastic deformation prior to fracture. On the other hand, applying a stress under conditions which lie to the left of the line will result in brittle fracture. The temperature range over which the rapid changes occurs is called the transition region.

If the curves of yield stress and brittle fracture do not intersect, there is no ductile to brittle transition. The yield stress curves for FCC materials generally lie below the brittle fracture stress curves and FCC materials do not experience this transition.

Figure 15.33 shows the energy consumed in an impact test as a function of temperature, for mild steel. A high strain rate can be achieved in impact testing machines by fast loading. Increasing the strain rate is equivalent to lowering the temperature, therefore materials that are ductile when strained slowly at a given temperature will behave in a brittle manner when subjected to a high strain rate.

From a design point of view, the ductile to brittle transition is very dangerous.

15.6.5 Methods of Protection Against Fracture

In the Griffith theory it was observed that surface cracks are more effective than internal cracks and cause brittle fracture. Fatigue cracks often propagate from the surface, inwards. Protection against fracture or failure by crack propagation, in many cases, involves the following methods to make the surface cracks ineffective.

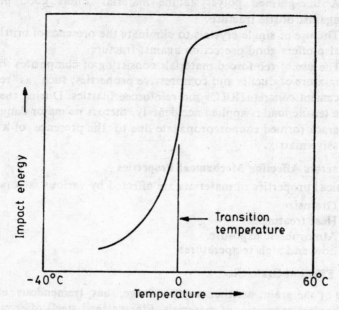

Fig. 15.33 *The variation of impact energy as a function of temperature in mild steel*

(i) Introducing compressive stresses on the surface to counteract the tensile stress causing cracks to propagate. This can be achieved by heating, say, silicate glass to its softening temperature and than cooling it rapidly to produce contraction or introduce compressive stresses. This makes the glass stronger and is used in producing tempered glass which is commonly used in the windshields of automobiles.

In order to increase surface resistance against crack propagation, case-hardening processes and shot peening shot blasting with small steel balls driven by a black of air of metals are the other methods which can be used.

(ii) Polishing the surface to a good finish offers better protection as it removes some of the cracks from the surface.

(iii) (a) Proper design considerations should be taken to avoid sharp corners and notches in the parts to eliminate points of stress concentration.

(b) Impurity free materials should be used for fabrication work as inclusions will produce brittle phases.

(c) The direction of elongation should be parallel to the direction of stress applied in a good design.

(iv) Failure along grain boundaries should be avoided by proper distribution of solute atoms in metals to prevent the formation of brittle phases at grain boundaries. This can be done by preventing the formation of films of impurities along the grain boundaries.

(v) A fine-grained polycrystalline material offers good protection against brittle fracture.

(vi) The use of single crystals to eliminate the presence of brittle phases also offers good protection against fracture.

(vii) The use of reinforced materials consisting of composites having a mixture of ductile and compressive properties, such as reinforced cement concrete (RCC) and reinforced plastics. During their use, if a tensile load is applied accidentaly, there is no major danger as the crack formed cannot propagate due to the presence of a compressive matrix.

15.7 Factors Affecting Mechanical Properties

Mechanical properties of materials are affected by various factors, such as:
 (i) Grainsize
 (ii) Heat treatment
 (iii) Atmospheric exposure
 (iv) Low and high temperatures

15.7.1 Effect of Grain Size

The size of the grain, whether small or large, has tremendous effects on the mechanical properties of materials. Fine-grained steels offer more resistance to cracking, produce fine surface finishes and can be drawn or plastically deformed better as campared to coarse-grained steels. Coarse-grained steels, which are indicated by small grain-size numbers, have lower strength than fine-grained structures but have higher hardenability, creep strength at high temperatures and better forging properties.

Grain size directly controls the extent of slip interference by adjacent grains and toughness, ductility and fatigue of metals. Figure 15.34 is a typical curve showing the effect of grain size and mechanical properties.

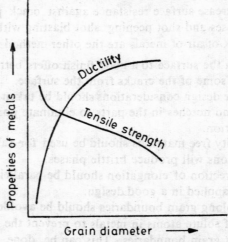

Fig. 15.34 *Effect of grain size of metals*

15.7.2 Effect of Heat Treatment

Mechanical properties like ductility hardness, tensile strength, toughness and wear resistance can be improved by heat treatment. Heat treatment is generally done for the following purposes:

   (i) To refine the grain and improve mechinability
  (ii) To relieve the internal stresses induced in the metals during forging castings and cold working of the metals.
 (iii) To improve resistance to corrosion
 (iv) To modify the structure, either coarse grained or fine grained
  (v) To improve chemical, magnetic, electrical and thermal properties.
 (vi) To produce a hard surface on a ductile core or interior.
(vii) To improve mechanical properties.

15.7.3 Effect of Atmospheric Exposure

Most metals get oxidized when exposed to the atmosphere. The effect on metals is indicated by film formation or film breakdown. A film is formed due to oxidation of the metal surface by air while the latter is due to electrochemical action requiring the presence of an electrolyte. In the presence of humid air iron rusts, whereas in dry air no visible change is observed. The presence of moisture, sulphur dioxide, hydrogen sulphide and various other corrosive conditions decrease the electrical resistivity of metals. The atmospheric effect on the metal depends on the followings:

  (i) Characteristic properties of the metal
 (ii) Value of the protective film on its surface
(iii) Electrolyte in contact with the metal
(iv) Presence of certain reducing agents
 (v) Local cells formed due to development of cracks and discontinuity on the protective film surface resulting in corrosion of the exposed part.

Metals like stainless steel, aluminium, nickel, etc. function in their highly passivated state under atmospheric conditions, providing an exceptional atmospheric resistance to their surface even when exposed to rain. Under very corrosive conditions, as in highly industrialized and urban atmospheres, in the presence of reducing agents, these metals have tarnished surfaces due to passivity destruction.

15.7.4 Effect of Low Temperatures

With decrease in temperature there is an increase in the tensile strength and yield strength of all metals in general.

1. Alloys of nickel, copper and aluminium retain most of their ductility and toughness at low temperatures even though there is increase in tensile strength.

2. For unnotched mild steel, the elongation and reduction in cross-sectional area is satisfactory upto $-130°$ C but after that it goes down to a large extent.
3. Near absolute zero temperature many metals exhibit the phenomenon of superconductivity.
4. Low temperatures cause low thermal vibrations and lattice parameters are stabilized.
5. Below $-100°$ C non-ferrous metals show better properties than ferrous metals.

5.7.5 Effect of High Temperatures

1. Yield stress and ultimate tensile strength decrease with rise in temperature.
2. Stiffness and fracture stress of many metals also decrease with increasing temperature.
3. At higher temperatures, embrittlement of steel occurs and toughness is reduced.
4. At higher temperatures, creep takes place and the material fails even at a very small stress.
5. Due to rise in temperature there is a corresponding rise in thermal vibration of atoms causing changes in structural properties.

15.8 TECHNOLOGICAL PROPERTIES OF METALS

These properties are highly desirable in the fabrication, forming and shaping materials for useful purposes. Important technological properties are weldability, machinability, castability and formability.

15.8.1 Weldability

This is defined as the capacity of the metal to be joined or welded under the fabrication conditions imposed in a specific suitably designed structure, and to perform satisfactorily in the intended service. In other words, a metal has good weldability if it can be welded readily in order to perform satisfactorily in the fabricated structure.

Weldability is the main factor upon which the various welding processes are based. The weldability of a metal is affected by the following factors:

(i) The composition of the metal
(ii) Heating and cooling effects
(iii) Affinity for oxygen (oxidation) and other gases before and at welding temperature
(iv) Design, selection and service conditions

15.8.2 Machinability

This is defined as the ease with which a given material can be cut, permitting the removal of material with a satisfactory finish at lowest cost. Common machining operations are turning, threading, sawing drilling, shaping, grinding, milling, etc.

This property of the material enables it to be machined by cutting tools. Machinability depends upon the following factors:

CHEMICAL COMPOSITION OF METALS Smaller amounts of lead, manganese and sulphur increase the machinability, whereas high alloy contents, presence of abrasive inclusions and more than 0.6% carbon content decrease machinability.

MICROSTRUCTURE Undistorted fine grains and uniform microstructures with very low carbon contents help increase machinability, whereas large distorted non-uniform structures with high carbon contents adversely affect machinability.

MECHANICAL PROPERTIES Machinability of soft materials is good whereas highly ductile and very hard materials are difficult to machine.

CUTTING CONDITIONS Heat-treatment processes affect the machinability of metals. Annealing and normalizing processes generally help make metal parts soft and increase their machinability, whereas quenching and hardening processes decrease their machinability.

To determine the machinability in any production process, the following points need consideration:

(i) Value of cutting forces and power required
(ii) Life of cutting tool and coefficient of friction
(iii) Quality of surface finish and velocity of cut (metal removal rate)
(iv) Type of work and type of tool
(v) Form and size of chips.

15.8.3 Castability

The ease at which a metal can be cast into forms is known as castability. It depends on the following factors:

SOLIDIFICATION RATE The ease with which a metal will continue to flow after it has been poured in the mould cavity depends on its analysis and pouring temperature. Some metals, such as grey cast iron, are very fluid and can be poured into intricate shapes (thin sections) of complicated castings.

GAS POROSITY Certain metals are easily affected by oxygen and nitrogen in the molten state. These gases are trapped during the cooling of castings, resulting in the formation of small holes and voids in the castings.

SHRINKAGE This is the reduction in volume of a metal on changing from a molten to a solid state. The amount of shrinkage produced during casting varies from metal to metal and some shrinkage allowance is taken into consideration in the design of a pattern and mould-making There are a few elements which may be added to the molten metal to increase and decrease the shrinkage rate.

SEGREGATION This is the non-uniform distribution of constituents in a metal. Usually a concentration of certain constituents and or impurities arising during freezing and generally persisting throughout subsequent operations.

15.8.4 Formability

The ability of metals to be formed into different shapes is known as formability. It is based on the flowability and ductility of the metal which, in turn, is based on its crystal structure. The factors affecting the formability of metals are grain size, hot and cold working alloying elements, and heat-treatment processes such as annealing and normalising.

Low-carbon steel has good metal-forming qualities and thus can be easily cold-worked. High-carbon steels are unsuitable for metal forming processes like rolling, spinning, drawing, extrusion, forging, etc. In order to prevent the distortion of crystals during cold working, hot-working processes are suitable for increasing the ductility as well as formability of crystals.

REVIEW QUESTIONS

1. Explain weldability and machinability of metal. (AMIE Winter-1985)
2. Describe in detail the purpose and procedure for testing of metals for
   (a) Tensile strength
   (b) Impact strength   (AMIE Winter 1985)
3. Draw a stress-strain diagram for a low-carbon steel specimen, indicating the proportionality limit, elastic limit, yield point and the point of maximum loading and rupture. Explain the above data. (AMIE Summer 1984)
4. Write short notes on ductility, malleability and machinability.
   (AMIE Summer 1984)
5. Describe briefly the radiographical method for non-destructive examination of engineering components. How will you decide the exact location of the flaw?
   (AMIE Summer 1984)
6. Write short notes on
   (a) Fracture   (Summer 1983)
   (b) Weldability   (Summer 1982)
   (c) Elastic strain   (Summer 1982)
7. (a) Describe briefly the important mechanical tests which give valuable information about metals.

(b) What does the creep test signify?
(c) With a neat sketch, explain the creep-testing method for metals
(AMIE Summer 1982)
8. Discuss the effect of heat treatment processes on the mechanical properties of steel. (AMIE Winter 1982)
9. Briefly explain the following:
   (a) Relation between hardness and strength of metals
   (b) Factors affecting fatigue strength of materials
   (c) Impact properties of materials as applied to design
   (d) Importance of creep testing (AMIE Winter 1982)
10. Discuss the principle of non-destructive testing by the following methods:
    (a) Magnetic dust method
    (b) Penetrating liquid test
    (c) Ultrasonic test method (AMIE Winter 1982)
11. What is creep? Draw a typical creep curve and explain the different stages of creep. Explain the effect of grain size, strain hardening, heat treatment and alloy addition on creep. Elaborate your answer with reference to steel.
(AMIE Winter 1983)
12. Mention the technological properties that control the weldability of metals. (AMIE Summer 1983)
13. What is fracture? What are the effects of cracks on fracture strength? Explain the characteristics of brittle fracture and ductile fracture. (AMIE Summer 1983)
14. How is impact strength measured? What is the SI unit for expressing impact strength? (AMIE 1982)
15. Discuss the factors affecting the impact properties of materials.
(AMIE Winter 1981)
16. What are the different types of fracture in metals? (AMIE Winter 1981)
17. (a) Explain the Brinell hardness test for mild steel specimens.
    (b) Explain the Charpy impact test.
    (c) Differentiate between the fracture of mild steel and cast iron specimens in tensile testing. (AMIE Summer 1980)
18. (a) Explain the terms weldability and machinability.
    (b) Differentiate between primary creep, secondary creep and tertiary creep.
(AMIE Summer 1980)
19. (a) Name five important mechanical tests which give valuable information about metals and alloys.
    (b) What does the impact test signify? Explain with the necessary formulations the procedures to be adopted in an impact test conducted on a pendulum-type impact-testing machine.
20. (a) Which are the different non-destructive tests? Explain briefly their fields of application.
    (b) Explain the Griffith theory of metals.
21. Sketch the conventional and actual stress and strain diagrams and state the significance of each.
22. What do you understand by machinability? What factors affect it and how can it be improved?
23. What are the different methods for determining the hardness of a metal? State their advantages and disadvantages.
24. Under what conditions may a ductile material fail in brittle structure?

16
PHOTO ELECTRIC EFFECT

16.1 INTRODUCTION

We have already discussed in Ch. 2 that electrons absorb energy, change orbits and sometimes become free electrons. The photoelectric effect is the emission of electrons from a metal surface when it is illuminated by high-frequency electromagnetic radiation. The photoelectric effect was discovered by Hertz in 1887 when he observed that the passage of an electric current through a gas discharge tube was made much easier when ultraviolet light was allowed to fall on the cathode. It was also noted that a high-voltage discharge jumped greater distances than when it was left in the dark. Hallwachs, in 1888, discovered that an insulated zinc plate charged negatively, lost its charge when a beam of ultraviolet light was directed on it. If the surface was charged positively, there was no loss of charge under the action of light. This phenomenon was explained by him by the theory that ultraviolet light ejects electrons from the metallic surface. This phenomenon is known as *photoelectric effect*, and the electrons emitted are called photoelectrons.

Metals like zinc, cadmium, selenium, etc., are sensitive only to ultraviolet light; whereas alkali metals like sodium, potassium, etc., are sensitive even to visible light. Various photosensitive devices like photo tubes or "electric eyes", photoconducting devices like automatic door opening and closing devices and photovoltaic cells like solar batteries or solar cells are common examples of photoelectric effect.

16.2 EXPERIMENTAL STUDY OF THE PHOTOELECTRIC EFFECT

Figure 16.1 (a) shows a typical arrangement for the study of the photoelectric effect. Photoelectric cells are usually made by depositing a thin layer of an alkali metal on the inner surface of a small vacuum tube. If the cell is to operate in ultraviolet light it is made of quartz, whereas if it is to be used in visible light it is made of common glass. The cell must be thoroughly evacuated as any oxygen in the air will combine chemically

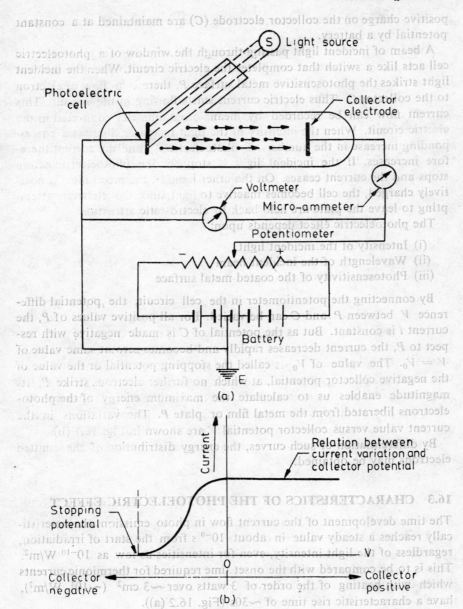

Fig. 16.1 *Photoelectric cell*

with the active metal layer, contaminating its surface and making it insensitive to visible light. A window or a small section is left clear to serve as an inlet for the incident light. Photoelectrons, upon leaving the metal surface, are attracted and collected by the positively charged collector or electrode C. The negative charge on the metal plate or film (P) and the positive charge on the collector electrode (C) are maintained at a constant potential by a battery.

A beam of incident light passing through the window of a photoelectric cell acts like a switch that completes the electric circuit. When the incident light strikes the photosensitive metal surface P, there is a flow of electron to the collector C. Thus electric current starts flowing in the circuit. This current flow can be recorded by means of an ammeter connected in the electric circuit. When the intensity of the light increases, there is a corresponding increase in the number of photoelectrons, and the current therefore increases. If the incident light is stopped, the photoelectric action stops and the current ceases. On the other hand, if the metal film is positively charged, the cell becomes inactive to light since the elrctrons attempting to leave the plate are held back by electrostatic attraction.

The photoelectric effect depends upon:

(i) Intensity of the incident light
(ii) Wavelength of the incident beam
(iii) Photosensitivity of the coated metal surface

By connecting the potentiometer in the cell circuit, the potential difference V between P and C can be varied. For all positive values of P, the current i is constant. But as the potential of C is made negative with respect to P, the current decreases rapidly and becomes zero at same value of $V = V_0$. The value of V_0 is called the stopping potential or the value of the negative collector potential, at which no further electrons strike P. Its magnitude enables us to calculate the maximum energy of the photoelectrons liberated from the metal film or plate P. The variations in the current value versus collector potential V are shown in Fig. 16.1 (b).

By differentiation of such curves, the energy distribution of the emitted electrons may be obtained.

16.3 CHARACTERISTICS OF THE PHOTOELECTRIC EFFECT

The time development of the current flow in photo emission characteristically reaches a steady value in about 10^{-9} s from the start of irradiation, regardless of the light intensity, even for intensities as low as 10^{-10} W/m². This is to be compared with the onset time required for thermionic currents which with heating of the order of 3 watts over $\sim$3 cm² ($\sim 10^4$ W/m²), have a characteristic rise time of $\sim$30s (Fig. 16.2 (a)).

For a given frequency of incident light, there is a well defined maximum kinetic energy of ejected electrons which is independent of the light intensity. For thermionic emission, the higher the power input the higher the temperature and therefore the higher the maximum electron kinetic energy (Fig. 16.2 (b)).

For photoemission, the maximum electron kinetic energy is found to be related linearly to the frequency of the incident light. This linear relation

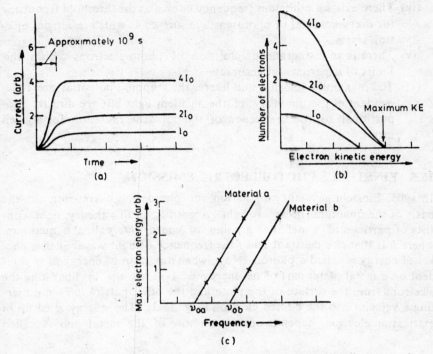

Fig 16.2 *Characteristics of photoelectric emission for different light intensities and emitting materials*

is the same for all materials but the frequency at which KE_{max} goes to zero (photo-current ceases) is different for different materials. Furthermore, below the *cut-off* frequency, no current is observed regardless of the light intensity (Fig. 16.3 (c)).

For a thermionic process, there should only be a relation between total power input and electron energy. The electron kinetic energy should not be related to the radiation frequency, so there should not be a cut-off frequency.

16.4 LAWS OF PHOTOELECTRIC EMISSION

(i) The velocity of the electrons emitted is independent of the intensity of light and depends only upon the frequency (or wavelength) of the light and the nature of the metal. It varies linearly with the frequency of the incident light.

(ii) The rate of emission of electrons is directly proportional to the intensity of the incident light.

(iii) Photoelectric emission is independent of the temperature of the photo-cathode.

(iv) There exists a minimum frequency known as the threshold frequency for every material of photosensitive surface at which electron ejection starts.
(v) There is an instantaneous emission of photo-electrons within the limits of experimental accuracy.
(vi) The maximum velocity, and hence, the stopping potential are independent of the intensity of the incident light but are directly proportional to the frequency of the incident radiation for a given metal.

16.5 EINSTEIN'S PHOTOELECTRIC EMISSION

In 1905, Einstein gave an explanation for photoelectric emission on the basis of the quantum theory of light. According to this theory, light consists of particles of a definite amount of energy hf, called a quantum, where h is Planck's constant and f the frequency of light wave. Such a packet of energy is called a photon. Now, when a photon of energy hf is incident on a metal plate, part of its energy ϕ_0 is used up in liberating the electrons from the surface of the plate and the other part ($\frac{1}{2} mv^2$) in imparting a velocity v to the ejected electron (Fig. 16.3). The energy used up in extricating electrons depends upon the nature of the metal and is called

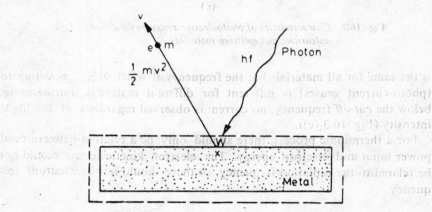

Fig. 16.3 *Photoelectric effect. A light quantum (photon) of energy hf incident on a metal surface, ejects an electron with a velocity v given by the Einstein equation*

the work function. When electrons leave the surface of the plate, the metal plate becomes positive to an equal extent and there will be attraction between the positive surface and the negative electrons, tending to pull the electrons back into the surface of the metal. Work will be done to overcome this pull in order to enable the electron to leave the surface. This work depends upon the nature of the element and accounts for the work function. The symbol ϕ is commonly used for work function. For elements it is found to be of the order of 10^{-12} ergs per electron.

Equating the energy,

$$hf = \tfrac{1}{2} mv^2 + \phi_0$$

or
$$\tfrac{1}{2} mv^2 = hf - \phi_0$$
$$= hf - hf_0 \quad \text{(taking } \phi_0 = hf_0)$$
$$= h(f - f_0)$$

This equation is called Einstein photoelectric equation. The term f_0 is called the *threshold frequency* and gives the minimum frequency of the incident light which will just be able to eject photoelectrons with practically zero velocity.

The velocity of the ejected electrons can be measured by the application of magnetic and electric fields. The use of X-rays and gamma rays from radioactive substances have confirmed the truth of Einstein equation.

In order that a photoelectron may be ejected, it is essential that the frequency of the incident radiation be greater than the threshold frequency. The velocity of the ejected electrons will depend only upon the frequency, and not upon the intensity (which does not add to the energy of the incident photon but only accounts for their number. Again, the number of electrons ejected will depend upon the number of photons falling, and hence upon the intensity of light. This explains the photoelectric laws.

Again, the velocity of electrons ejected from near the surface will be larger than those coming from the interior of the metal, and hence the velocity of the electrons varies with a certain maximum for all ejected electrons.

The number of electrons emitted per unit area per second determines the magnitude of the photoelectric current. For a given wavelength, this is proportional to the intensity of the incident light. The emission varies with wavelength also. Alkali metals show the maximum effect in or near the visible part of the spectrum.

16.6 MILLIKAN'S VERIFICATION OF EINSTEIN'S EQUATION

Millikan, in 1916, verified the Einstein photoelectric equation. His experiment was based on the determination of Planck's constant h and the stopping potential V_0 from the photoelectric effect. Millikan's experiment was very suitable for low-frequency radiations like ultraviolet and visible light.

The apparatus used for this purpose was a vaccum tube to prevent the possibility of a gas film being formed (Fig. 16.4). Cylinder blocks of lithium, sodium and potassium are used as cathodes or emitters. Cylindrical blocks are attached to a wheel (W) which can be rotated by means of an electromagnet (M). One of the cylindrical blocks is brought opposite

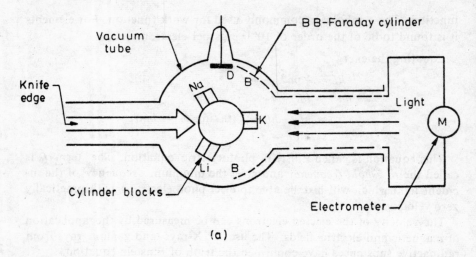

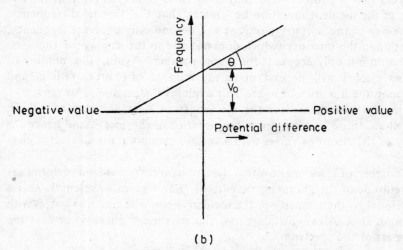

(a) Verification of Einstein's photoelectric equation
(b) Determination of h and w

Fig. 16.4 Millikan's experiment

the knife edge K, which can be advanced towards the cylinder and rotated so that a clean surface of the metal is cut. Since a high vacuum is maintained, the cut surface will not be corroded in the time required for the experiment. Monochromatic light is passed through the window E. The cylinder BB (Faraday cylinder) and disc D are made of oxidised copper to reduce reflection of light to a small quantity so photoemission is not produced by the light used for lithium, sodium and potassium.

The stopping potentials of photoelectrons were measured by increasing the positive potential on the emitting surface or cathode. This can be done by passing the light from a mercury vapour lamp through the window and illuminating the surface at K. Radiations of six different wavelengths, in turn, were used and a curve of current-potential difference obtained, from which the limiting voltage (stopping voltage) V_0 for zero current could be found.

Then
$$e.V_0 = \tfrac{1}{2} mv^2 = hf - p,$$

where p is the energy required to release the electron without kinetic energy.

The observed value of the stopping potential V_0 must be corrected so that there may be a contact difference of potential between the metal and the cylinder BB. This is measured by turning the wheel W until the metal under examination faces the disc D which is connected to an electrometer. By raising or lowering the disc, the potential difference is maintained between the disc and the metal. This was taken as the value of the contact difference of potential and used as a corrective to find the proper value of the stopping potential V_0.

Millikan found that, on plotting the stopping potentials V_0 against a selected range of frequencies f, a straight line was obtained.

Also, $dV_0/df = h/e$. Since e is known, the experiment gives the value of h, which was found to be 6.56×10^{-34} joules-s. The linear relation (Fig. 16.4 (b)) between V_0 and f and the correctness of the value for h may be considered to establish the validity of Einstein's equation on an experimental basis. The currently accepted value of h is 6.625×10^{-34} joules-s.

From the intercept on the frequency (f) axis, the threshold frequency f_0 can be calculated and hence the work function w.

16.7 PHOTOELECTRIC CELLS

The device which converts light energy into electric energy is known as a photoelectric cell.

Photoelectric cells, based upon the different types of photoelectric effects employed, can be classified as follows:

(i) Photo-emissive cells
(ii) Photo-voltaic cells
(iii) Photo-conductive cells

16.7.1 Photoemissive Cells

In photo-emissive cells or phototubes, emission of electrons is caused by the energy of a light beam striking a metallic surface having a low work function. This sensitive surface is enclosed in a vacuum, a gas-filled envelope or quartz and the emitted electrons are captured by a positive anode. Photo-emissive cells are of two types.

(i) Vacuum photo-emissive cell or phototube
(ii) Gas-filled photo-emissive cell or gas photo tube.

In the vaccum photo-tube, the cathodic surface is treated with a low work-function material and a small wire is used as anode (Fig. 16.5 (a)). The cathode is semicircular in appearance. The anode collects the electrons emitted when the light strikes the light-sensitive cathode. Anode voltages are large enough to achieve current saturations, i.e., all electrons emitted by voltages over about 40 volts reach the anode.

The current output of a vacuum photo tube is only a few micro-amperes, which is very small. To increase the current output and sensitivity for a given luminous flux, an inert gas, usually argon or neon, is introduced at low pressure to cause ionization. Ionization by collision of electrons with gas molecules increases the current five to ten times. This effect is also known as *thermionic effect*.

When photo-tubes are used at extremely low light levels, the output of the tube may be fraction of an ampere. For many applications, such as for the reproduction of sound recorded on film in television, etc., its is necessary to amplify the current output of photo-tubes. These currents can be directly amplified in a device known as photomultiplier tube or electron multiplier. These are based on the principle of secondary emission.

The cathode K emits electrons when light strikes it. These electrons are accelerated towards the positive electrode D (Fig. 16.5 (a)). They bombard the sensitive surface of D_1 and several secondary electrons are emitted.

Suppose the sensitive surface has a secondary emission coefficient x above unity, then x electrons leave D_1 for every electron striking it on the average. The electrons may again be accelerated towards another positive electrode D_2 at a potential above that of D_1. This surface may also have a secondary emission coefficient x, so that x^2 electrons leave D_2 for every electron originally leaving the photoelectric electrode K.

This process may be repeated n times on n electrodes, giving an overall current gain x^n. In practice, x may range between 5 and 10. This type of tube is used in modulated light applications where conventional tubes cannot be used. Figure 16.5 (c) shows photomultiplier tubes used in TVs and other devices,

16.7.2 Photovoltaic Cells

In photovoltaic cells, the radiant energy striking the cell surface generates an emf. This emf can be applied to an external circuit. The external circuit may be designed so that the external current flow is a linear function of the light intensity. This type of photoelectric device consists of oxides or compounds coated on metallic plates. Another name for this is barrier-layer cell.

Photovoltaic cells can be classified as:

Photoelectric Effect 339

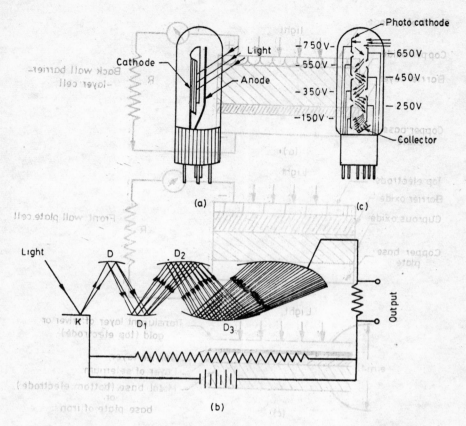

Fig. 16.5 *Photo-emissive cells (a) vaccuum tube (c) photo-multiplier*

(i) Copper-copper oxide photovoltaic cells
(ii) Iron-selenium photovoltaic cells

Copper-copper oxide cells are either back-wall cells or frontwall cells.

In the back-wall cell, the semiconductor cuprous oxide is deposited on a copper-based plate. The blocking layer or barrier existing between the copper-based plate and the cuprous oxide is the point of origin of photo electrons. A semi-transparent metallic layer is placed over the oxide layer. The copper-based plate constitutes one electrode and a thin copper ring in contact with the semi-transparent metallic layer another. When illuminated, the photo-electrons travel from the cuprous oxide to the copper, i.e., from the semiconductor to the metal through the barrier layer (Fig. 16.6 (a)). As the barrier layer is remote from the incident light, this cell is known as a back-wall cell.

When the barrier layer lies between the top electrode and the cuprous oxide, the cell is called a front-wall cell. The current direction is from the semiconducting cuprous oxide to the metal electrode through the barrier

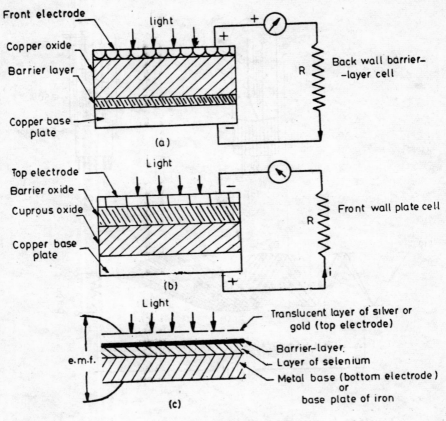

Fig. 16.6 *Photovoltaic cells*

layer. Consequently, the direction of electron flow in the external circuit is opposite to the back-wall cell (Fig. 16.6 (b)).

The front-wall and back-wall cells are also known as photon cells. The sensitivity of the front-wall cell is greater than that of the back-wall cell because light reaches the barrier without being partially absorbed by the ineffective cuprous oxide.

The iron-selenium cell is a front-wall cell and it (Fig. 16.6 c) consists of a base plate of iron. In this cell a thin layer of iron selenide is deposited on the iron-based plate. The iron selenide layer is then covered with a silver layer. The base plate is connected to a copper ring and forms one electrode. Another copper ring in contact with the silver layer forms the second electrode.

The main characteristics of photovoltaic cells are:

(i) No external voltage is required as in photoconductive or photoemissive cells.

(ii) A small dc voltage is generated so amplification directly by vacuum tube is not possible.
(iii) The response falls rapidly with modulated light because of high barrier-layer capacities.
(iv) Voltage sensitivity can be increased if cells are connected in series.
(v) Highly sensitive to high temperatures. The selenide cell is damaged if operated above 65 °C.

The photo-voltaic cell is used in devices like portable exposure-meters, direct-reading illumination meters, solar cells, solar batteries and other monitoring operations in industries.

THE SOLAR BATTERY When the primary purpose of the photovoltaic cell is to serve as an electric energy supply, we consider them solar cells. They function as a battery to supply voltage, but the energy source is the light of the sun instead of the chemicals of dry batteries. Of course, extended periods of good weather are a requirement.

The solar battery is a type of photovoltaic cell. These units consist of carefully refined wafer-thin strips of silicon, about the size of razor blades, into the faces of which an impurity has been diffused in the form of a microscopic layer near the surface. These formed strips constitute a p-n junction. The energy in sunlight moves electrons to one face and holes to the other, thus giving polarity to the battery.

Among the advantages of the solar cells are
(i) Long life
(ii) Relatively high efficiency
(iii) More power developed than in other photovoltaic devices.

These batteries have been used as a source of energy on remote repeater stations, wire-communication circuits, and space satelites.

16.7.3 Photoconductive Cell

These are based on the principle of conductivity of semiconductor materials whose resistances change according to the radiant energy received. The resistivity of semiconductor materials like selenium, lead sulphide, cadmium sulphide, etc., is decreased when irradiated. Therefore, these materials have dark resistance and low irradiated resistance. It was shown that, for photo-emission to occur, relatively large photon energies were necessary and, as a result, there are no known photo-emitters for wavelengths beyond the near infrared region of the spectrum. In semiconductors, however, relatively small photon energies are capable of producing electron-hole pairs internally, and hence of increasing the carrier concentrations and conductivity of the material. This photo-induced conductivity can be used for light detection from the ultraviolet through the infrared region.

We know that the electrons in the valence band can be excited into the conduction band by visible light when the energy gap of the semiconductor is in the right range. The additional holes and electrons created by the incident light can lead to an increase in the current in an approximate external circuit. The current is a direct measure of the light intensity and the devices used for detecting and measuring the light energy are called photoconductors.

Figure 16.7 (a) shows the simplest form of a photoconductive cell using selenium. There are two electrodes provided with the semiconductor material. When the cell is unilluminated or dark, its resistance is always high and hence current through the circuit is low. On the other hand, the current through the circuit will become large due to a decrease in its resistance as soon as the cell is illuminated. The shape of the semiconductor is so fabricated as to get a large ratio of dark to light resistance.

In Fig. 16.7 (b) is shown another commonly used cadmium sulphide cell. Due to its shape, it provides very high dark-to-light ratio as well as a maximum response at 5000 Å. In order to increase the contact area with the sensitive material; the two electrodes are generally extended in the interdigital pattern.

Photoconductive cells are generally used for detecting ships and aircrafts by the radiations given out by their exhausts or fuel gases, as well as for telecommunication by modulated infrared light.

In Fig. 16.7 (a) is shown a circuit for automatic door opening. In this, a photoconductor is arranged with a light beam falling on it continuously.

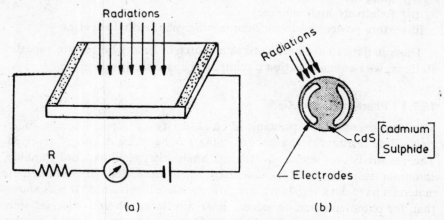

Fig. 16.7 *Photoconductive cell*

When an object approaches the door, this beam is interrupted, reducing the current flow in the photoconductor which then activates the circuit to open the door. Similarly, this cell can be used for low-speed counting or sorting work, in fire or burglar alarms and street lights can be automatically switched on when the sun goes down. With an invisible infrared beam, photoconducting devices can be used to detect a distant object in the sky.

16.8 APPLICATIONS OF PHOTOELECTRIC CELLS

Photoelectric cells are widely used in various fields of industry these days. Some important applications are given below:
- (i) Street lighting
- (ii) Television or cathode ray tubes
- (iii) Photometry
- (iv) Counting machines
- (v) Burglar alarms
- (vi) Fire alarms
- (vii) Reproduction of sound in motion pictures
- (viii) Cell-systems in hotels
- (ix) Solar batteries
- (x) Remote sensing in defence work
- (xi) Automatic relays
- (xii) Contrast and brightness control in TVs
- (xiii) Automatic timing devices
- (xiv) Sorting machines
- (xv) Photocopiers or photostat work

Example

The work function of sodium is 2.3 ev. Calculate the threshold frequency and the corresponding wavelength. (AMIE Summer 1979)

Solution

Work function $W = hf_0$

$$= 2.3 \text{ eV}$$

$$f_0 = \frac{W}{h} = \frac{2.3 \times 1.602 \times 10^{-19}}{6.625 \times 10^{-34}} \text{ Hz}$$

Now, $W = hf_0$

$$= h\frac{C}{\lambda_0} = \frac{6.625 \times 10^{-34} \times 3 \times 10^8}{\lambda_0} \text{ Joules}$$

when λ_0 is in meters

$$= \frac{6.625 \times 10^{-34} \times 3 \times 10^8}{1.602 \times 10^{-19} \lambda_0} \text{ eV} \quad \text{when } \lambda_0 \text{ is in Å}$$

$$= \frac{12400}{\lambda_0} \text{ eV}$$

The work function of sodium is 2.3 eV

or $\quad 2.3 \text{ eV} = \dfrac{12400}{\lambda_0}$

or $\quad \lambda_0 = \dfrac{12400}{2.3} = 5390 \text{ Å}$

Example

A blue lamp emits light of a mean wavelength of 4500 Å. The lamp is rated at 150 watts and 8% of the energy appears as emitted light. How many photons are emitted by the lamp per second? (AMIE Winter 1976)

Solution

Light energy emitted per second

$$= 150 \times \dfrac{8}{100} \text{ J}$$
$$= 12 \text{ J}$$

where frequency $f = C/\lambda$.

$$C = 3 \times 10^8 \text{ m/s}$$
$$h = 6.625 \times 10^{-34} \text{ J/s}$$

and $\quad \lambda = 10^{-10}$ m

Now, energy carried by the electron $= hf$

$$= h \cdot \dfrac{C}{\lambda}$$
$$= \dfrac{6.625 \times 10^{-34} \times 3 \times 10^8}{4500 \times 10^{-10}}$$
$$= 44 \times 10^{-20} \text{ J}$$

Number of photons emitted by the lamp per second

$$= \dfrac{12}{44 \times 10^{-20}}$$
$$= 26.98 \times 10^{18}$$

Example

Electrons are emitted with zero velocity from a certain metal surface when it is exposed to radiations of 6800 Å. Calculate the threshold frequency and work function of the metal. (AMIE Summer 1978)

Solution

Using the Einstein equation for photoelectric effect, we get

$$hf = w + \tfrac{1}{2} mv^2$$
$$= hf_0 + \tfrac{1}{2} mv^2$$

where $W = hf_0$

f_0 = Threshold frequency
f = Frequency of incident light
h = Planck's constant = $6.625 \, 10^{-34}$ J-s

As $\quad V = 0$
so $\quad \frac{1}{2} mv^2 = 0$
Now $\quad hf = hf_0$

or $\quad f = f_0 = \dfrac{C}{\lambda} = \dfrac{3 \times 10^8}{6800 \times 10^{-10}} = 4.416 \times 10^{14}$ Hz

Work function $\quad W = hf_0 = 6.625 \times 10^{-34} \times 4.416 \times 10^{14}$
$\qquad\qquad = 2.918 \times 10^{-10}$ J

Example

A radio receiver operates at a frequency of 440 kHz and a power of 10 kW. How many photons per second does it emit?

Solution

Power of radio receiver = 10 kW = 10,000 watts
$\qquad\qquad\qquad\qquad\quad$ = 10,000 joules per second
$\qquad\qquad\qquad\qquad\quad$ = 10^4 joules per seconds $\quad$ (1 watt = 1 J/s)

Now the energy of each photon will be given by the relation,

$$E = hf$$
$$E = 6.625 \times 10^{-34} \times 440 \times 10^3 \text{ joules}$$

Substituting values of h and f

$$= 291.28 \times 10^{-30} \text{ J}$$

Number of photons emitted per second

$$= \dfrac{\text{Power}}{\text{Energy}} = \dfrac{10^4}{291.28 \times 10^{-34}}$$

$$= 34.32 \times 10^{30}$$

REVIEW QUESTIONS

1. What is the photoelectric effect? Explain the characteristics of the photoelectric effect.
2. Explain the laws of photoelectric emission.
3. Derive einstein's photo electric equation and show that the frequency of the incident radiation affects the photoelectric emission.
4. Define the following terms:
   (a) Stopping potential
   (b) Thresho d frequency

     (c) Photoelectric work function
     (d) Planck's quantum theory of radiation
5. Explain the experimental study of the photoelectric effect. What are the fundamental laws of photoelectric emission?
6. Describe the different types of photoelectric cells. List the important applications of photoelectric cells.
7. Write short notes on the following:
     (a) Photomultipliers
     (b) Solar batteries
     (c) Phototubes
     (d) Photo-sensitive surfaces
8. Briefly describe the origin of photoelectrons and the effect of temperature on their emission.
9. What is the photoelectric effect? How will you demonstrate it experimentally? (AMIE May 1978)
10. State and explain Einstein's photoelectric equation. (AMIE May 1978)
11. State Einstein's law for the photoelectric effect. Describe an experiment to verify this law. (AMIE Summer 1980)
12. Give a brief account of Millikan's experimental verification of Einstein's photoelectric equation.
13. Describe some of the devices operating on the photoelectric effect. (AMIE 1975)

17
POWDER METALLURGY

17.1 INTRODUCTION

Powder metallurgy is likely to grow in importance as a branch of fabrication because of the growing need for more and more complex alloys and components demanding special properties which are obtainable only by these means.

The technique of agglomerating metal powders into engineering components is called powder metallurgy. This process involves the preparation and processing of metal powders and then making objects from a single metal or alloyed metal powders with or without the inclusion of non-metallic powders. In general, the ability of the process to change these powders into solid metals depends first upon the principles of solid-phase welding and second upon the transportation of matter by diffusion. This process is very useful in producing dimensionally accurate ready-to-use parts without waste and directly without any further processing. Powder metallurgy principles were used by the Egyptians for producing implements as early as 3000 BC. In recent years, the powder metallurgy technique has acquired a very important place in the modern technology for producing a wide variety of parts which otherwise could not be made by any other manufacturing process.

Two conditions must be fulfilled before a powder metallurgy product can be made:

(i) It must be possible to form a continuously bonded matrix. The metal in the powder form must be able to respond to solid-phase welding.

(ii) The powders in which the basic materials are available must be capable of sufficiently close packing under pressure to permit welding to take place and, in the case of alloying, be capable of being sufficiently intimately mixed.

17.2 APPLICATIONS OF POWDER METALLURGY

(i) *For producing parts* Certain hard-material powders like tungsten, titanium, boron, etc. are used in the fabrication of cemented carbide tools for high-speed processing of CI, bakelite, hard rubber, etc.

(ii) *In porous products like filters and bearings* Brass and bronze powders are used for producing filters, and in the manufacture of porous bearings, iron, copper and tin powders are used.

(iii) *In lamination of high-precision parts such as magnets and self-lubricating bearings* Powders of nickel, cobalt, aluminium and iron are used for producing permanent magnets. Mixtures of copper, tin and iron powders with or without graphite are used in the manufacture of self-lubricating bearings.

(iv) *Where combined properties of two metals or of metals and non-metals are desired* Powder mixtures of tin, copper, graphite, lead, silica, etc. are used for fabricating brake linings, clutch facings, electric motor brushes and aircraft landing gears.

(v) Grinding wheels, emery and polishing papers are manufactured by powder metallurgy techniques.

(vi) Tungsten powder is used for making filament wire for electric lamps.

(vii) *For high-temperature service* Nozzles for rockets, missiles and parts used in plasma applications are made from highly refractory metals (mixture of tungsten and silver, etc.).

(viii) In making armaments, atomic energy applications, calculators, adding machine, automatic timmers and bimetallic strips used in temperature-control devices.

(ix) Products of complicated shapes like gears and rotors that require a lot of machining in the conventional methods of manufacture can be easily obtained in this method. Iron and graphite powders are commonly used for pump rotors and gears.

17.3 ADVANTAGES OF POWDER METALLURGY

Following are the advantages of the powder metallurgy technique:

(i) Machining operations are almost eliminated and, whenever machining is required, it is very little.

(ii) Close dimensional accuracy can be maintained, thus articles of any intricate shape can be manufactured

(iii) Non-metallic substances like, graphite can be mixed with metals in any proportion to obtain the desired properties.

(iv) Products can be obtained from alloys of poor casting qualities.

(v) Alloys not readily machinable in their finished form and yet required to be produced to their finished shape are more cheaply produced by this method.

(vi) Although the initial cost of metal powder is high, there is no wastage of material throughout the entire process of manufacturing.

(vii) Highly skilled labour is not required.

(viii) A high rate of production may be achieved by cutting down production time.
(ix) It is possible to ensure uniformity of composition and structure by the controlled conditions of working; thus products of high purity are obtained without the risk of contamination.
(x) Parts from highly refractory metals can be produced easily.
(xi) Perhaps the greatest advantage possesed is the ease with which a wide range of properties like porosity, density, purity, particle size, etc., can be obtained.
(xii) Free from limitations imposed by phase diagrams. For example, powders of copper and lead are successfully shaped by powder metallurgy, whereas it is difficult to produce an alloy of these metals which are mutually insoluble in the liquid state.
(xiii) The application of diamonds in processing industries has become possible by this method.
(xiv) In the production of magnetic cores of special properties.

17.4 DISADVANTAGES AND LIMITATIONS

(i) The size of the product is bound to be limited both by handling strength considerations during manufacture and by economic limitations upon equipment size, e.g., high cost of metal powders and dies.
(ii) The size of the product is also limited as it depends upon the capacity of the press and compression ratio of the powders.
(iii) Due to low flowability of metal powders, complicated shapes are difficult to attain.
(iv) Powder metallurgy products do not have as good physical properties (strength) as cast or wrought parts.
(v) This process is not economical for small-scale production.
(vi) Die design limitations limit the shape of the product.
(vii) Metal powders require special storing and handling facilities, as there is a risk of fires, explosions and health hazards with toxic powders.
(viii) There is difficulty in obtaining particular alloy powders—of steels, bronzes, brasses, etc.

17.5 DESIGN CONSIDERATIONS FOR POWDER METALLURGY

In designing parts for powder metallurgy, the following points should be considered:

(i) Abrupt changes in section thickness should be avoidep as far as possible.
(ii) Deep and narrow sections should be avoided.

(iii) Holes should not be made in the direction of pressure.
(iv) Very close dimensional tolerances should be avoided in the direction of pressing.
(v) Small holes, knurlings, undercuts and threads should be produced by machining.

17.6 PROCESS DESCRIPTION

The process of powder metallurgy can be divided roughly into four stages:
(i) Manufacture or preparation of the powder.
(ii) Blending or mixing of powders in desired proportions.
(iii) Compacting the powders into the desired shapes in suitable dies at specified pressures.
(iv) Sintering the compacted components in furnaces provided with controlled atmospheres.

17.6.1 Powder Manufacture

Many methods are available for the production of metal powders. Unfortunately, most are expensive and this is one, possibly the main reason why powder metallurgy is not more widely used. There are four methods of producing powders (Fig. 17.1):

(a) and (b) *Mechanical pulversisation*
(c) *Atomisation*
(d) Chemical reduction process.
(e) Electrolytic process

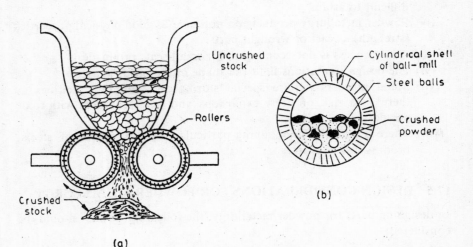

Mechanical pulverising methods
Fig. 17.1 (Contd.)

(Contd.)

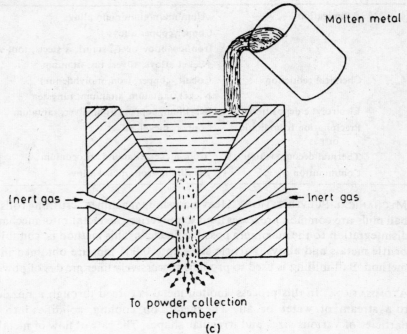

(c)
Atomisation technique

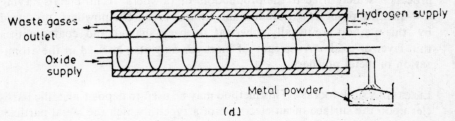

(d)
Reduction process

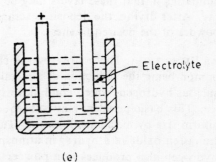

(e)
Electrolytic process
Fig. 17.1 Power forming methods

TABLE 17.1 *Processes Used for Manufacturing Metal Powders*

Atomization	Aluminium/aluminium alloys
	Copper/copper alloys
	Iron/low alloy steels, stainless steels, tool steels
	Nickel alloys, silver, tin, titanium
Chemical reduction	Cobalt, copper, iron, molybdenum
	Nickel, tantalum, titanium, tungsten
Electrolytic deposition	Beryllium, copper, iron, silver, tantalum
Precipitation from liquid or gas	Copper, nickel, silver
Thermal decomposition	Iron, nickel, titanium, ziroconium
Comminution	Beryllium, iron, nickel alloys

MECHANICAL PULVERISATION   Heavy crushing machines, crushing rolls and ball mills are common processes used to pulverise the metal into mechanical disintegration to the maximum possible fineness. This method is suitable for brittle metals and alloys. Particle sizes upto 0.001 mm are obtained in this method. Ball-milling is used to produce successively finer grades of powders.

ATOMISATION   In this process, molten metal is forced through a nozzle into a stream of water or air. The metal, on cooling, solidifies into tiny particles of various sizes and irregular shapes. The rate of flow of metal, air pressure, and size of nozzle are the controlling factors for particle size. This process is widely used in the production of powders from metals having low melting points, e.g., aluminium, tin and zinc. The powders produced by this method are usually spherical or pear shaped and coated with a thin layer of oxide. A stream of inert gas may also be used in the atomisation of metal powders.

ELECTROLYTIC PROCESS   This method may be used to deposit metallic particles upon the surface of an electrode of a type to which the metal particles will not bond. It is used to produce stainless powders from solid metal. The metal is first treated to cause precipitation of chromium carbide at the grain boundaries so that these layers may be subsequently electrolysed out of the way. After drying, the deposit is scraped off and pulverised to produce a powder of the desired fineness.

CHEMICAL REDUCTION   There are several chemical methods, one of the most common being the reduction of a metal oxide in an appropriate atmosphere such as hydrogen, which combines with the oxygen and leaves the metal free. This basic powder may subsequently be finished to a finer and more uniform size by mechanical means. Tungsten powder is obtained by heating tungsten oxide in a hydrogen atmosphere. Reduction of iron chloride in hydrogen also produces iron powders. Copper powder is also produced by this method.

17.6.2 Characteristics of Metal Powders

The properties of powder-metallurgy products are mainly dependent upon the characteristics of metal powders. Important characteristics are given below:

- (i) Particle size and distribution
- (ii) Shape of particle
- (iii) Apparent density
- (iv) Purity
- (v) Sintering ability
- (vi) Flowability
- (vii) Compressibility
- (viii) Chemical composition

17.6.3 Blending or Mixing

Mixing of powders preparatory to compacting is a very important aspect of the main process. If the material is all of one type, then the influence of mixing is not great, but if there is more than one kind of metal powder, mixing must be done very carefully to ensure that the powders are uniformly distributed. For this purpose, electrically operated double-cone apex mixers are used. Correct proportioning of the powders is extremely important, since the ratios of the quantities will govern the nature of the final product as regards both mechanical efficiency and physical properties.

The blending or mixing may be dry or wet. In wet mixing, water or some organic solvent is used to ensure better mixing. In order to reduce friction and ease ejection of pressed parts, lubricants such as graphite, stearic acid and lithium stearate are added during blending.

17.6.4 Compacting

The next stage in powder metallurgical manufacture is the compacting or pressing of the powders into their semi-finished form, preparatory to sintering. It is also called briquetting of loose powder into a green compact of accurate size and shape. This process is carried out at room temperature in steel dies and punches. High pressures are required. The degree of pressure required will depend on (a) the required density of the final product and (b) the ease with which the particles will weld together in the green form. Compacting pressure may be applied in several ways, e.g., die pressing, roll pressing and extrusion.

DIE-PRESSING This is done in special presses which include a feed hopper for the powder, some form of shaping die to form the product, and a ram to apply the correct pressure in the right direction. Die pressing is essentially a limited process in relation to the size and continuity of product (Fig. 17.2 (a)).

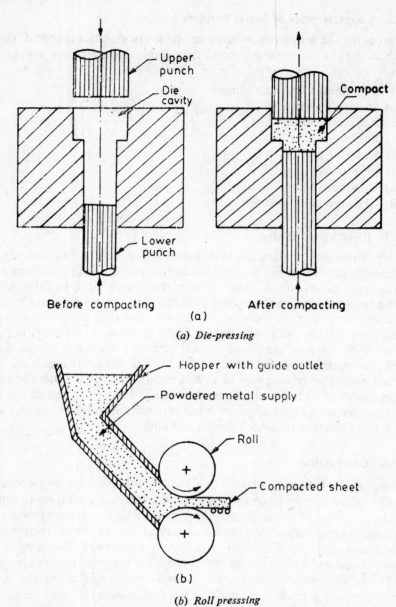

(a) Die-pressing

(b) Roll presssing

Fig. 17.2 *Compacting the metal powders*

ROLL-PRESSING This is often used for production of continuous strip sections, using the system shown in Fig. 17.2 (b). The purpose is to replace the die and ram with two rolls of the appropriate size, into which a regulated stream of powder is guided. The rolls apply the necessary compacting pressure in a continuous sequence.

EXTRUSION METHOD This does not give such efficient control as given by die or roll pressing. It is difficult to obtain high densities and some porosity is always left.

17.6.5 Sintering

The process of heating the green compact to a temperature below the sintering temperature is called *pre-sintering*. This is necessary to remove the lubricants and binders added during blending and to increase the strength of the green compact. Pre-sintering is required for a few metal powders. This process is carried out in the case of tungsten carbide powder to ease machining of the part which otherwise becomes very hard after sintering. Afrer this process, the compacted part accquires sufficient strength to be handled and machined without difficulty. Pre-sintering is carried out on hand-to-machine parts in which holes are to be drilled. Pre-sintering can be altogether eliminated if no machining of the final product is required.

Sintering is the process of heating the compacted component at an elevated temperature, preferably below the melting point of the major components, in a furnace with controlled atmospheric conditions. Sintering helps in bonding the individual powder particles, either by melting of minor constituents, diffusion or by metallurgical reactions between the constituents to achieve some particular type of structure. The duration of the sintering treatment is an important factor, particularly in materials which sinter readily. Atmospheric control is necessary to avoid oxidation and other undesirable reactions.

Sintering may be considered as the bonding of adjacent surfaces in a mass of powder or, compacting by heating. It results in numerous structural changes within a powder mass and significant improvement of mechanical properties. Densification or removal of porosity is often associated with sintering.

Conventional sintering involves heating of the green compact to some temperature below its melting point, or solidus, in a protective or reducing atmosphere at normal pressure.

In general, the following changes occur in the powder compact when it is subjected to sintering.

(i) Particles begin to bond together, increasing the strength of the compact as well as its electrical and thermal conductivity.
(ii) As sintering time is extended (or sintering temperature raised), the strength continues to increase.
(iii) The increase in strength is often accompanied by a decrease in porosity (increase in compact density).
(iv) Grain growth may occur and the final grain size may be much larger than the original powder particle size.
(v) Pores become smoother and more spherical in shape as sintering progresses.

(vi) In a suitable sintering atmosphere, absorbed gases are removed from the compact and particle surface oxides are reduced.

SINTERING MECHANISM The various structural changes that take place during sintering depend on mass transport to and from various regions within the powder compact. A net decrease in free energy must be associated with sintering if these structural changes are to occur spontaneously. Probably the most dominant mass transport machanism involved in normal sintering is solid state diffusion through the crystalline lattice and via short-circuit paths such as particle surfaces and grain boundaries. The rate at which diffusion occurs in a given structure depends primarily on temperature. Therefore, sintering rates are strongly influenced by the temperature at which the process is carried out.

Sintering is growing in importance because of the growing use of powder metallurgy. It is a method by which powdered particles are bonded in a compact unit of powdered metal. It is, in fact, a form of solid-phase welding. Its purpose is to cause a compact of the powdered particles to change into a coherent unit by solid-phase welding. Figures 17.3 (a) and (b) show the mechanism and effect of sintering.

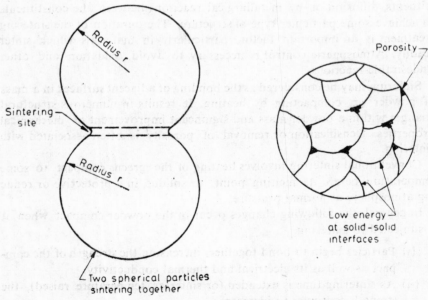

Fig. 17.3 *Effect of sintering*

After careful preparation and mixing of metal powders the required shape is produced by compacting and then heating to a sufficient intensity to give conditions suitable for solid-phase welding. Heating for sintering is invariably done in a controlled atmosphere to asssit the diffusion and bonding of powder particles and to prevent oxidation. Most sintering furnaces operate in a hydrogen atmosphere. Many sintering furnaces are of the electric-resistance type because they are most readily adaptable to heating in a controlled atmosphere.

Sintering furnaces (Fig. 17.4) are not usually of large size. The product enters at one end and is fed through continuously or in a series of steps until it passes out at the other end into a cooling zone or to the next part of the process.

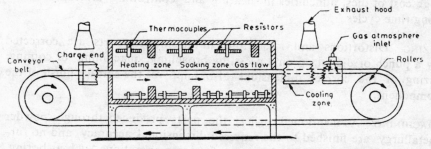

Fig. 17.4 *Sintering furnace (continuous type) used in powder metallurgy*

For many powder metallurgy products the sintering process can take place in one simple stage, but there are materials which require complicated sintering treatment in as many as five or six stages, according to the desired metallurgical results.

Sintering is done,

(i) to achieve strength and bonding
(ii) to produce a particular type of structure
(iii) to cause a metallurgical reaction between the constituents to achieve alloying
(iv) to produce a dense structure.

The duration of the sintering treatment is also an important factor. Prolonged treatment time helps tungsten carbide crystals to merge together in one particular area, giving an excessively enlarged grain.

A controlled atmosphere has a marked influence upon the mechanical properties which are finally achieved.

Sintering is done to achieve strength and hardness in the finished product. The sintering time varies from one hour to several hours. Sintering drives off volatile materials from the compacted part, leaving a porous object. Thus this powder matallurgy technique is suitable for the production of self-lubricating bearings, filters, etc.

17.7 SECONDARY OPERATIONS

Many products can be used in the 'as-sintered state' but where better surface finish and close tolerances are desired, further processing is required.

The various operations subsequent to sintering are called secondary operations. The common ones are discussed below.

HOT PRESSING   Hot pressing is a combination of pressing and sintering. For improved strength, hardness, greater accuracy and higher densities of parts obtained by the usual methods—hot pressing is used, for example, for the fabrication of cemented carbide parts. Moulding and sintering takes place at the same time in the die. The use of the process is limited by the high cost of dies, difficulties in heating and controlling atmosphere and long time cycles.

SIZING   Distortions and alterations in the sintered parts are corrected by a punch or die set up on a press, by broaching, machining or coining. Sizing of the correct dimensions also improves the surface finish of the component.

MACHINING   As mentioned earlier, the parts produced through powder metallurgy are finished to the required dimensional accuracy and no further machining is required. However, for some operations such as boring, turning, drilling, tapping, etc., on the fabricated parts to produce small holes, threads and undercuts which cannot be produced through the powder metallurgy technique, machining has to be done.

COINING   It is similar to sizing but the sintered parts are repressed in the die to reduce blow holes and give additional density and strength to the component.

IMPREGNATION   When self-lubricating properties are desired, the sintered parts may be impregnated with oil, wax grease or other lubricants. By capillary action the lubricant is retained in the part and comes to the surface by heat of friction and by external pressure. In order to improve machinability, corrosion resistance and sealing prior to plating, sintered parts may be impregnated with plastics or other polymers.

PLATING   Plating is done for protection from corrosion and for a pleasing appearance. The procedure for plating powdered-metal parts is different from those used for metal parts. The plating operation consists of first impregnating the part with a plastic resin to eliminate voids before the regular plating operation is adopted. Plating also improves wear-resistance and electrical conductivity.

INFILTRATION   Infiltration is required to give the increased strength, hardness and density not obtainable by straight sintering. The process consists of introducing the molten metal to close the pores in the sintered parts by capillary action. Low melting point metals are used for infiltration. The operation should be done as quickly as possible in the furnace.

HEAT TREATMENT Sintered parts are heat-treated to improve strength, hardness, wear resistance and grain structure.

17.8 POWDER METALLURGY AND FABRICATION

Although the field of powder metallurgy is so widespread in application and varied in its nature, it is possible to classify the types of structure, in a general sort of way:

(i) Porous metals
(ii) Alloys with excessively high melting temperatures
(iii) Alloys with poor casting qualities
(iv) Alloys containing insoluble constituents required in a fine dispersion
(v) Alloys not readily machinable in their finished form and yet required to be more dimensionally accurate than they would be in a casting or, alternatively, alloys which can be produced in their finished shape more cheaply by powder metallurgy.

A few powder metallurgy fabrications are discussed below.

17.8.1 Porous Parts

Porosity is generally an undesirable characteristic of parts for structural applications. This same characteristic, however, when properly controlled, is the key to sintered porous media. Porous parts are used to retain liquid as in bearings, filtering, metering of liquids or gas and sound attenuation in telephones, microphones and hearing aids.

Porous metal sheets are rolled from metal powders like copper, brass, bronze and stainless steel. Two processes are used to make porous metal sheets. One is gravity sintering and the other is rolling. In gravity sintering, a uniform layer of powder is placed on ceramic trays. Sintering of this layer is done for 45 hours at high temperatures in an atmosphere of dissociated ammonia. To obtain a better surface finish and uniformity, sheets are rolled and fabricated into suitable shapes. In another process, the metal powders are fed through a hopper between two rollers which compress the powders into a metallic sheet. It is further conveyed through the sintering furnace and rolled once again to the final size. Rolling helps in producing sufficient strength by compression and interlocking. Rolling also results in uniform mechanical properties and controlled porosity.

17.8.2 Self-Lubricating Porous Bronze Bearings

Self-lubricating or permanently lubricated porous powdered-metallurgy bearings and bushings are the two best examples to illustrate the uniqueness of the powder metallurgy process. Sacrifices in some mechanical properties result from the use of a less than fully dense part, but the fact that the pores can be used as oil reservoirs more than compensates for this loss. During use, as the bearing gets hot, the oil expands and flows to the bearing surface. On cooling it is drawn back into the pores by capillary action. These bearings are made by pressing elemental powder blends of copper-tin, followed by sintering for short periods of time (15 min. is typical). These bearing can run quietly and may be used in vertical positions for which solid bearings would normally be impractical because of lubricant run-out. They are useful if it is difficult to lubricate the part, such as in refrigerator motors or where oil splashing may interfere with the operation of the machine. These bearings are widely used in the automotive industry, in home appliances and in the leisure industry, e.g., starters, dishwashers, washing machines, tape recorders, business machines, electric fans and textile machines.

17.8.3 Non-Porous Bearings

Main bearings and connecting rod bearings for automobiles are produced by powder metallurgy. These are non-porous fluid-lubricated bearings made by this technique and are based on copper-lead alloys, copper-tin-lead alloys and lead or tin based white or babbitt metals. They offer some technical advantages over bearings made by casting. Depending on the specific composition and structure, these bearings are recommended for various combinations of load, velocity, fatigue and corrosion resistance. They are used in passenger-car engines, compressors, gear boxes, etc.

Other non-porous, fully dense bearings, such as the so called dry-metal graphite bearings, are used under poor lubrication conditions at intermediate temperatures of 250-350°C. These materials are based on brass and bronze (also on iron and nickel), and contain about 10-50% graphite. They are made mainly by hot pressing and can operate under adverse conditions of dirt and dust as in hot-water pumps, coal-cutting mechinery and the textile industry.

17.8.4 Powdered-Metal Friction Material

Powdered-metal friction materials are composed of a sintered mixture of powdered metals and known-metals developed to achieve the desirable friction characteristics. The most commonly used metal powders are copper, iron, lead and tin. The most commonly used non-metal powders are graphite and silica. Sometimes metallic filters are also used in combination with asbestos and rubber. Metal-based friction materials are stronger and more heat resistant and were developed in response to energy inputs and temperatures which exceeded the endurance of organic-based friction materials. Generally, powdered-metal friction materials are classified as either copper or iron based mixes depending on which of the two is the predominant constituent.

As sintered mixtures of all powdered metal friction materials are weak in tension and shear, they are supported by core or backing plates which are bonded directly to the sintered powdered metal and which take most of the mechanical stress required of the friction elements.

The core of backing member is usually of low-carbon steel but can be of other materials such as copper, bronze or cast iron.

Dry powders of the constituent elements are blended in the correct proportions, thoroughly mixed and moulded into compacts. All moulding of metal powders for friction purposes is done on flat sections only. This is required in order to achieve a constant density which has a definite effect on the frictional characteristics of the sintered material. Large production quantities are moulded in automatic molding machines, whereas small product volumes are hand moulded. Sintering in the present case refers to heat treating powdered-metal compacts to convert them into a solid, coalesced metallic mass. Bonding refers specifically to welding or brazing the powdered metal compacts to the steel cores.

The greatest single property which justifies the use of powdered-metal friction materials economically is their great wear-resistance. Other outstanding properties are:

(i) They absorb energy at higher rates and can be used at higher temperatures, greater pressures and speeds, such as in brakes and clutches.
(ii) They absorb a greater portion of the heat of the friction application than friction materials of other types.
(iii) They are not appreciably affected by climatic or other conditions such as heat, cold, dampness and fungi. This property is important in many applications because it means more consistant results through many adverse conditions.
(iv) Because of the greater wear resistance and physical strength of the bimetallic powdered-metal friction material, it is often possible to design a smaller, lighter and more compact friction assembly. The steel backing member also can be used as a structural part of the brake of clutch mechanism, saving both weight and assembly costs.

Powdered-metal friction materials are mainly used for wet and dry clutches and brakes. Figure 17.5 shows applications of powdered-metal friction material. When bonded to only one side of the backing member, it is known as single face bonding, as used in brake linings, cylindrical block linings and single-face clutch facings. The most popular powdered-metal friction materials is the double-face or sandwich facings type. The powdered-metal material is bonded to both sides of the steel core, forming a sandwich, as in splined disks. The most popular spline for this type of friction element is the standard, involute gear tooth.

17.8.5 Copper-Infiltrated Parts

In copper infiltration, the pores of a sintered part are filled with liquid copper. Because of a poor surface appearance and lack of dimensional

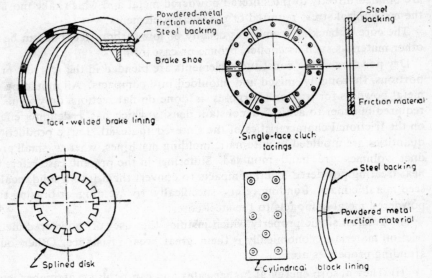

Fig. 17.5 *Powdered-metal friction materials*

control, these attempts to obtain full density in a porous powdered-metal steel part through copper infiltration did not lead to commercial success. In 1946, these problems were solved through the addition of graphite to P/M steel part and addition of iron to the copper infiltrant, which prevented erosion of the farmer.

The major developments in this field since World War II may be summarized as follows:

(i) New copper infiltrants offering improved performance have been developed.
(ii) Larger parts are being infiltrated.
(iii) The average matrix density of the steel part is higher because of the availability of inexpensive high-compressibility iron powders.
(iv) Water-atomized hardenable low-alloy steel powders may be infiltrated.

The one-step infiltration process is widely used, where sintering and infiltration of the base part are combined into one processing step. The infiltrant slug may rest either on top of the base part (top-infiltration) or beneath it (bottom-infiltration). Typical infiltrating conditions are 30 minutes at 1120°C in endo gas with few points anywhere from 20 to 50°C. Dissociated ammonia or nitrogen-based atmospheres are also used. There are so-called residue-free and residue-type infiltrants; the latter leave a small non-adhering residue which can be disposed of easily.

Copper-infiltrated parts are widely used in load-bearing applications in many types of machines and engines. P/M gears are particularly well suited to copper infiltration because infiltration adds strength to the teeth of the gear which often suffer from lower density.

The single most important reason for infiltrating a part with copper is to enhance its mechanical porperties.

Producers of parts usually provide complete sets of data on the properties of infiltrated parts, including dimensional tolerances.

Successful infiltration requires that the surface free energy of the system be smaller after infiltration than before.

17.8.6 Electrical Contact Materials

Flaked, ball-milled and other forms of copper may be used in combination with graphited materials to form carbon brushes used extensively in electrical motor units. The high conductivity of copper is combined with the light-weight and lubricating properties of graphite. The sintered product of such a combination gives efficient, long-lived service in the transfer of electric current between moving and stationary parts of electrical equipment.

Other composite contact materials with good current-carrying capacity and thermal conductivity are silver or copper, with tungsten and molybdenum which have good refractory and hardness properties. These sintered materials are used in relays, circuit breakers and welding electrode.

17.8.7 Diamond Impregnated Tools

The powder metallurgy technique is also used to make diamond tool bits by blending iron powder and 30% diamond dust. The mixture is pressed into compacts and sintered at 1000°C. These tools are used for cutting hard materials like glass, porcelain and tungsten carbide.

17.8.8 Magnetic Materials

Alnicos, the alloys of aluminium, nickel, cobalt and iron, used as magnetic materials, are produced by the powder metallurgy technique. These alloys are difficult to cast and not easily machinable. Ferrites are also produced by powder metallurgy because they possess good magnetic characteristics.

17.8.9 Tungsten Wires

Wires for filaments in the lamp industry are made from pure tungsten powder. Tungsten powder is produced by reduction of tungsten oxide with hydrogen. The powder is compacted, pre-sintered and re-sintered in a number of times to get a ductile tungsten bar. This bar is hot swagged to a diameter of about 2 mm or less and is then drawn through diamond or tungsten dies to about 0.2 mm diameter. Further finishing of the drawn bar is done at 1000°C, in different dies, till it is finished to the desired diameter. Tungsten wire has more than twice the strength of the hardest steel. It is commonly used as a filament in electric lamps.

17.8.10 Cemented Carbides

Hard, abrasive materials such as tungsten carbide tend to be excessively brittle and therefore, relatively useless for machining operations or other applications where any shock loading is likely. To overcome this difficulty, a range of hard metals, known as cemented carbides, have been developed. These are alloys with finely dispersed, insoluble constituents embedded in a continuous matrix.

The method of fabrication adopted for cemented carbides is to incorporate in the matrix some relatively soft material, in sufficient amounts to form a ductile matrix in which the hard particles can be located. This matrix has to be able to withstand relatively high temperatures and so, for cutting metal applications, cobalt is used as the bonding material. The abrasive particles embedded in the cobalt may be composed entirely of tungsten carbide or may be mixtures of different carbides with the necessary characteristics.

The carbides most often used, in addition to tungsten carbide, are titanium carbide and, possibly, tantalum carbide. Of these carbides, titanium carbide is harder than tungsten carbide while tantalum carbide is rather softer. These carbides may be incorporated in powder metallurgical products, in different particle sizes, to give the required characteristics to the finished metal.

Powders of tungsten carbide (and other carbides if added) and cobalt are properly mixed and then compacted in alloy steel dies at pressures of about 48 kg per square mm. The sintering is done in two stages. In the pre-sintering, cobalt is fused at about 900°C, whereas carbides of tungsten and of other metals remain intact. The pre-sintered products are machined to give the exact shape and size required for the final product. The final sintering is carried out at a temperature of 1300°C, for two hours. The products are allowed to cool gradually in the sintering furnace. Cemented carbides may be used in cutting tools and wire-drawing dies.

17.8.11 Metal-Bonded Ceramics

Other applications of powder metallurgy are in the fabrication of metal-bonded ceramics or cermets. These are made from certain metals with high melting points and are incorporated in a metal matrix. This class of material is most commonly prepared for turbine blades in gas turbines where the temperature of operation is in the range of 900-1300° C. To make cermets, materials like tungsten carbide, magnesium oxide, graphite, chromium oxide, platinum, tantalum and silica may be used. Sometimes cermets are also known as refractory material composites.

REVIEW QUESTIONS

1. What is the powder metallurgy technique? Describe briefly the methods of powder forming suitable for this technique.

2. What are the applications, advantages, disadvantages and limitations of the powder metallurgy technique?
3. What are the main characteristics of metal powders?
4. Describe briefly the process of powder metallurgy.
5. Outline the advantages of secondary processes in powder metallurgy processes.
6. Explain the objectives of pre-sintering and sintering.
7. List the important products of powder metallurgy and explain the objectives of powder blending.
8. (a) Briefly explain the methods of manufacturing metal powders.
   (b) Explain the advantages and disadvantages of using powder metallurgy.
   (AMIE Summer 1982)
9. Write short notes on the following:
   (a) Sintering   (Summer 1983)
   (b) Self-lubricating bearings   (Winter 1982)
10. Describe the method of forming a powdered-metal compact by compaction, giving sketches.   (AMIE Winter 1980)

18
ELECTRON THEORY OF METALS

18.1 INTRODUCTION

Many fundamental properties of solids, like electrical and thermal conductivities, magnetic and optical properties, depend upon their electronic structure. The development of the electron theory of solids, started in the earlier part of the 20th century. At present it is the basis for the classification of all solids. When applied to metals, it explains forces of cohesion and repulsion, binding the energy levels and the behaviour of conductors and insulators and magnetic materials.

The theory was developed in three stages from a band model of free or valence electrons.

(i) Drude-Lorentz free electron theory
(ii) Sommerfield free electron theory
(iii) Zone theory

18.2 METALLIC BONDING

Metallic bonding has already been discussed in Sec. 4.6. The valence electrons in a metal are free to move in different directions in the metal. These electrons are called free electrons and move inside the metal. The binding forces in the metals are due to the electrostatic attraction between the positive ions and negative cloud or gas of electrons. Such a free electron model of metals was actually proposed by Drude and developed by Lorentz.

The modern picture of metallic bonding assumes that the metallic bond is more closely related to the covalent or electron pair bond and resembles the ionic bond. Essentially, the metallic bond is an unsaturated covalent bond allowing a large number of atoms to be held together by mutual sharing of free electrons. Also, in metallic bonding the density of electrons between the atoms is much lower than allowed by the Pauli exclusion principle. This allows the electrons to move freely from point to point without significant increase in energy.

18.2.1 Cohesive and Repulsive Forces

The bonds between atoms in solids are made of cohesive and repulsive forces which hold the atoms at definite distances from each other. It is desirable to understand these forces upon which the electron structure of atoms depends. Primarily, because the close proximity of two atoms places too many electrons into interacting locations, mutual repulsion results. An equilibrium position is reached at which cohesive and repulsive forces are equal. Large cohesive forces produce high melting points and fairly large elastic values of materials and higher mechanical strength. For further reading refer Sec. 4.3.

18.3 DRUDE AND LORENTZ THEORY

The free electron theory was first proposed by Drude and further developed by Lorentz. It was assumed that the metal crystal consists of positive metal ions whose valence electrons are free to move throughout the volume of the metal, forming an electron gas.

The movement of electrons obeys the laws of the classical kinetic theory of gases. The Drude and Lorentz theory is also known as the *classical free electron theory*.

The basis of the classical free electron theory is that when a voltage is applied to the metal, the negatively charged free electrons are attracted towards the positive field direction and produce an electronic current within the metal. In this theory it is assumed that

(i) the mutual repulsion between valence electrons is ignored,
(ii) a completely uniform potential field exists due to its ions.

The main deficiencies in the classical electron theory are:

(i) Over-estimation of specific heat by considering the total energy of electrons.
(ii) Electrons in the electron gas do not absorb heat like a gas obeying classical gas laws.

18.4 SOMMERFIELD FREE ELECTRON THEORY

This theory is a modification of the Drude and Lorentz theory, including the concept of quantum mechanics which takes into account the fact that a moving electron behaves as if it is a system of waves. The following are the basic assumptions of this theory:

(i) The valence electrons in a metal are free.
(ii) Valence electrons are confined to move within the boundaries of a crystal. It shows that electrons have a lower potential energy inside the crystal than outside.

(iii) The potential energy of the electron is uniform (constant) within the crystal, as in the Drude theory.
(iv) The electronic specific heat of metals is very low.

Sommerfield, in his theory, supposed that the potential of an electron in a metal is uniform. He applied the Schrödinger equation to calculate the total energy E. Schroedinger suggested that the quantum states of electrons in an atom might obey an equation similar to that of a vibrating string clamped at both ends.

Considering the quantum free electron theory in one dimension, the classical relation for the kinetic energy is

$$E = \tfrac{1}{2} mv^2 \tag{1}$$

Since the potential energy of the electron within the crystal is assumed to be constant, it may be zero and the total energy is then equal to the kinetic energy. Now applying the De-Broglie relation

$$\lambda = \frac{h}{mv}$$

where λ is the wavelength that describes the motion of the electron
mv is the momentum of the electron
h is Planck's constant

and $$K = \frac{2\pi}{\lambda} \tag{2}$$

where K = wave number and is extensively used in electron theories.
Combining the above equations,

$$E = \frac{K^2 h^2}{8m \pi^2} \tag{3}$$

i.e., the relation between energy and wave number obtained is parabolic, as shown in Fig. 18.1 (a).

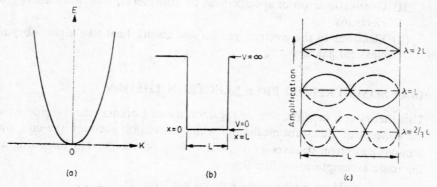

(a) Parabolic relationship between energy and wave number
(b) Potential well (one-dimensional model)
(c) Standing waves in a string with fixed ends

Fig. 18.1 Free electron model

One very important model of the free electron theory in the present case, particularly in a metal, is that the electrons inside it are confined by a potential well of the same volume as the solid as shown in Fig. 18.1(b). Here it is assumed that an electron is confined within it, and its potential energy is zero within the well and infinite outside the well. The quantum states of a single electron in the potential well are found by solving the Schroedinger equation with appropriate boundary conditions,

$$\nabla^2 \psi + \frac{8\pi^2 m}{h^2}(E - V)\psi = 0 \qquad (4)$$

or
$$\nabla^2 \psi + K^2 \psi = 0 \qquad (5)$$

where ψ is the wave function.

The boundary conditions are $\psi = 0$ at $x = 0$ and $x = L$ (and similar conditions for the y and z directions). The solution of the above equation gives the wave function,

$$\psi = A \sin K x \qquad (6)$$

which is a sine wave with amplitude A. Three wave functions that satisfy the condition that controls the motion of electrons in one dimension, and the vibration of waves travelling in a string of length L (the ends of which are fixed, are shown in Fig. 18.1 (c).

$$\therefore \quad \lambda = \frac{2L}{n}$$

$$n = \pm 1, \pm 2, \pm 3, \ldots \qquad (7)$$

$$K = \frac{n\pi}{L} \qquad (8)$$

Putting the value of K in the energy equation,

$$E = \frac{h^2 n^2}{8\, mL^2} \qquad (9)$$

This is the condition for the one-dimensional problem. The corresponding 3-d case would be one in which an electron moves in all directions so that quantum numbers n_x, n_y and n_z are required corresponding to the three coordinate axes. In terms of a cubically shaped block of metal of side L, the allowed energies are given by

$$E = \frac{h^2}{8\, mL^2}(n_x^2 + n_y^2 + n_z^2)$$

$$= \frac{h^2}{8\, mv^{2/3}}(n_x^2 + n_y^2 + n_z^2) \qquad (10)$$

where $v = L^3$ is the volume of the assumed metal cube.

18.5 ELECTRON ENERGIES IN A METAL

Equation 10 gives the total values of energies of the valence electrons in a metal. This also shows that among E_1, E_2, E_3, etc., each energy value is greater than the preceding value by the same amount E_1. For each different set of (n_x, n_y, n_z), there is an energy state. The number of states that have a given energy increases rapidly with increase in the number of different values of n_x, n_y, n_z. As any change in these values influences the change in E by the square of their values (n_x^2, n_y^2, n_z^2), so the number of states in the metal is vary large. If a plot is made for the number of states per interval of energy $N(E)$ or density of energy states V_s the total energy E, $N(E)$ increases parabolically with increasing E as shown in Fig. 18.2 (a).

The valence electrons have a tendency to occupy the lowest available energy states. It is, however, essential to consider all the electrons in the single system because of the mutual interactions among all the electrons forming the electron gas. This is possible only when the Pauli exclusion principle is applied. According to this, only two electrons can occupy a given state, specified by the three quantum numbers (n_x, n_y, n_z), one with spin up and the other with spin down or opposite spin. By this principle, at 0 °K the electrons fill all the states up to a certain maximum energy level (E_{max}) called the *Fermi level* and all quantum states in the energy levels above this level are empty (Fig. 18.2 (b)). The value of E_{max} depends on how many free electrons there are. The level at which the probability of occupation is

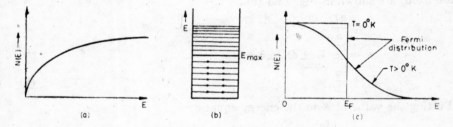

Fig. 18.2 *Electron energy levels*
(a) *Density of energy levels vs energy*
(b) *Filling of energy levels by electrons*
(c) *Probability of occupancy by an electron at various temperatures*

50% is the Fermi level E_F. If an electron is to be removed from the Fermi level and taken out of the metal, some energy is required. This is called the *work function* (ϕ), which is equal to the energy normally measured when an electron is removed from the surface of the metal. At 0 °K an electron in the Fermi level possesses the highest energy of the electrons in the metal and is the easiest to remove.

Now we shall consider the effect of temperature on the electron energy levels. At temperatures above 0 °K, the electrons near the top of the distribution are excited and move into higher quantum states. The probability that a state at a level of energy E is occupied by an electron at $T = 0$ °K is constant and equal to unity up to the Fermi level E_F and zero above it. When the temperature is raised, excited electrons move into the new levels, as shown in Fig. 18.2 (c). The change in electron energy distribution can be described by the following expression:

$$\int N(E) = \frac{1}{[1 + \exp(E - E_F)/kT]}$$

where $\int N(E)$ represents the probability that a state of energy E is occupied at temperature T. Thus at $T = 0\,°K$ when $E < E_F$, the exponential term becomes zero and

$$\int N(E) = 1$$

when $E > E_F$, the exponential term becomes infinite and

$$\int N(E) = 0$$

By putting

$$\int N(E) = \tfrac{1}{2}$$

it can readily be seen that $E = E_F$. So at any temperature we can define the Fermi level as that level for which the occupation probability is 1/2. The function $\int N(E)$ is called the *Fermi distribution function*.

18.6 ZONE THEORY OF SOLIDS

The free-electron theory of Sommerfeld described electrical conductivity, electron emission, as well as certain magnetic properties satisfactorily, but it failed to explain the behaviour of free electrons in semiconductors and insulators. The zone theory explains this with the concept of a periodic potential field by assuming that the electrons move in a region of constant potential. The electrons move in a periodic field provided by the lattice. It was explained in the discussion on the periodic table of elements that the atomic arrangement is periodic in all solids. In other words, within any real metal there is a periodic arrangement of positively charged ions through which the electrons move.

Figure 18.3 (a) is a one-dimensional representation of the actual potential in which a free electron moves in a crystal. The deep attractive wells mark the positions of the positive ions. These ions are regularly spaced at the lattice spacings. A deep potential well is located at each ion due to the coulomb forces when the potential V varies periodically with x.

Considering the wave motion of electrons, when the wavelength of an electron travelling through the crystal is such that it will be diffracted by a particular set of atomic planes (Fig. 18.3 (b)) the Bragg diffraction condition is satisfied:

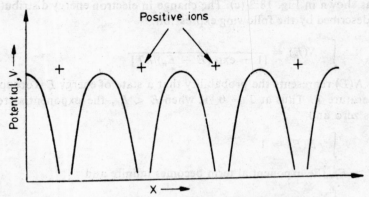

Fig. 18.3 (a) Periodic potential inside a crystal

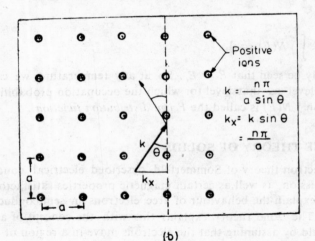

Fig. 18.3 (b) Bragg reflection showing the movement of electrons through the lattice

$$n\lambda = 2d \sin \theta$$

where n is any integral

λ the wavelength of electrons

d the lattice spacing and

θ the angle of incidence of the electrons.

Now, $\dfrac{2\pi}{\lambda} = K$

where K is the wave number

Substituting the value of K in the Bragg equation,

$$K = \frac{n\pi}{d \sin \theta}$$

K is a vector in the same direction in which the electron is moving and its magnitude is proportional to the magnitude of its momentum.

According to the above equation, there is one series of values corresponding to the integral n for which electrons are diffracted and do not enter freely through the crystal.

18.6.1 Brillouin Zone

In the equation

$$K = \frac{n\pi}{d \sin \theta}$$

K represents the wave number. It is a vector in the direction in which the electron is moving and its magnitude is proportional to the magnitude of its momentum. The components of the vector are K_x, K_y and K_z along the x, y and z directions, representing a K-space. When the Bragg condition is satisfied, the moving electron suffers diffraction by the lattice plane. For this to happen

$$K_x = \frac{n\pi}{d}$$

for a one-dimensional lattice where $n = \pm 1, \pm 2$ for the critical value of K.

Consider a simple cubic lattice of two dimensions. Bragg's condition is first satisfied by the two ((100) and (010)) planes of this lattice. As the energy of the electrons increases, a point is reached when Bragg's condition for another set of planes (110) is satisfied. The area in wave-number space enclosed by a line corresponding to K-vectors is known as the *Brillouin zone*. For the first zone, one integer is ± 1 and the other integer is 0. For the second zone, each integer is ± 1.

Figure 18.4 (a) shows the first two Brillouin zones for a square lattice.

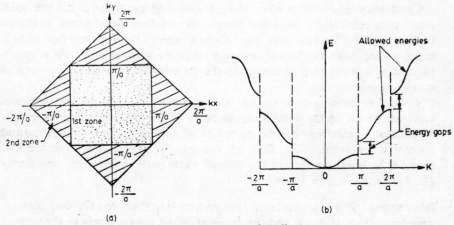

Fig. 18.4 *A two-dimensional Brillouin zone*

The *K*-space is divided into zones. Within each zone, when the electron touches the boundary, it encounters an energy barrier preventing further increase of wave number unless some energy is available to excite the electron over the barrier. The barrier is called the *energy gap*. Figure 18.4 (b) shows energy gaps and Brillouin zones for a simple cubic lattice.

The different Brillouin zones for a given crystal have the same area. Thus, the areas of the first zone and the second zone are equal. This is a very important characteristic of Brillouin zones. With the help of Brillouin zones, solids can be classified easily into conductors and insulators.

18.6.2 Zones in Conductors, Insulators, and Semiconductors

The characteristics of conductors, insulators and semiconductors can now be explained with the help of the zone theory.

CONDUCTORS To determine the electrical conductivity of a solid, it is necessary to know the size of the energy gap between zones and how completely the Brillouin zones are filled. When an external field is applied, there is a flow of electrons in the direction of the field. Electrons so excited are capable of going to the second zone and filling the vacant energy states. In a metal there are always unfilled states at energy levels very slightly higher than the highest energy of the filled states and electrons can be raised by applying the external field.

Consider Fig. 18.5(a). There is an energy gap between the first and second zones as in a monovalent metal, in which theoretically at the most two electrons per atom are present. Since the atoms of the metal have only a single valence electron per atom, only half the first zone could be filled with the available electrons. The second zone is at a comparatively higher energy level. Since half the energy levels are vacant, any small applied electric field can accelerate electrons to move into the second zone or higher energy levels.

Conduction also occurs when there are enough electrons to fill the Brillouin zone completely, provided there is an overlapping of zones as shown in Fig. 18.5 (b). In this case the highest energy level in the first zone is much greater than the lowest energy level in the second zone. As a result, there is an overlapping of energy levels. In metallic crystals like silver and copper, zones overlap.

In the overlapping-zone system, it is assumed that there are sufficient electrons to fill up the first zone. Because of the overlap in energy levels, higher energy electrons can be moved to other levels in the second zone and there will be conduction. Due to the transfer of electrons from the first zone to the lower portion of the second zone, both zones are now partly filled with the energy levels.

INSULATORS For an insulator, the energy gap between the two zones is very large (Fig. 18.5 (c)) and the energies of the lowest levels in the second

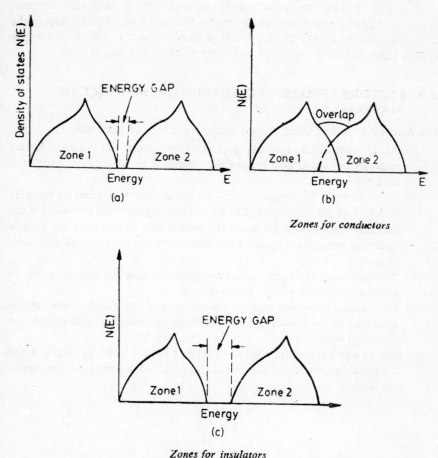

Fig. 18.5 *Density of states as a function of energy*

zone should be much higher than that of the upper levels in the first zone. Again, consider a monovalent metal that has just enough electrons to fill the first zone completely. Each zone is capable of holding a maximum of two electrons per atom. Thus the first zone contains two electrons per atom and as the energy gap between the first and second zone is wide, the charge cannot flow to the second zone in the solid. The electrons in the first zone cannot move to the higher energy levels because all the states are filled there and a normal electric field cannot excite them to rise to the next allowed energy level. Diamond is an excellent insulator having four valence electrons per atom in each zone, just sufficient to completely fill the first zone. The second zone is separated from the first by a wide energy gap.

SEMICONDUCTORS Semiconductors are a group of materials in which the energy gap between the filled and unfilled zones is very small and zones do not overlap. They can be made to conduct electricity if sufficient thermal energy is supplied to the system, to enable the electron to jump across the small gap into the empty zone. At $0°K$, a semiconductor behaves as an insulator since the thermal energy available to cross the gap is zero.

18.7 FACTORS AFFECTING ELECTRICAL RESISTANCE OF MATERIALS

The following factors affect the electrical resistance of materials:
  (i) As discussed earlier, alloying elements or impurities increase lattice imperfections and also reduce the electrical conductivity compared to a pure metal.
 (ii) With increase of temperature, the electrical resistance of metallic conductors also increases. The higher the temperature the wider is the thermal vibration of atoms in the lattice and this results in atomic spacings being less regular, thus decreasing the mobility of electrons in metals.
(iii) The nature of the material, its composition and the condition of its surface, influence electrical resistance.
(iv) Mechanical processes like cold working and increased strain hardening can cause a number of dislocations, thereby increasing the electrical resistance.
 (v) Due to age hardening, electron mobility is restricted. Therefore the electrical resistance of the metal increases. Annealing has an opposite effect.

APPENDICES

A   Mechanical Properties of Important Mevals and Alloys   *379*
B   Physical Properties of Some Elements   *380*
C   Physical Properties of Common Metals   *382*
D   Physical Constants   *382*
E   Selected SI Units   *383*
F   Conversion Factors   *384*
G   Properties of Some Important Semiconducting Materials   *385*
H   Periodic Table   *386*

APPENDIX A

Mechanical Properties of Important Metals and Alloys

Metals	Brinell hardness	Modulus of elasticity kg/cm² × 10⁶	Modulus of rigidity kg/cm² × 10⁶	Ultimate strength kg/cm²		Shear
				Tension	Compression	
Mild steel	120	2	0.8	4000-6000	4300	3500
Wrought iron	100	1.8-2.0	0.75-0.95	2400-4500	2700	2700
Cast iron	150-225	1-1.5	0.6-0.8	700-2000	4000-9500	1100
Copper	47	0.7-1.2	0.42	2400	3000	2000
Brass	70	0.6-1	0.35-0.4	1750	750	1100
Aluminium	21-24	0.8	—	1250	2500	900

APPENDIX B

Physical Properties of Some Elements

Metal	Symbol	Atomic no.	Crystal structure of solid	Melting Point °C	Density g/cm³
Aluminium	Al	13	FCC	660	2.7
Antimony	Sb	51	Rhombic	630	6.67
Arsenic	As	33	Rhombic	814	5.73
Beryllium	Be	4	CPH	1280	1.85
Bismuth	Bi	83	Rhombic	271.3	9.8
Boron	B	5	Orthorhombic	2300	2.3
Cadmium	Cd	48	HCP	321	8.65
Chromium	Cr	24	BCC	1888	7.19
Cobalt	Co	27	α CPH / β FCC	1492	8.7
Copper	Cu	29	FCC	1083	8.9
Titanium	Ti	22	HCP	1820	4.54
Vanadium	V	23	BCC	1735	6.0
Manganese	Mn	25	Cubic Comp	1244	7.4
Iron	Fe	26	BCC	1539	7.87
Nickel	Ni	28	FCC	1453	8.90
Zinc	Zn	30	CPH	419	7.133
Germanium	Ge	32	Diamond cubic	958	5.36
Zirconium	Zr	40	α CPH / β BCC	1750	6.5
Molybdenum	Mo	42	BCC	2620	10.2
Silver	Ag	47	FCC	961	10.49
Tin	Sn	50	BC tetra	232	7.31
Tungsten	W	74	BCC	3380	19.3
Platinum	Pt	78	FCC	1769	21.45
Lead	Pb	82	FCC	327	11.3
Magnesium	Mg	12	CPH	649	1.74
Uranium	U	92	Orthorhombic	1130	18.7

APPENDIX C

Physical Properties of Common Metals

Metal	Symbol	Colour	Melting point °C	Thermal conductivity cgs units	Eelectrical resistivity microhm-cm	Specific heat Cal/g/°C	Co-efficient of linear expansion $\times 10^{-6}$ at 20 °C	Density g/cc
Iron	Fe	Greyish	1539	0.18	9.71	0.109	11.7	7.9
Aluminium	Al	White with bluish tinge	660	0.53	2.69	0.215	23.0	2.7
Copper	Cu	Yellowish red	1083	0.94	1.55	0.092	16.7	8.9
Silver	Ag	White	961	1.00	1.6	0.056	18.9	10.5
Zinc	Zn	Bluish white	419	0.27	5.9	0.092	33.0	7.1
Lead	Pb	Bluish grey	327	0.08	20.60	0.031	29.1	11.3
Tin	Sn	Silvery white lustrous with bluish tinge	232	0.16	12.8	0.054	21.4	7.3
Nickel	Ni	Silvery white with yellowish tinge	1453	0.151	6.84	0.105	12.8	8.9
Magnesium	Mg	Silvery white	650	0.38	4.4	0.245	26.0	1.7

APPENDIX D
Physical Constants

Avogadro's number	$N = 6.023 \times 10^{23}$/g mole
Velocity of light in Vacuum	$C = 3 \times 10^8$ m/s
Boltzmann's constant	$K = 1.38 \times 10^{-23}$ joules/°K
	$= 1.380 \times 10^{-16}$ ergs/°K
Gas constant	$R = 8.314 \times 10^3$ joule/kg mole °K
	$= 1.987$ cal/g mole °K
Planck's constant	$h = 6.63 \times 10^{-34}$ joules-s
	$= 6.63 \times 10^{-27}$ erg-s
Electron charge	$e = 1.602 \times 10^{-19}$ coulomb
Electron rest mass	$m_0 = 9.109 \times 10^{-31}$ kg
Electron charge-to-mass ratio	$e/m_0 = 1.76 \times 10^{11}$ coulomb/kg
Permeability constant	$\mu_0 = 4\pi \times 10^{-7}$ henry/m
Faraday's constant	$F = 9.649 \times 10^7$ coulomb/kg equivalent
Acceleration due to gravity	$g = 9.81$ m/s^2
Standard temperature and pressure	stp $= 1.013 \times 10^5$ N/m^2

APPENDIX E
Selected SI Units

Length L or l	metre	m
Mass m	kilogram	kg
Force F	newton	kg m/s^2
Energy E	joule	kg m^2/s^2
Power W	watt	kg m^2/s^3
Resistance R	ohm	kg m^2/s coulomb^2
Resistivity ρ	ohm m	kg m^3/s coulomb^2
Current I	ampere	coulomb/s
Current density J_i	ampere/m^2	coulomb/s-m^2
Potential V	volt	kg m^2/coulomb s^2
Potential gradient or field strength E	volt/m	kg m/coulomb s^2
Inductance L	henry	kg m^2/coulomb^2
Capacitance C	farad	coulomb^2 s^2/kg m^2
Permittivity or dielectric constant ϵ	farad/m	Coulomb^2 s^2/kg m^3
Electric dipole moment	coulomb metre	coulomb m
Electric polarization P	coulomb per square metre	coulomb/m^2
Magnetic flux	weber	kg m^2/coulomb s
Magnetic flux density B	weber/m^2	kg/coulomb s
Magnetic permeability	henry/m	kg m/coulomb^2
Magnetic field strength H	ampere/metre	coulomb/m s
Magnetic dipole moment m	weber metre	coulomb m^2/s
Entropy S	joule per kelvin	kg m^2/s^2/$°$K

APPENDIX F
Conversion Factors

Length

1 m = 10^2 cm = 10^6 microns = 10^9 nanometres = 10^{10} Å = 39.4 in = 3.28 ft
= 1.094 yd

1 cm = 10^4 microns = 10^7 nanometres = 10^8 Å

1 inch = 2.54 cm = 0.0254 m

1 ft = 12 in = 30.5 cm = 0.305 m

1 micron (μ) = 1 μ m = 10^{-6} metre = 10^{-3} millimetre

Force

1 newton = 10^5 dynes = 0.102 kgf = 0.225 lbf

1 kgf = 10^3 g = 2.21 lbf = 9.81 newtons = 9.81 × 10^5 dynes

1 dyne = 10^{-5} newtons = 0.102 × 10^{-5} kg = 0.225 × 10^{-5} lb

1 lbf = 0.454 kgf = 454 gf = 4.45 newtons = 4.45 × 10^5 dynes

Pressure—Stress

1 newton/m² = 1.02 × 10^{-7} kg/mm² = 1.45 × 10^{-4} psi = 10 dyne/cm²

1 kg/mm²/9.81 × 10^6 newton/m² = 1.426 × 10^3 psi = 9.81 × 10^7 dyne/cm²

1 psi = 6.89 × 10^3 newton/m² = 7.02 × 10^{-4} kg/mm² = 6.89 × 10^4 dyne/cm²

1 atmosphere = 760 mm of Hg = 1.01 × 10^5 newton/m² = 1.03 × 10^{-2} kg/mm²

Energy

1 joule = 10^7 erg = 0.239 cal = 0.625 × 10^{19} eV

1 erg = 10^{-7} joule = 0.239 × 10^{-7} cal = 0.625 × 10^{12} eV

1 cal = 4.18 joule = 4.18 × 10^7 erg = 2.62 × 10^{19} eV

1 eV = 1.602 × 10^{-19} joule = 1.602 × 10^{-12} erg = 23,000 cal/mole

1 hph = 27 × 10^4 kgfm = 632 kcal = 2.65 × 10^6 joules

1 kwh = 3671 × 10^2 kgfm = 860 kcal

Others

1 gauss = 10^{-1} weber/m²

1 debye = 0.33 × 10^{-29} coulomb-m

1 oersted = 79.6 ampere-turn/m

1 gauss/oersted = 79.6 × 10^{-4} henry/m

1 radian = 57.3 degrees

APPENDIX G

Properties of Some Important Semi-conducting Materials

Material	Energy gap	Melting pt
	eV/m/°C	
Aluminium antimonide (AlSb)	1.60	1050
Bismuth telluride (BiTe)	0.16	585
Cadmium sulphide (CdS)	2.40	350
Gallium arsenide (GaAs)	1.53	1238
Gallium phosphide (GaP)	2.40	1350
Gallium antimonide (GaSb)	0.80	705
Germanium (Ge)	0.75	940
Indium phosphide (InP)	1.34	1070
Indium arsenide (InAs)	0.45	940
Indium antimonide (InSb)	0.25	525
Lead Sulphide (PbS)	0.40	1114
Lead Selenide (PbSe)	0.26	1065
Lead telluride (PbTe)	0.30	904
Silicon (Si)	0.10	1420
Tin (Sn)	0.10	232

APPENDIX H

Periodic Table of the Elements

I	II		III	IV	V	VI	VII	VIII
1 H 1.0079								2 He 4.0026
3 Li 6.939	4 Be 9.0122		5 B 10.811	6 C 12.0111	7 N 14.0067	8 O 15.9994	9 F 18.9984	10 Ne 20.179
11 Na 22.089	12 Mg 24.305		13 Al 26.9815	14 Si 28.086	15 P 30.9738	16 S 32.064	17 Cl 35.453	18 Ar 39.948
19 K 39.102	20 Ca 40.080	21 Sc 44.956						
		22 Ti 47.900	23 V 50.942	24 Cr 51.996	25 Mn 54.938	26 Fe 55.847	27 Co 58.993	28 Ni 58.710
29 Cu 63.540	30 Zn 65.370	31 Ga 69.720	32 Ge 72.590	33 As 74.922	34 Se 78.960	35 Br 79.909	36 Kr 83.800	
37 Rb 85.470	38 Sr 87.620	39 Y 88.905	40 Zr 91.220	41 Nb 92.906	42 Mo 95.940	43 Tc 99	44 Ru 101.070	45 Rh 102.905
46 Pd 106.400	47 Ag 107.870	48 Cd 112.400	49 In 114.820	50 Sn 118.690	51 Sb 121.790	52 Te 127.750	53 I 126.9044	54 Xe 131.300
55 Cs 132.905	56 Ba 137.340	57 La 138.92	72 Hf 178.490	73 Ta 180.948	74 W 183.85	75 Re 186.20	76 Os 190.2	77 Ir 192.2
78 Pt 195.09	79 Au 196.967	80 Hg 200.590	81 Tl 204.370	82 Pb 207.190	83 Bi 208.980	84 Po 210	85 At 210	86 Rn 222
87 Fr 223	88 Ra 227	89 Ac 227						

TRANSITION ELEMENTS

LANTHANIDE SERIES

| 58 Ce 140.13 | 59 Pr 140.92 | 60 Nd 144.24 | 61 Pm 145 | 62 Sm 150.35 | 63 Eu 152.0 | 64 Gd 157.25 | 65 Tb 158.92 | 66 Dy 162.50 | 67 Ho 164.93 | 68 Er 167.3 | 69 Tm 169 | 70 Yb 173.04 | 71 Lu 174.98 |

ACTINIDE SERIES

| 90 Th 232.038 | 91 Pa 231 | 92 U 238.030 | 93 Np 237 | 94 Pu 242 | 95 Am 243 | 96 Cm 247 | 97 Bk 249 | 98 Cf 251 | 99 Es 254 | 100 Fm 257 | 101 Md 257 | 102 No 255 | 103 Lr 256 |

INDEX

Abrasive 1, 223, 364
Absorption 94
Activation energy 99, 102, 106
Addition polymers 210
Agglomerated structures 222
Allotropic transformation 164
Allotropy 163
Alloy 134
Alloying elements 137
Alpha-iron 34
Amorphous material 28, 282
Anisotropy 282
Annealing 177
Anode 265, 268
Anodic reaction 267
Anti-ferromagnetic 255
Atomic
  movements 97
  models 9
  number 8
  packing 31, 38, 222
  packing factor 34
  radius 36
  structure 5
  weight 8
Austempering 188
Austenite 162, 170
Avogadro's number 9

Bainite 163
Band
  conduction 231
  energy 230
  model 230
Baushinger effect 122
Binary diagram 139, 144
Blue brittleness 124
Body centred cubic 34
Bohr's theory 12
Bonds
  chemical 61
  covalent 48, 54

  ionic 51
  metallic 56, 366
  mixed 60
  primary 48
  secondary 48, 60
Boundries
  grain 69, 105
  tilt 69
  twin 70
Bragg's diffraction 371
  law 82
Bravais lattice 31
Breaking strength 300
Brittle fracture 317
Burger's vector 67, 72

C curve 172
Carbide 274, 364
  cemented 364
  iron 164
  tungsten 369
Carburizing 104
  gas 194
  liquid 194
  pack 193
Case hardening 104, 192
Castability 327
Cast iron 168
Cathode protection 278, 280
Cell 265
  galvanic 269
  photoelectric 337
  unit 30
Cement 223
Cementite 162
Ceramic 1, 221, 223
Charpy test 309
Chemical
  bonding 62
  reduction 352
Cleavage 318
Climb dislocation 73

Close-packed hexagonal 35
Coatings 223-225
  inorganic 223
  metallic 276
  organic 224
  protective 223
  surface 276
  type of 235
Coercive force 367
Coining 358
Cold work 126
Collector 331
Collision of electrons 80
Composite materials 220, 221
Compressive stresses 323
Concrete 209, 222
Conductivity 229
  electrical 229
  metal 227
  thermal 366
Conductor 374
  semi 375
Constitution diagram 134
Continuous cooling transformation 174
Cooling curves 141
Coordination number 34
Copolymer 210
Coring 145
Corrosion 262
  atmospheric 271
  contact 274
  direct 263
  electrochemical 264
  environment 275
  erosion 273
  fatigue 272
  fretting 274
  galvanic 268
  high temperature 269
  intergranular 271
  mechanism of 266
  pitting 270
  prevention of 274
  selective 274
  stress 272
  type of 263, 270
Covalent solids 56
Creep 284, 292
  curves 293
  fracture 295
  mechanism 294
  test 312
Critical
  point 141
  shear stress 113
Cross
  linking 214
  slip 73
Crystal 30
  deformation 111
  direction 46
  geometry 28
  imperfection 64
  planes 40
  single 222
  structure 4, 28
  symmetry 35
  systems 31
Cubic crystals 34
Curie temperature 245
Cyaniding 196

De-Broglie 25, 368
Decarburization 205
Defects
  line 66
  point 65
  schottky 65
  surface 69
Deformation 104
  elastic 109
  slip 111
  twinning 118
  of materials or metals 119
  plastic 110
  single crystal 111
Degree of freedom 140
Delta-iron 166
Dendritic structure 158
Diagram
  equilibrium 139
  isothermal 172
  phase 139
Diamagnetic 252
Diamond 375
  structure 57
  tools 363
Dielectric 240
  constant 240
  loss 242
  strength 241
  types of 244
  uses of 248
Diffraction 81
  pattern 82
  technique of X-rays 81

Diffusion 97
  application of 97
  coating 277
  coefficient 100
  couple 103
  mechanism of 98
  vacancy 295
  self 103
  volume 105
Dipole 60, 241, 244, 249
Dislocation 66
  climb 73, 295
  edge 66
  pile up 120
  screw 67
  theory 116
Domains, magnetic 254
Donor level 236
Drude and Lorentz theory 367
Ductile
  brittle transition 321
  fracture 318
Ductility 300

Edge dislocation 66
Einstein's equation 325, 334
Elastic 109
  deformation 109
  limit 2'4, 299
Elasticity 283
Elastomers 217
Electro-chemical corrosion 264
Electrode potential 265, 266
Electrolyte 261
Electron 5, 80
  charge 7
  configuration 4
  energy 369
  free 57, 227, 330, 366
  theory of metals 366-376
  valence 56
Electroplating 225
Electrostriction 246
Elongation 281
Embryo 156
Enamel 224
End-quench 185
Endurance limit 312
Energy
  activation 106
  band 231
  gap 231-237, 374
  levels 18
  thermal 234
  states 231
Equilibrium diagrams 139, 164
Erosion 206
Etching 163
Eutectic 146
  reaction 166
  temperature 166
Eutectoid 166, 169
  hyper 169
  hypo 169
  point 166
Extension 283
Extrinsic semiconductors 234

Face centred cubic 35
Fatigue 289
  failure 291
  fracture 290
  limit 290
  test 311
Fe–C diagram 165
Fermi-level 233, 236, 370
Ferri-magnetism 253, 255
Ferrite 162, 258
Ferro-electricity 244
Ferro-magnetism 253
Fibres 219
Fick's law
  first 100
  second 102
Filters 348
Flame-hardening 199
Flow, viscous 217
Forbidden energy gap 231
Force
  coercive 367
  Vander waals 56, 60
Formability 328
Fracture 316
  brittle 317
  cleavage 318
  cup and cone 317
  ductile 318
  fatigue 290, 292
  models of 316
  protection against 322
  types of 316
Frank-Reed source 71
Frankel defect 66
Free energy 155
Freezing 158, 161
Furnace, heat treatment 200

Index

Fusion 158

Galvanic
 cell 269
 corrosion 263, 269
 protection 276
 series 266
Galvanised coating 276
Gamma iron 163
Gas carburizing 194
Germanium 227, 233
Glass
 fibre 222
 transition 161
Grain
 boundry 105, 295
 growth 127, 130, 160
 refining 178
 size 171
Graphite 60
Griffith's theory 319

Hardenability 184
Hard magnetic materials 259
Hardening 182
 age 376
 case 192
 flame 199
 induction 197
 methods 187
 strain 121
 surface 197
 work 121
Hardness 288
 Brinell 302
 Rockwell 305
 Vickers 303
Heat treatment 176
 furnaces 200
 processes 177
Hexagonal close-packed 34, 119
Hot pressing 358
Hot working 131
Hooke's law 110, 282
Hume Rothery rules 37
Hydrogen
 bond 61
 embrittleness 272
Hysteresis
 loop ferroelectric 245
 magnetic 254

Impact
 strength 289
 test 308
Imperfection
 in crystal 64
 line 64, 66
 point 64
 structure 64
Inclusions 185
Induction hardening 197
Inhibitors 274-278
Insulators 277, 240, 374
Interatomic
 attraction 106
 distances 106
Interstitials 99
 diffusion 99
 impurities 105
Intrinsic semiconductors 233
Ionic
 bonds 51
 solids 52
Ionization 85
Iron-carbon
 alloys 164
 diagram 164
Isothermal transformations 172
Isotope 8
Isotropy 281
Izod test 309

Jog dislocation 74
Jominy
 bar 186
 end quench test 185
Junction, pn type 237

Kα radiation 86
Kinetic energy 80, 337
Kirkendall effect 100

Lattice 40
 Bravais 31
 space 29
Laue method 90
Ledeburite 168
Lever rule 142
Limited solubility 149
Line imperfection 64, 66
Linear
 density 44
 polymers 213
Limit, elastic 284, 299
Liquids 141

Liquidus 142
Low angle boundaries 70, 129, 319
Luder's band 124

Machinability 326
Machining 358
Magnetic
  domains 254
  dust method 314
  materials 250, 363
  moments 249
  properties 248
Magnetism 248
Magnetostriction 257
Martempering 188
Martensite 163
  tempered 163
  transformation 175
Materials 1
  agglomerated 222
  ceramic 222
  composite 220
  multiphase 221
  organic 2
  reinforced 221
  selection of 2
Matrix 163, 193
Mechanical
  properties 281
  tests 296
  working 110
Melting point 355, 379
Metallic bond 56, 366
Metals 1, 2
Microconstituents 162
Microstructure 4
Millikan's experiment 335
Miller indices 40, 222
Molecules 48
Moment 249
Monomer 209
Mosley's law 87
Muffle 202
Multiphase 221

n-type crystal 235
NaCl structure 52, 53
Natural rubber 217
Necking 284, 318
Neoprene 218
Neutrons 7
Nitriding 195
Non-destructive testing 313

Normalising 181
Nominal stress 284
Nucleation 157
Nucleus 5
Number
  atomic 8
  orbital 22
  magnetic 23
  quantum 22
  spin 23
Nylon 210, 216

Orange peel effect 124
Orbit 10-20
Ordered solid solution 136
Organic coatings 223
Organic materials 2
Orientation
  preferred 126
  random 127
Ortho-rhombic 31
Oxidation 97, 205, 263
Oxygen 99

p-type crystal 235
Pack-carburizing 193
Paramagnetic 253
Passivity 268
Patterns, diffraction 81
Pauli's exclusion principle 24, 370
Pearlite 162
Penetrating liquid test 314
Periodic table 25, 385
Peritectic 150
Permeability 248
Phase
  amorphous 161
  crystalline 161
  diagram 139
  rule 140
  metastable 175
  transformation 154-175
Phenol formaldehyde 216
Photoconductive cell 341
Photo-electric
  cell 337
  effect 330
Piezo-electricity 247
Pitting 261, 270
Plastic deformation 110
Plastics 214
pn junction 237
Point defects 65

Poisson's ratio 283
Polar and non-polar materials 242
Polarization 242
Polymer 209
  addition 210, 212
  chain 213
  condensation 212
  cross-linked 214
  crystalline 213
  linear 214
  structure 213
Polymerization 210
Porous parts 359
Potential
  difference 336
  electrode 266
  energy 50
Powder
  crystal method 92
  manufacture 350
  metallurgy 347
Powder factor 242
Preferred orientation 126
Primary
  bonds 48
  solidification 166
Process annealing 180
Properties of
  alloys 1
  cathode rays 6
  metals 1
  X-rays 78
Proportional limit 284
Protective coatings 223, 276
Protons 7
Pyrometry 204

Quantum number 22
Quenching 182
  media 184
  rate 184

Radiation 82
Radii, atomic 3
Radiography 315
Random solid solution 136
Rare earths 26
Rebound test 306
Recovery 128
Recrystallization 129
  temperature 126
Refractory 2
Reinforced materials 221

Repulsive forces 367
Resins 210, 216
Resistivity 228
Rockwell hardness 305
Rotating crystal method 91
Rubber 217

Schottky defect 66
Scratch test 306
Screw dislocations 67
Secondary bonds 48, 60
Self diffusion 103
Semiconductors 227, 233
  extrinsic 234
  intrinsic 233
Shear
  strain 283
  stress critical 113, 283
Shell 18
  sub 24
Single crystal 222
Sintering 223, 355
Sizing 358
Slip
  band 115
  direction 114
  planes 111, 114
  processes 112
  systems 115
Soft magnetic materials 258
Solid solutions 135
Sorbite 163
Space lattice 29
Spectral series 16
Stacking faults 71
Strain aging 124
Stress and strain 283
  concentration 290
  diagram 284

Tempering 190
Tensile strength 288
Ternary diagram 152
Tests
  fatigue 312
  hardness 300, 301
  impact 308
  jominy 185
  tensile 297
Thermoplastic resins 216
Thermosetting resins 216
Tilt boundary 69
Toughness 300

Transformations
  isothermal 172
  secondary 168
Troosite 163
T-T-T diagram 172
Twinning 111

Ultimate tensile stress 299
Ultra-sonic test 315
Unit cell 30

Vacancies 65
Valence
  bands 232
  electrons 56, 370
Vander Waal's forces 49, 60, 240
Vander's hardness test 303
Viscoelastic 109
Vitrous
  coatings 209, 217, 223
  solid 161
Vulcanization 217

Wavelength 16
Wavemechanics 25

Weight, atomic 8
Weldability 326
Whiskers 222
Wood 220
Work
  function 370
  hardening 121
Working
  cold 126
  hot 132

X-rays 76
  applications of 93
  characteristic 79
  origin of 80
  properties of 78
Yield
  point 123, 284
  strength 299
Young's modulus 283, 299

Zinc 276
Zirconium 352, 379
Zone theory 367